ROUTLEDGE LIBRARY EDITIONS:
INTERNATIONAL TRADE POLICY

Volume 19

THE POLITICAL ECONOMY OF MANUFACTURING PROTECTION

THE POLITICAL ECONOMY OF MANUFACTURING PROTECTION

Experiences of ASEAN and Australia

Edited by
CHRISTOPHER FINDLAY AND ROSS GARNAUT

Routledge
Taylor & Francis Group
LONDON AND NEW YORK

First published in 1986 by George Allen & Unwin (Publishers) Ltd

This edition first published in 2018
by Routledge
2 Park Square, Milton Park, Abingdon, Oxon OX14 4RN

and by Routledge
711 Third Avenue, New York, NY 10017

Routledge is an imprint of the Taylor & Francis Group, an informa business

British Library Cataloguing in Publication Data
A catalogue record for this book is available from the British Library

ISBN: 978-1-138-06323-5 (Set)
ISBN: 978-1-315-14339-2 (Set) (ebk)
ISBN: 978-1-138-29771-5 (Volume 19) (hbk)
ISBN: 978-1-315-09835-7 (Volume 19) (ebk)

Publisher's Note
The publisher has gone to great lengths to ensure the quality of this reprint but points out that some imperfections in the original copies may be apparent.

THE POLITICAL ECONOMY OF MANUFACTURING PROTECTION: EXPERIENCES OF ASEAN AND AUSTRALIA

Edited by
CHRISTOPHER FINDLAY
AND ROSS GARNAUT

ALLEN AND UNWIN
Sydney London Boston

First published in 1986
Allen & Unwin Australia Pty Ltd
8 Napier Street, North Sydney, NSW 2060, Australia

Allen & Unwin New Zealand Limited
60 Cambridge Terrace, Wellington, New Zealand

George Allen & Unwin (Publishers) Ltd
18 Park Lane, Hemel Hempstead, Herts HP2 4TE England

Allen & Unwin Inc.
8 Winchester Place, Winchester, Mass. 01890 USA

National Library of Australia
Cataloguing-in-Publication entry:

The Political economy of manufacturing protection.

Includes index.
ISBN 0 04 330378 1 (pbk.).

1. ASEAN. 2. Free trade and protection – Addresses, essays, lectures. 3. Asia, Southeastern – Commercial policy – Addresses, essays, lectures. 4. Australia – Commercial policy – Addresses, essays, lectures. I. Findlay, Christopher. II. Garnaut, Ross. (Series: ASEAN series).

382.7

Set in 10/11 pt Times by Graphicraft Typesetters Ltd, Hong Kong
Printed by Singapore National Printers (Pte) Ltd.

Contents

Acknowledgements

This project was sponsored by the ASEAN-Australia Economic Relations Research Project. Thanks are due to Hal Hill, Mohd Ariff and the Research Directors' Steering Committee. The editors are also grateful for the advice of a number of readers of the country chapters.

The text was expertly processed by Kerry Clift and edited by Maria Sobolewski and Rosemary Thomson. Helen Wickens drew the diagrams and Rosemary Thomson prepared the index.

Tables

Abbreviations

ACT	Action Committee on Tariffs
ACTID	Action Committee on Tariff and Industrial Development
ADO Tax	Automatic allocation of foreign exchange tax
AID	Agency for International Development
ASEAN	Association of South East Asian Nations
ASIC	Australian Standard Industrial Classification
BE	*Bonus expor*: export bonus certificates
BKPM	Badan Koordinasi Penanaman Modal: Investment Coordination Board
BOI	Board of Investment
BOT	Bank of Thailand
BTN	Brussels Tariff Nomenclature
CCCN	Customs Cooperation Council Nomenclature
CES	constant elasticity of substitution
CIC	Capital Investment Committee
c.i.f.	carriage, insurance and freight
c.k.d.	completely knocked down
COTT	Committee on Trade and Tourism
CPI	consumer price index
DP	*devisa perlengkap*: complementary foreign exchange
DSP	*Daftar Skala Prioritas*: investment priority list
EC	European Community
EDB	Economic Development Board
EEC	European Economic Community
EFTA	European Free Trade Area
ELES	extended linear expenditure system
EPP	Export Priorities Plan
EPZA	Export Processing Zone Authority

ERA	effective rate of assistance
ERP	effective rates of protection
f.o.b.	free on board
FIDA	Federal Industrial Development Authority
GATT	General Agreement on Tariffs and Trade
GDP	gross domestic product
GE	general equilibrium
GNP	gross national product
HICOM	Heavy Industries Corporation of Malaysia
HIID	Harvard International Institute for Development
HO	Heckscher-Ohlin
IAC	Industries Assistance Commission, Australia
IBRD	International Bank for Reconstruction and Development
IFC	International Finance Corporation
IFS	International Financial Statistics (of IMF)
IMF	International Monetary Fund
I–O	input-output
IPP	Investment Priorities Plan
ISIC	International Standard Industrial Classification
L/C	letter of credit
LDC	less developed country
MDC	more developed country
MIDA	Malaysian Industrial Development Authority
MPO	*Menghitung Pajak Orang*: withholding tax
n.e.c.	not elsewhere classified
NIC	newly industrialising country
NRP	nominal rates of protection
NTB	non-tariff barrier
OECD	Organisation for Economic Co-operation and Development
OLS	ordinary least squares
pdf	probability density function
PPH	*Pajak Penghasilan*: income tax
P3DN	*Peningkatan Pendayagunaan Produksi Dalam Negeri*: the Junior Minister for the Promotion of the Use of Domestically Produced Goods
PTA	preferential trading arrangements
PTC	Pitt's trade classification
SACT	Special Advisory Committee on Tariffs
SE	*Sertifikat Ekspor*: export certificate
SITC	Standard International Trade Classification
TAB	Tariff Advisory Board
TAC	Tariff Advisory Committee
TCF	textiles, clothing and footwear

TDB	Trade Development Board
TNI	*Tata Niaga Impor*: approved importers system
TOT	terms of trade
UNIDO	United Nations Industrial Development Organisation
VAO	value added in output
VAT	value added tax
VER	voluntary export restraint
WVA	wages as share of value added

Introduction

PROTECTION is a persistent feature of economic policy in developed and developing countries alike. Australia and all of the original ASEAN countries except Singapore provide high protection for their manufacturing sectors.

The evidence that high protection for import-competing industries and large biases in incentives against exporting industries seriously damage growth performance is now strong. Ariff and Hill (1985) review the literature and evidence with particular application to the ASEAN countries.

It is now widely accepted beyond the community of academic economists even, for example, by ministers with economic responsibilities in Australia and the ASEAN countries, that high protection holds back economic growth. In response to this belief, there have been attempts at import liberalisation in Australia and all of the ASEAN countries. They have been only partially successful. The systems of incentives in Australia and four of the original ASEAN countries remain strongly biased against expansion of the internationally competitive, export-oriented industries which must play key roles in future economic growth.

This collection of papers presents new material on a range of matters related to protection in Australia and the ASEAN countries.

One major focus is the extent and structure of manufacturing protection in Australia and the ASEAN countries. The historical material and statistical analysis of the structure of protection are examined for the light that they can throw on reasons why protection is so pervasive when it is so widely recognised to be against the national interest of the countries which impose it. This examination provides a useful backdrop to the analysis of recent attempts at trade liberalisation.

Other papers in the volume examine the effects of protection on ASEAN-Australian trade in manufactures, present the results of research on the effects of new forms of protection, and apply new techniques to the analysis of the effects of protection in the ASEAN countries and Australia.

The first six papers in the volume describe the origin, current structure and rationale of manufacturing protection in Indonesia, the Philippines, Thailand, Malaysia, Singapore and Australia. These papers, by economists writing about their own countries, represent a major advance in our knowledge about manufacturing protection in this region. The material presented on the current levels and structure of protection is summarised later in this chapter, and the influences on the historical development of protection are summarised in the final chapter of the volume (Chapter 11).

A number of these chapters attempt to draw conclusions on the causes of protection from statistical analysis of its interindustry structure. This line of analysis emerges as not very powerful, so that in examining the causes of protection we are thrown back onto more conventional historical and political techniques.

The recent history of attempts at import liberalisation in five of the six countries is far from encouraging given the high importance of these issues for the continuation of strong economic growth. However, we believe that the material presented in the country chapters provides insights into the causes of protection that could be useful to governments seeking to promote import liberalisation in future. Our thoughts on these matters are presented in Chapter 11.

Following the six country papers about protection in the ASEAN countries and Australia, Chapter 7, by Harrison and Rutström, applies general equilibrium modelling techniques to the analysis of the effects of protection in the region as a whole.

Chapters 8 and 9, by Hamilton, break new ground in the analysis of the effects of import quotas and voluntary export restraints on importing and exporting countries. The analysis of ASEAN systems for allocation of export licences under voluntary export restraints is of considerable policy significance and an area in which there has been little application of rigorous economic analysis.

In Chapter 10, Warr applies another general equilibrium model, the ORANI model of the Australian economy, to the detailed analysis of the effects of Australian protection on the scale and composition of Australian imports from ASEAN.

Finally, Chapter 11 applies the insights of the volume as a whole, and particularly of the six country chapters, to answering the questions: How did high protection come to be installed? Why does it persist in the face of evidence that it detracts from economic performance? And how can we do better in the promotion of trade liberalisation and economic growth?

The volume makes considerable use of a number of concepts and techniques for defining, measuring and analysing the effects of protection. Some readers may not be familiar with them, so it is as well to introduce them now.

Extent of industry assistance

The nominal rate of protection

The question of the extent of industry assistance has a number of dimensions and each requires a different empirical measure. For example, if questions are being asked about the extent to which domestic prices have been raised above world prices by tariffs and other restrictions on imports, then the appropriate indicator is the nominal rate of protection (NRP). The country studies reported here all examine the effects of the tariff, but they have not all been able to cover the impact of quantitative restrictions. For example, the work on Indonesia considers tariffs and the discriminatory impact of sales taxes, but does not cover the quantitative restrictions and government control of values for duty. The exclusions are not so important in the consideration of other countries, either because a wider range of import restrictions is examined, or because the excluded instruments are less significant.

There are other problems in the calculation of NRPs. Competition amongst domestic producers may keep product prices below the 'international prices plus the tariff' which is conventionally assumed in the analysis of the effects of protection. (In Australia this is known as 'water in the tariff'). Alternatively, the potential protection available may not be fully used because of various methods of avoidance, for example, smuggling, understatement of import values or corruption in the customs system.

Further, it may not be easy to take fully into account in the calculation of protection level the effects of fiscal devices which are not imposed at the border, but which nevertheless affect the level of assistance to domestic industries. The discriminatory sales taxes that are applied in Indonesia, Thailand and the Philippines are examples of nonborder levies which have protective effects.

These problems would be overcome if we based estimates of protection on comparisons between domestic and world prices, rather than relying on published tariff rates, or on estimates of tariff rates based on calculations of tariff revenue as a proportion of the value of imports. These comparisons would capture the effects of quantitative restrictions, of discriminatory sales taxes and of 'water in the tariff'.

However, there are considerable empirical problems in the application of this method. For example, it is often difficult to obtain prices for sufficiently similar goods.

It would be useful for further work to focus on extension of the NRP estimates to incorporate the effects of quantitative restrictions and other policies.

Effective rate of assistance

For many purposes related to analysing the economic effects of assistance it is more important to estimate how restrictions on imports have changed the incomes of people supplying factors to an industry than it is to measure the effects on domestic prices. In this case, it is the effect of government policy on *value added* in the industry which is of interest. Information is required on how both prices of output and of material inputs have changed as a result of government policy. These effects can be measured by the *effective* rate of protection, rather than the *nominal* rate.

It is most useful if the effective rate takes into account more than just the impact of policies at the border. It should measure the effects of all forms of government assistance to industry, including subsidies, which raise the price at which output is sold,or reduce the prices that producers pay for inputs into production. It should also incorporate the effects of other preferential policies such as government procurement, which could be analysed in the same manner as a subsidy.

Estimates of *effective rates of assistance* (ERAs) are available for Australia. In the Philippines chapter, investment incentives were found to have minor protective effects compared to tariffs and the indirect tax system. In other country studies in this volume, the coverage of the effective rate measures is more limited and must be defined as effective rates of *protection* (ERP), covering only policies applied at the border and not of assistance, which would incorporate the effects of all policies which change relative prices.

The extension of ERPs to include other forms of assistance, or at least calculation of the price equivalents of these forms of assistance, also warrants further work.

The effective rate of assistance refers to factor rewards in a single industry taken as a whole. Some other policies affect relative prices and factor rewards across a number of industries, or for only parts of an industry. These considerations can become very important where many policies are operating simultaneously in an uncoordinated manner and with conflicting objectives. These situations are identified by the authors of some of the ASEAN country papers in this volume. For example, there may be a policy to increase the income of economically weak or indigenous groups, but this policy could be contradicted by commercial policy elsewhere. There also have been conflicts reported, in the volume, between the rhetoric of industry policy, favouring export promotion, and its actual performance, which introduces the question of the relative levels of assistance. A single simple measure such as the ERP is unlikely to be able to represent all of the conflicting influences in situations like these.

Ranking of effective rates

In public debate the claim may be made that a particular sector is receiving assistance, for example, as a result of a direct subsidy. The calculation of the ERA takes into account the reality that the net effect of government assistance depends on the impact of those policies on the input prices paid by the industry. The agricultural sector for example may receive more direct, and therefore more obvious, subsidies from government than manufacturing, but agriculture could still suffer discrimination because of the effects of manufacturing sector protection on the prices of its material inputs. If the policy interest is with the relative impacts of government intervention, then a ranking of ERAs will be required. All the country chapters include tables of effective rates at the industry level, subject to the qualifications about the varying degrees of coverage, from which these rankings can be derived. The country studies in this volume reveal that there is commonly an extremely strong bias against export activities, either in the whole economy, or within the manufacturing sector. This is evident in the summary data presented in the next section.

Comparative Data on NRPs and ERPs

Table Intro. 1 shows levels of nominal and effective rates of protection (or assistance in the Australian case) for the manufacturing sector in the six countries of interest. The industries within the sector have been combined into two groups—importables (net imports positive) and exportables.

A couple of points are evident from the Table. Firstly, in terms of degree of effective protection for importables, the highest ranking country in this group is Indonesia, followed by Thailand, the Philippines, Malaysia, Australia and finally, the outlier, Singapore. Secondly, there is a general bias against the exportable goods sectors; even though industries in that sector may receive positive effective rates of

Table Introduction 1: Average effective rates of protection, ASEAN and Australia[a], selected years, 1974–1980

Country	Year	Exportables	Importables	Total Manufacturing
Indonesia	1975	−6	98	39
Philippines	1974	4	61	44
Thailand	1980	24	89	65
Malaysia	1978	13	53	34
Singapore	1979	4	1	3
Australia	1978	4	30	26

Note: Effective rates of protection weighted by value added. Industries are classified by whether net exports are positive or negative. Coverage of policies varies between countries: see sources for details.

Sources:
Indonesia: as for Table 1.5
Philippines: as for Table 2.8
Thailand: as for Table 3.9
Malaysia: as for Table 4.3
Singapore: as for Table 5.6
Australia: Industries Assistance Commission, *Annual Report*, 1978–79, Table 1.3.3.

assistance or protection, effective rates in the import competing sectors are generally higher. Singapore is the exception.

Effects of industry assistance

The question of the effect of assistance also has a number of dimensions. These can be grouped under the headings of consumption and production impacts.

Consumption effects

On the consumption side, the protection policy will change relative prices and distort choices. For example, Australian protection for the motor vehicle industry or the clothing industry has raised prices and led consumers to buy fewer motor vehicles and clothes. The extent of this effect can be measured by combining the estimate of NRP with estimates of the elasticity of demand. A summay statistic which incorporates the welfare effect, that is the loss of consumer surplus due to the price increase, can be calculated in a similar manner.

Production effects

On the production side, the policy interest will be in the impact of protection or assistance policy on the structure of output. The estimation of output effects would be relatively simple if there were no intermediate inputs. Consider the case of two broad groups of goods, exportables and importables. A tariff at a common rate raises the domestic price of importables compared to exportables and therefore leads to a shift of resources in the production of import competing goods. (The tariff is therefore equivalent in many of its economic effects to an export tax). The production effects can be inferred by examining in this case the nominal rate of protection.

However, in the presence of many goods and the use of intermediate products, the estimation of output effects is more complicated. In the first place, an effective rate rather than a nominal rate is required. Rankings of effective rates of protection (or assistance) indicate how outputs of different industries change as a result of the government policies where effects on inputs are taken into account. The analysis can be extended by examining the variation of effective rates within a sector to estimate the extent to which the pattern of output has been distorted within the sector as a result of government policy. Given the same average level of effective rate for two sectors, the economic inefficiency and loss due to the production effects would be greater in the sector with the higher degree of variation, because the welfare cost increases more than proportionately with the level of assistance. The policy implication is that effective rates should be equalised within sectors to reduce the size of the production cost. Discussion of the patterns of protection in the country chapters includes some information on the degree of variation in effective rates for this reason.

General equilibrium analysis

There has been some criticism of the use of effective rates for the purposes outlined above, along the lines that this concept is not a reliable indicator of output changes because it does not take into account the possible effects of policy changes on major variables affecting costs and relative prices right across the economy, such as wage levels, and the exchange rate. Hence attempts are now being made to address these wider issues through the application of general equilibrium (GE) models, which incorporate more flexible assumptions about production than does the ERP or ERA formula.

Harrison and Rutstrom in this volume compare ERP estimates derived using the standard procedure, which they call basic ERPs, with estimates of changes in value added from the results of a GE model in which global trade liberalisation is simulated.

An examination of the effects of protection need not be restricted to the country imposing the protection policy; also of interest is the effect on trading partners. These effects are incorporated in the Harrison and Rutström type of model and, if of interest, the bilateral trade implications of liberalisation could be highlighted.

The chapter by Peter Warr in this volume also applies a GE model, in this case the Australian ORANI model, to the analysis of the effects of protection. Warr uses the model to measure the impact of Australian protection on its imports from the ASEAN countries.

Analysis of the causes of protection

Underlying the analysis of the causes of protection in Chapters 1 to 6 is the idea that policies may be introduced either because they are believed to be in the national interest (generally the national interest in economic growth, but sometimes in more equitable income distribution), or because they reflect the pressures that vested interests apply to the political process.

In reality, of course, policy outcomes reflect the interaction of national interest and private interest pressures.

The influence of national interest considerations can be assessed through analysis of the economic effects of protection (both on growth and on distribution), and through examination of government and wider community beliefs about the effects of protection. The gap that has often existed between beliefs about the effects of protection, and what economic analysis declares to be its effects, requires explanation. The role of vested interests in nurturing misperceptions cannot be ignored. Nevertheless, substantial progress can be made by looking at what was believed about the effects of protection at the time decisions were made, without resort to unfamiliar analytic techniques.

The analysis of the role of private interest groups is more challenging. There is no point in asking either representatives of interest groups or their political contacts about the role of pressures from these sources in the development of policy. Indirect means must be used to establish their importance.

If vested interest groups are of major importance in the policy making process, it is likely that they will be more effective in industries with certain characteristics, than in others. This differential power of interest groups could be expected to generate a differentiated structure of protection.

The highly differentiated protection structures in Australia and the four ASEAN countries suggest a major role for vested interest groups, since the main national interest arguments that have been influential would suggest more uniform rates across industries. If we could establish a close correlation between the rates of effective protection (presumed to be a dependent variable) and the strength of certain characteristics which could reasonably be presumed to reflect industries' power in the political process, we would have achieved two results: the confirmation of a powerful causal role for vested interests in the policy-making process; and an understanding of the characteristics that make some vested interest groups more and others less politically influential.

Several of the country chapters present the results of detailed statistical analysis along these lines. The identification of industry characteristics that might be associated with the effective application of political pressure takes into account the likely weakness of diffuse groups with many members, such as consumers and primary producers. Such groups face large costs of co-ordination and a small individual benefit from liberalisation which reduces the incentive to become involved in political activity. This tendency is reinforced by the reality that the outcome of political pressure by the group is available to all, whether or not a particular member has contributed to the campaign. This encourages some members to free ride on others and makes it even less likely that the coalition will be effective.

There are major difficulties in statistical analysis along these lines, in defining both dependent and independent variables effectively. Economic analysis suggests that the effective rate of assistance is the appropriate dependent variable, but a case could be made for the use of nominal rates of protection in some circumstances. The choice of independent variables is even more complex, often involving some arbitrariness, since the relevant theory of political economy is too little developed to identify definitively the appropriate industry characteristics.

For these and other reasons, the statistical analysis of the inter-industry structure of protection should be seen as a tentative and, for the ASEAN countries, preliminary entry into new territory.

Choice of policy instrument

A further question is what determines the choice of policy instrument? The various measures that are used to provide assistance for industry are catalogued in the country papers. However, in this volume the focus was on levels of protection, or assistance where possible, rather than on the choice of instrument so there are few empirical studies of this problem to report. Extension of the empirical work to this question would be useful, for example, by relating industry characteristics to the use of various policy instruments.

The standard theoretical results are that, from an efficiency point of view, subsidies are preferred to tariffs which are preferred to quotas as instruments of achieving a domestic output target. However, this sort of result is typically ignored in the actual decision making on industry policy. There are two explanations. One is that policy makers do not believe theoretical rankings. Implicitly they may be arguing that subsidies also have to be financed and this will also create distortions or that subsidies are more likely to attract retaliation by other countries. The other explanation is a political one, that the effects of quotas are more obscure than tariffs and therefore less likely to generate opposition. In the absence of any opposition, producers will prefer a subsidy, because they can receive larger increments to their profits at a smaller burden on consumers, and thereby with smaller political resistance. However, given that consumers would oppose such a measure, a group lobbying for protection may favour tariffs which place a bigger, but less obvious, burden on consumers. Over time of course, the effects of tariffs and quotas become more widely understood and this explanation of the choice of policy instrument becomes less powerful.

The uncertainties associated with world markets may also influence the choice of instrument for protection. For example, producers may demand protection to isolate them from the world market and may prefer a quota instead of a tariff, depending on whether the world price fluctuates more than domestic demand. Producers may also argue for anti-dumping forms of protection policy to act as an insurance policy against a price fall.

An important contribution to this general topic of the choice of instrument of protection is Carl Hamilton's work on voluntary export restraints and preferential import quotas. Hamilton in this volume examines the analytical questions about the impact of these policies, compares them to global import quotas and looks at the actual practice of the administration of VERs in the ASEAN countries.

1 The structure and causes of manufacturing sector protection in Indonesia

MARI PANGESTU AND BOEDIONO*
November 1984

FURNIVAL stated in 1944, in reference to the import restriction policies adopted by the Colonial Dutch government in the 1930s that '... it has further been found necessary for the State to intervene in protecting both the local market for home products and also the new industrial developments.'[1] In 1984, according to an official statement,[2] 'an acceptable system of protection will be continually developed with a two-fold aim: to protect domestic producers from excessive import competition and to ensure that consumers will not pay too high a price for domestically produced goods.'

It can be seen, then, that the theme of protecting domestic industries in Indonesia has not changed since the 1930s. However, while the main theme has remained unchanged, the justification for protection and the instruments used have changed. In the 1930s import restrictions were in the form of tariffs, import quotas and licences directed to curb the flow of cheap Japanese imports, as well as to promote domestic industrialisation. In the 1980s there is a whole set of instruments—ranging from tariffs to credit rationings—which have protective effects and were set up not just to protect domestic industries from import competition but also to protect certain groups in the economy, to redistribute income, to promote non-oil exports and to promote the use of domestically produced components.

The aim of this paper is to examine the evolution of the system of protection in Indonesia since the 1930s. The levels of protection are empirically estimated by using the standard effective rates of protection measures. In addition, to enhance the understanding of the

* Mari Pangestu is a graduate student in the Economics Department, University of California. Boediono is Lecturer in the Faculty of Economics at Gajah Mada University in Jogyakarta.

observed structure of incentives, we attempt to identify and analyse the economic-political forces underlying the system of protection. An understanding of the rationale for protection in the Indonesian context may help in understanding why the liberalisation of the trade regime according to laissez-faire principles would be difficult to attain or only possible at a gradual pace, despite growing awareness of the adverse effects of protection policies.

The system of protection

Considering protection policies have existed since the 1930s, it is relevant to provide a brief historical overview of the main objectives and instruments of protection in the manufacturing sector in the latter part of the pre-independence period, beginning around 1930, and in the Old Order period (1949–1965). This overview will then be followed by a more comprehensive discussion of the system of protection in the New Order period (1966-present).

The pre-independence period[3]

The aims of protection in this period were to limit the influx of cheap Japanese goods, which threatened to displace imports from the Netherlands and to encourage domestic industrialisation, especially on Java. By 1935, the share of imports from Japan had reached 30 percent of total imports whereas Japan only purchased 5 percent of total exports.[4] The main policy instruments used were discriminatory tariffs, import quotas and licences. Quantitative restrictions were used instead of tariffs because of conditions in earlier Anglo-Dutch treaties. There were two types of quotas. Under the free quota system, the origin of the imports was not restricted,[5] while in the case of national quotas, the origin of imports was linked to the destination of exports. Furthermore, the import of certain goods, mainly cheaper consumer goods, was restricted to holders of an import licence.

The main effect of these policies was that the share of imports from the West rose. By 1938, the share of imports from Japan had fallen to 15 percent, while from Holland it was 22 percent.[6]

The Old Order period, 1950–1966

The period was characterised by complex and continually changing regulation on exchange rates, import and export duties, import prepayment schemes and quantitative restrictions. The following two sections provide a brief description[7] of the first phase (1950–1957), which was of a more pragmatic orientation and the second phase (1958–1966), which was of a more radical orientation under the Guided Economy principle.

1950–1957

Although the foreign trade regulations and policies that evolved during this period were aimed at increasing tax revenues and allocating scarce foreign exchange, especially after the end of the commodity boom due to the Korean War, the system led to high levels of nominal protection. An important principle that was practised throughout this period was taxation of imports by degree of 'necessity'.

The administrative ease of collecting trade taxes, compared with collecting income and corporate taxes, made such a system a logical choice for increasing the tax revenues of the newly independent country. Table 1.1 shows government revenue from taxes on trade for the period 1951–1960. It can be observed that the revenue from trade taxes made up a substantial proportion of total revenue at 66 percent in 1951, the year of the Korean boom, and then fluctuated between 32–55 percent of total tax revenue between 1952 to 1957. The contribution of the different types of taxes has also varied over time. Revenue from foreign exchange certificates and import surcharges and, from 1957, the 20 percent tax on export certificates appears to be the most significant. During this period the government deficit also rose rapidly and led to inflationary pressure.

Several types of export inducement certificates,[8] and exchange taxes on export proceeds and on import payments, underlie the multiple exchange rate system that developed during this period. Under the various export inducement certificates, exporters[9] (except in the petroleum sector) surrendered all or part of their receipts and received marketable certificates in exchange. Private importers had to purchase export inducement certificates of an equivalent foreign exchange value and pay exchange taxes on imports on top of the official exchange rate. The export inducement certificate market was supposed to operate freely, but the government intervened frequently by issuing certificates, or more directly, by pegging the price of the certificates.

The exchange taxes on all export receipts were usually levied at a basic rate, while the exchange taxes on import payments, or import surcharges, varied according to the principle of 'necessity'. Other instruments used to earn revenue as well as to limit imports were a simple tariff schedule with two or three rates and import prepay-

Table 1.1 Government revenue from taxes on trade: Indonesia (Rupiahs, million and ( ) indicates percentage of total taxes on trade)

	1951	1952	1953	1954	1955	1956	1957	1958	1959	1960
Import duties	1,404	1,396	1,283	995	1,105	1,872	1,785	1,644	1,850	2,460
	(20.5)	(26.4)	(30.5)	(36.6)	(29.9)	(25.7)	(22.4)	(9.4)	(21.7)	(17.2)
Export duties (normal and "other")	1,309	1,819	1,013	544	752	424	199	185	170	450
	(19.2)	(34.2)	(24.1)	(20.0)	(20.3)	(5.8)	(2.5)	(2.2)	(2.0)	(3.1)
Foreign exchange certificates and	4,122	2,082	1,914	1,179	1,843	4,985	4,411	2,925	3,500	5,800
import surcharges	(60.3)	(39.3)	(45.5)	(43.4)	(49.8)	(68.5)	(55.3)	(34.6)	(41.1)	(40.5)
20% tax on B.E.s	—	—	—	—	—	—	1,587	3,706	3,000	5,600
							(19.9)	(43.8)	(35.2)	(39.1)
Total taxes on trade	6,835	5,297	4,210	2,718	3,700	7,281	7,982	8,460	8,520	14,310
Total net revenue	10,303	9,684	9,707	8,467	10,307	15,733	16,873	19,656	24,163	44,039
Trade taxes as percentage of net revenue	66.3	54.7	43.4	32.1	35.9	46.3	47.3	43.0	35.3	32.5

Sources: Figures for 1951 to 1958 from *Annual Reports* of the Bank Indonesia.
Figures for 1952 differ from those given by Paauw (1960), *Financing Economic Development*, p. 434.
Figures for 1959 and 1960 are estimates only and come from *Nota Keuangan Negara*, 1960.
(This table is taken from Corden & Mackie (1962), Table 1).

ments. Import prepayments required the importer to advance a deposit equal to a certain percentage of the estimated rupiah cost of his imports with the bank.

As a consequence of the high import prices resulting from the import regulation system and owing also to political motivations, illegal barter trade increased substantially between 1957–1958 and probably reached a peak in the early 1960s. Barter trade was the practice of accepting payment for exports in terms of goods in order to avoid the various taxes and restrictions on trade and profit from the high import prices.

During parts of this period the government controlled imports of industrial raw materials to prevent speculative activities and delegated the task to national indigenous importers and distributors. The preference for national indigenous entrepeneurs was part of the *Benteng* or protection policy aimed at protecting Indonesian entrepeneurs. As a result, the allocation of import licences came under the discretion of officials. It was also widely known that many indigenous importers who received licences sold their licences to, or became the legitimate 'front' for, established importers.

The main results were that government control over raw materials, despite officially fixed prices, in fact increased speculation and market price fluctuations through manipulation of available stocks. This added to inflationary pressure from quantitative restrictions and continual government deficits. Direct controls over the allocation of scarce foreign exchange also led to corruption, smuggling and other malpractices.

The Guided Democracy period 1958–1965

The main objectives of trade policy during this period continued to be a combination of revenue raising and exchange control, but with a much stronger emphasis on the establishment of indigenous Indonesian control over all aspects of economic activity. The first major step in implementing this objective was the nationalisation of all Dutch enterprises in December 1957, including the five big trading houses.[10] The period began with a severe foreign exchange shortage,[11] and the recurring problems of rapid inflation due to the scarcity of goods and a government budget deficit.

A plethora of government controls followed the nationalisation move. First, in April 1959, government trading houses obtained monopoly rights to import nine categories of goods which comprised 75 percent of imports. Private importers could only import nonessential goods and the number of importers was restricted to 400.[12] Second, beginning in early 1960, the government also began to control the traditional retail distribution channels previously handled by Chinese and Indian traders. The main result was a disruption of the distribution of goods and a scarcity of most consumer goods.

The disruption of imports of raw materials, in turn, seriously affected the infant domestic industries—mainly the textile and kretek industries—that had been set up behind the protective effects of import licensing and multiple exchange rates. The government did not actively encourage the private sector and perceived state owned industries as forming the base for industrial development.

There were many changes of the basic instruments[13] of foreign trade policy already in use during this period. The main problem with the multiple exchange rate system in such an inflationary period was that the rates tended to move sharply with inflation and necessitated government intervention to peg the prices of foreign exchange certificates to prevent too rapid a depreciation of the rupiah. This problem explains the many changes in regulations and explains why the government did not allow the rupiah to float freely.

The reliance on foreign trade taxes as a source of revenue was still substantial during this period. In 1963 and 1964, about 70 percent of government revenue came from foreign trade taxes, especially import duties and surcharges.[14] The high percentage coming from the foreign trade sector compared with other years reflects the low rates of collection from other sources, such as income and corporation taxes, resulting from the economic disruption during that period.

By the end of the period, the country experienced high rates of inflation,[15] very low levels of trade and almost zero foreign exchange reserves, a domestic manufacturing sector operating at only an estimated 20–30 percent of capacity, and badly damaged economic and physical infrastructure. There was also widespread smuggling due to the overvalued exchange rate, caused by high inflation rates and unserviceable debts.

The New Order period, 1966—present

Up to 1970, the foreign trade system was relatively liberal compared to later years. A brief description of the period between 1966–1971 will serve as an introduction to a more detailed description of the protection system that evolved in the 1970s.

1966–1971

The main objectives of foreign trade and exchange policies underlying the multiple exchange rate system were to ration foreign exchange and to influence the level and composition of imports, as part of the economic stabilisation and rehabilitation program of the government. Protection of domestic industries was not a central issue at this stage. Exporters surrendered their receipts and received in exchange, net of the 10 percent ADO (Automatic Allocation of Foreign Exchange) tax, which was allocated to the outer islands[16] and net of the 15 percent foreign exchange tax on category A (major) exports,[17] *Bonus Ekspor* (BE) or Export Bonus Certificates equal to the government set check prices.[18] They also received the *Devisa Perlengkap (DP)* or Complementary Foreign Exchange equal to the excess of the actual f.o.b. price of the commodity over the check price or overprice.

For most of this period, each type of foreign exchange had its own free market since the BE rate applied to imports on the BE list (mostly essential goods) and the DP rate applied to all other imports. The government influenced the market BE rates by the selling of foreign exchange and foreign credit (Aid-BE) through the Central Bank at set rates. Since the Aid-BE rate applied to imports from the donor country under tied aid agreements, Aid-BE rates were usually set below the market BE rate to encourage imports from donor countries. The DP rates became increasingly higher than the BE rates over the period and thus the various exchange rates influenced the composition of imports.

Tariffs did not appear to be an important tool in influencing the volume and composition of imports. Other than the premium on the DP rate, the evaluation base rate and limitation on credit for imports also limited the flow of imports.[19] The evaluation base rate, which is the exchange rate used for customs valuation and was set at a different level from the other exchange rates, was revised upwards a few times during this period.[20]

As concern began to switch towards increasing domestic production, credit restrictions were relaxed and tariffs on some products were increased in 1968. Production improved, notably textiles, partly because of these policies and partly owing to increased foreign and domestic investment resulting from the 'open door' investment policy of the government.[21]

Despite the efforts by the government to increase revenues from taxes outside the foreign trade sector, a substantial portion of government revenue continued to come from the foreign trade sector. From Table 1.2, it can be seen that in 1968, 38 percent of government domestic revenue originated from taxes on international trade, and import duties were by far the most important trade tax category.

The rice crises of late 1967 led to large increases in prices and precipitated an increase in the prices of both BE and DP exchange.

Table 1.2 Central government receipts—1968–1982: Indonesia (In billions of rupiahs and percentage of total domestic revenue in ( ))

	1968	1972/3	1975/6	1978/9	1981/2
Taxes on income	*51.0*	*301.1*	*1558.3*	*2996.3*	*10,100.3*
out of which:					
Corporate tax on oil	25.6	197.9	1205.2	2308.7	8627.8
	(17.1)	(33.8)	(54.8)	(54.1)	(70.6)
Taxes on domestic consumption	*36.7*	*125.9*	*234.4*	*491.4*	*888.0*
Taxes on international trade	*57.3*	*141.4*	*309.5*	*587.0*	*887.9*
	(38.2)	(24.2)	(14.1)	(13.8)	(7.3)
Import duties	37.3	76.8	175.1	295.3	536.2
	(24.9)	(13.1)	(8.0)	(6.9)	(4.4)
Import sales tax	6.1	29.9	73.4	125.5	223.3
	(4.1)	(5.1)	(3.3)	(2.9)	(1.8)
Export tax	13.9	34.7	61.0	166.2	128.4
	(9.3)	(5.9)	(2.8)	(3.9)	(1.0)
Non tax receipt	*4.8*	*16.7*	*98.6*	*191.4*	*336.4*
Domestic revenue	149.8	585.1	2200.8	4266.1	12,212.0
Development funds	n.a.[1]	*149.5*	*491.9*	*1035.5*	*1709.0*
TOTAL REVENUE		734.6	2692.7	6301.6	13,921.6

Note:
1 Foreign credit played an important role in balancing the budget in the first few years of the new order period. In 1967, foreign credits amounted to 36.5 percent of government revenue (Table IV-1 in Panglaykim & Thomas 1973)).

Source: 1968, from Table VII-1 in Panglaykim & Thomas (1973); 1972–1982, Ministry of Finance.

The government intervened[22] by stopping the sale of general BEs in July 1968 to encourage the sale of Aid-BEs which were then sold through the general BEs system, but sales of BEs had to be resumed three weeks later because of rapid increases in the DP rate. However, the spread between the BE and DP rates continued to be substantial owing primarily to speculation, and led to such malpractice as obtaining the BE exchange for fraudulent imports and selling it on the DP market at a higher rate.[23]

The multiple exchange rate system was finally abolished in 1970 and the exchange rate unified. Exporters had to surrender 100 percent of their receipts and foreign exchange and ADO taxes were replaced by a 10 percent export tax and direct rupiah subsidies to provincial governments. Henceforth, the government demonstrated its commitment to allowing the exchange rate to be more in line with the market value. The policies adopted were in part a reaction to the problems experienced with the multiple exchange rate system, which tended to exaggerate movements in the exchange rate and was open to abuse. The recovery of the economy from the post coup state of disrepair and hyperinflation coupled with the influence of Western trained economists in economic policy making, also contributed to the change in policy approach. In August 1971, pressure from balance of payments resurfaced and coupled with the floating of the US dollar, Indonesia devalued the rupiah by 10 percent to Rp 415 per US dollar.

1972—present

The main forms of protection in use are still tariffs or customs duties, import restrictions and import licences. However, in recent years many other instruments have come into existence under different government departments which have direct or indirect 'protective' effects. The revenue motive of foreign trade regulations is no longer important, as can be observed by the decline in the percentage contribution of taxes on international trade (see Table 1.2) between 1973–1982. It is clear that at present, revenue from taxes on oil companies provides a substantial proportion of total revenue.

Table 1.3 provides a list of the main policy instruments which are perceived to have protective effects. In what follows, a description of the main policy instruments is undertaken where the main details, changes over time, recent developments, the institutional set up of the policy, and the stated objectives are discussed. Examples of the goods affected are provided whenever possible. It should be noted that 'protection' has been interpreted in the broad sense to mean protection of the manufacturing sector in general as well as protection of certain types of, and certain groups involved in, the manufacturing activity. The priority sectors have changed over time, with import substituting, final goods industries being encouraged in the early 1970s and, more recently, export oriented and upstream industries. The economically weak sector, at least until 1984, usually refers to *pribumi* or indigenous entrepeneurs and/or small scale industries and cooperatives.

Tariffs

The basic tariff schedule can only be changed by the Parliament. However, surcharges and exemptions in addition to the basic scheduled rate are allowed after being evaluated by a 'team' comprised of representatives from the Departments of Industry, Trade and Finance, and, since 1983, the Junior Minister for the Promotion of the Use of Domestically Produced Goods (*Peningkatan Pendayagunaan Produksi Dalam Negeri*) or P3DN. Once the changes are agreed upon, they are formalised by a decree from the Minister of Finance. There have been three types of classifications of the scheduled rates up to the present time. These were the Geneva Nomenclature (1965–1973), the Brussels Tariff Nomenclature (1973–1980), and the

Table 1.3 Policy instruments having potential protective effects on domestic manufacturing: Indonesia

	Policy instruments	Principal decision makers
1	Customs duties	Department of Finance
2	Import sales taxes	Department of Finance
3	Domestic sales taxes	Department of Finance
4	MPO (withholding tax)	Department of Finance
5	Import prefinancing requirements	Bank Indonesia/Department of Finance
6	Prohibition or quota on certain imports/exports	Department of Trade on recommendation from Department of Industry or other relevant Departments
7	Government monopoly on certain products	Cabinet
8	Checkprices:	
	Exports	Department of Trade
	Imports	Customs Office/Department of Finance in coordination with Department of Industry & Trade
9	Export taxes	Department of Finance
10	Special export taxes	Department of Finance
11	Various levies on specific products (e.g., timber exports)	Individual Departments
12	Exchange rate	Cabinet/Bank Indonesia
13	Special exchange rates	Department of Finance
14	Export certificates:	
	Percentages	Department of Finance
	Checkprice	Department of Trade
15	Investment licensing	BKPM (Investment Coordination Board)/ Department of Industry
16	Special tax/customs duties facilities	Department of Finance on Recommendation from BKPM
17	Preferential credit policies	Bank Indonesia/Department of Finance
18	Compulsory use of domestic components by domestic industries	Department of Industry, BKPM
19	Compulsory use of domestic products in government projects[1]	Cabinet/Minister for the promotion of the domestically produced goods
20	Maximum wholesale prices for certain imported goods (billet/ ingot, newsprint, etc.)	Department of Trade
21	Counter trade policy	Department of Trade
22	Import licensing (Tata Niaga Impor)	Department of Trade

Note:
1 There is an informal rule of thumb that one should use domestic components if the price differential with respect to imports is only 15 percent.

Customs Cooperation Council Nomenclature (CCCN) from 1 April 1980. Under the CCCN the basic rates applicable are: 0, 5, 10, 25, 30, 40, 60, 90 and 100 percent. The structure is still cascading with the ranges of basic tariffs being 0–20 percent for basic raw materials, 20–40 percent for intermediate goods and 50–100 percent for finished goods. The range of tariffs is 0 to 270 percent after considering all changes, with the following structure: luxury and finished goods (40–270 percent), intermediate goods (15–30 percent), and capital goods and raw materials (0–10 percent).

One major change outside of the scheduled rates occurred in April 1979, in the wake of the November 1978 devaluation. The tariffs on one thousand goods and services were significantly reduced by 50 percent for most goods, and the number of imports subject to a specific tariff was increased by four hundred goods. The specific tariff had existed since 1973, but the number of goods affected was still small until 1979. The new specific tariff ranged from Rp 300/kg for concentrated orange juice to Rp 272,500/unit for video casette recorders, with many variations within the range. The aims of the policy were to reduce the consumption of imported goods in line with the austere lifestyle that the government was advocating, and to simplify and therefore increase the effectiveness of tariff revenue collecting since the value or price of the good was no longer a criterion in calculating the amount payable. Many of the goods experiencing a tariff reduction have subsequently experienced an increase.

Taxes and restrictions affecting imports

There are several regulations that can affect the domestic price of the imported good and therefore can be considered to have protective effects. Each will be considered briefly in the following sections.

The import sales tax and the Menghitung Pajak Orang (MPO) tax

The import sales tax is collected at the same time as the tariff. The base is the c.i.f. value of imports plus tariffs plus a 5 percent trade margin. The import sales tax has a protective effect since the import sales tax rate is usually higher than the domestic sales taxes on similar goods.

As part of the April 1979 regulations, one thousand types of goods experienced a decrease of 50 percent in the domestic sales tax and this change appears to have increased the harmonisation between the import and domestic sales taxes and, thus, there is a fall in the protective effect of the import sales tax.[24] The specific import sales tax, ranging from Rp. 321/unit for inner tyre tubes to Rp. 152,600/ unit for videos, was also extended in April 1979 for similar reasoning as that applied to the extensions in specific tariffs.

The *Menghitung Pajak Orang (MPO)* or withholding tax amounts to a prepayment of the corporate profit tax and has a protective structure, since the tax rate is usually higher for imports.[25] Pitt has pointed out that the MPO will only have a minor protective effect if it is seen as a prepayment of corporate profit taxes. However, if firms choose not to claim an MPO tax credit so that their profits appear lower,[26] the MPO tax can be considered like a sales tax.

The December 1983 tax reforms which came into effect on 1 January 1984, are aimed mainly at reducing the country's dependence on oil revenues by increasing government revenue from the non oil and non traded sectors. Thus, it is not expected to have significant effects on present protection policy except with regard to the introduction of the Value Added Tax (VAT)[27] and the treatment of the MPO tax. The replacement of the import sales tax and domestic sales tax by the VAT removes the protective effect of the difference between the two sales taxes. The base VAT rate is 10 percent on all manufactured goods, whether domestically produced or imported, and the basic rate can be changed by the government within the 5–30 percent range. There is also a separate sales or excise tax of 10 or 20 percent, depending on the category, on luxury goods as part of the government's austerity drive. In line with the promotion of non oil exports, exports are not subject to VAT and exports of luxuries are also not subject to the excise taxes.

The inclusion of the MPO tax in the *Pajak Penghasilan* (PPh) or income tax will remove the protective effects of the MPO tax being considered like a sales tax. Under the new PPh there is preferential treatment for licensed importers with the PPh for licensed importers equal to 2.5 percent of the c.i.f. value of imports, while for unlicensed importers it is 7.5 percent.

Import check prices

Since 1976, an interdepartmental team has been formed to set import check prices which represent the minimum price that can be declared at customs. As a result, this could mean an undervaluation of import statistics.[28] The stated aim of the check price system is to reduce negotiations regarding the valuation base upon which the duties are levied at customs. In this sense, by speeding up the clearing of goods through customs, it offsets in part the protection arising out of lengthy customs procedure.

In the past, check prices have also been used to change levels of protection without changing the tariff. An example of such a change is the 75 percent increase in the check prices on some textiles in 1975. This policy appears to be no longer in use, although occasional clamours for protection often include demands for an increase in check prices.[29] Some producers have also complained that the check prices of the products used as inputs in their production were increased at a more rapid rate than the price of the products in the country of origin. Other producers have reported discriminatory practices where the check price of a certain brand is not increased at the same rate as another brand of a similar good.[30] At present the check prices have not been used a protective device and the inconsistencies that have arisen probably only reflect slowness to respond and/or lack of coordination on the part of the team monitoring check prices.

Quantitative restrictions and licensing

Quantitative restrictions on imports can take place through outright bans and implicit or explicit quotas. There are at present five groups of commodities that are banned. These groups include printed matter, radio and television receiving sets, completely built up automobiles and motorcycles, as well as various types of drugs and pharmaceuticals. The bans are partly for political reasons (for instance, on some Chinese printed matter); are partly health or safety oriented (for instance, drugs or chemical products regarded as dangerous), and also partly to promote assembling industries in the electronic and automotive fields.

There are only a few explicit import quotas in the Indonesian trade system, such as import quotas on the import of foreign films and on the import of powdered milk by milk processing plants. The latter was aimed at protecting small scale dairy cooperatives producing fresh milk. However, implicit quotas do appear to exist through the new approved importers system called the *Tata Niaga Impor* (TNI), which was introduced in November 1982. Prior to this system, importers who obtained a licence from the Department of Trade were either general importers, who have to be Indonesian nationals and who could import most goods, or importer-producers who import the raw materials and intermediate goods necessary for use in their production. Under the TNI system, goods listed under the nine categories[31] can only be imported by approved importers, except for imports under the foreign and domestic investment laws, foreign aid agreements, government projects, service contracts and production sharing contracts.

There are two types of licences issued under the system. First, there are general licences under which an importer is approved to import goods falling under certain categories; an additional import licence is not necessary and the amount, type, and country of origin of the goods are not specified. However, the government can limit the number of approved importers to *bona fide* ones who satisfy certain requirements such as past performance and the hiring of skilled labour. The importers must report their realised imports every three to six months and the approval is granted for a two year period.

Second, there are discretionary or specific licences. Only approved importers can import certain goods and they are then eligible to apply for a licence to import the goods which may, or may not, be automatically approved. The type and/or amount of goods imported are specified and as such there is a built in quota under this system. It is also possible for the government to impose a '0' amount or not approve the import licence if it does not want any imports of the good. The aim of the policy seems to be to control both the amount and type of goods entering the domestic economy. The list of such goods and amount of imports allowed and sometimes the fixed selling price of the good (e.g. steel products) are determined after discussions between the interdepartmental team, business and industry groups. Examples of such goods in the manufacturing sector are Kraft liner paper, piston rings, automotive components, milk products, steel, white cement, inner tyre tubes, some dyes, heavy equipment, PVC, motor vehicle tyres, threshers, hand sprayers, batiks, textiles, and glass sheets.

In theory, a domestic industry group which can show that it can produce the good in sufficient quantity, of acceptable quality and at a price competitive with imports, can petition for a limitation of competing imports through the interdepartmental team.[32] The interdepartmental team can consider import restrictions if the customs duty does not exceed 20 percent, agreement between consumers and producers is reached, and the restrictions are of a temporary nature and can be lifted any time when there are supply shortages or price instability.

A final point is that the import of certain goods is delegated to state owned enterprises for various reasons. Prior to the TNI system, imports of cement, fertilisers, and some processed agricultural goods, such as wheat flour and cooking oil, were already on this list. The main motivation behind the policy appears to be price stabilisation of strategic commodites by assurance of adequate supplies. Another reason appears to be the protection of government owned corporations producing cement, flour, and fertilisers and, more recently, steel, Kraft paper, tyres, and scrap metal. The reasons for the addition appear to be to protect some domestic industries which are considered already capable of satisfying domestic demand, to protect state owned corporations producing the goods, and to conserve foreign exchange. A related policy is the appointment of state trading corporations to import machinery on behalf of small scale industries, and they are also obliged to give technical and economic assistance to these industries.

Discretionary changes in regulations affecting imports

Changes in the scheduled tariffs, import sales taxes, MPO taxes, and other regulations affecting imports occur periodically at the discretion of different government departments, presumably after discussions with other relevant departments. A cursory survey[33] of such changes for the 1981–1983 period reveals several motives for discretionary changes, not including the investment incentives under BKPM: to stabilise prices by ensuring adequate domestic supplies of certain essential goods, to subsidise industrial inputs used in priority sectors, to sudsidise imports by government enterprises and the economically weak groups, to limit imports of goods that can already be produced domestically, to earn revenue, and to conserve foreign exchange. An example of import subsidisation is the 50 percent exemption of tariffs and import sales taxes on imports of equipment used by national contractors to enable them to compete with foreign contractors. The subsidisation of imported industrial inputs, through partial exemptions on tariffs and taxes for various industries—especially various domestic components industries—and the increase of tariffs and import sales taxes or the elimination of exemptions for goods that can already be produced domestically, appear to have occurred most frequently.

Export policy

The protection of domestic producers under an import substitution policy has an inherent bias against exportable industries by the lack of protection, if there are no accompanying export subsidies, or through outright taxation. This is reflected in the higher price of imports of raw materials and inputs caused by protection policies borne to some extent by the producers of exportable goods. Thus, the higher cost has the same result as a tax on exports. Export promotion policies potentially can offset some of these negative effects, while any policy that limits exports will increase the bias.

Export taxes and subsidies

Between 1969–1976 the export tax was a flat rate of 10 percent on most nonoil exports. In addition, there were specific taxes (*cesses*) on certain goods such as leather, and regional levies on most agricultural exports. The additional taxes were eliminated in 1976 and, in line with the promotion of manufactured exports, as well as to aid traditional exports, the export tax was reduced for many goods. In 1981 export taxes were either eliminated or further reduced for a number of goods such as rubber, tea, coffee, pepper, and sawn wood.

Since then, export taxes have remained low except for additional export taxes that can be levied by the Departments of Finance and Trade (since 1978), as long as the sum of the export and additional export tax does not exceed 50 percent. Additional taxes were usually used to restrict exports when the government was concerned with adequate domestic supply, such as with palm oil and coconut oil, and to earn revenue from 'strong' exports. Export taxes, by increasing domestic consumption, can also be interpreted as a type of subsidy to domestic industries using the exportables as raw materials, for example, as cloves and lumber which are inputs into the cigarette and plywood industries.[34]

Export promotion policies

Other than the export tax reduction, the 1976 export promotion package comprised: a reduction of the interest rate on credit for exporters and producers of exportables from 25 to 12 percent, the issuing of regional export licences, and the lowering of warehouse and handling fees. More recently, the 1982 export promotion package comprised: removal or substantial reduction of export taxes, the lowering of the interest rate on credit for export related activities to 6 percent for 'weak' exports and to 9 percent for 'strong' exports, introduction of an export credit guarantee, and insurance schemes, and the introduction of the counter purchase agreements. Under the counter purchase agreement, foreign contractors involved in government sponsored contracts must buy or arrange to buy Indonesian exports of equal value to the imports brought into Indonesia for the projects. In January 1984, the value of exports realised under the counter purchase agreement was US $238.5 million, out of which the industrial goods were: crumb rubber, plywood, sawmilled wood, aluminium ingots, garments and textiles.[35]

One other major policy change outside of the export package needs to be mentioned—the export certificate or *Sertifikat Ekspor* (SE) system which, substituted for the duty drawback scheme, was introduced in 1978. The main purpose of this scheme was to refund to producer-exporters the tariffs, sales taxes, and MPO taxes paid on imports of raw materials and other imports used in producing the product exported. It has since developed to become more of an export incentive system. The SE percentage is set by the Department of Finance on recommendation from the Department of Industry, which in turn bases the percentage on reports by the firm requesting the SE and on its own estimates. The percentage is supposed to reflect the average levies on imported inputs used. The Department of Trade is assigned the task of setting the applicable export check price. The Department of Finance implements the policy from the inspection point to the rebate point. At the present time about 1800 goods, out of which 1500 were textile goods, have been granted SE. The percentage range of the SE is around 0.5 percent for canned vegetables to 20 percent for pyjamas.

The two main problems that have been associated with the scheme were, first, that the tendency to use imported inputs was increased, since using similarly higher priced domestically produced inputs did not entail obtaining a rebate. The other problem associated with the scheme was of cases[36] of export falsification undertaken in order to earn the rebate. To rectify this, the inspection of exports to certain destinations has been tightened. There are also the recurrent problems associated with lengthy procedures in obtaining the SE.

The 1982 package also included the abolition of the requirement by exporters to sell their foreign exchange proceeds to Bank Indonesia. They can now use the foreign exchange to buy raw materials or capital goods used in the production of their exports or sell the foreign exchange to foreign exchange banks, importers, or foreign exchange dealers. Importers can buy foreign exchange from foreign exchange banks, exporters, or other sources and not just from Bank Indonesia.

A more recent policy to promote exports of processed goods is that of quantitative controls or bans on exports of raw materials, as will be discussed in the next section.

Export restrictions and division of export quotas

The amount of exports leaving the country is often regulated by domestic policy or by quotas from the country of destination. Restrictions arising out of domestic policy are usually motivated by concern to ensure sufficient domestic supply, such as with palm oil and cement, or to encourage the processing of raw materials, as in the case of timber and rattan. Exports of logs have been restricted since 1981 and the plan is to phase out all log exports by 1985. The aim is to encourage investment in the domestic timber processing industry and also to conserve forestry resources. A certain percentage of logs is supposed to be channelled to domestic timber processing industries. The recent growth of the plywood industry appears to have been largely due to the quantitative restrictions on the export of logs.[37]

The allocation of quotas imposed by countries of destination of the exports often includes provisions for the participation of new exporters. For instance, in the allocation of garment export quotas, 85 percent is allocated to existing exporters and the rest to new exporters to those countries. Promotion of indigenous exporters has not been a dominant motive in this system, accounting for only 5 percent of the total quota allocation.

Investment incentives

The investment incentives given by the government through the *Badan Koordinasi Penanaman Modal* (BKPM) or Investment Coordination Board have protective effects for several reasons. First, there are direct protective effects because of partial or full exemptions of tariffs and import sales taxes on imports posted on the BKPM master list to approved foreign and domestic investors.

In the early 1970s when import substitution in final consumer goods was the main thrust of industrialisation, a substantial proportion of inputs used was still imported and with the exemptions in force, the final goods industries enjoyed high effective rates of protection since there were already high tariffs and other restrictions on their final products. However, in the last few years, as the focus of industrialisation shifted upstream to the production of inputs, the 'negative' list, comprising goods deleted from the master list, was introduced in 1981. Goods on the negative list were goods deemed to be satisfactorily produced domestically, as determined by the Department of Industry, BKPM, and the Minister For The Promotion of The Use of Domestically Produced Goods.

Another investment incentive concerns the protection of certain groups in the economy by the reservation of certain activities and/or by the granting of tax holidays and other incentives, for groups such as the 'weak' economic sector or cooperatives, and domestic national indigenous operations. Traditional sectors, such as meat and poultry processing, are reserved for the weak economic sector. Furthermore, new investments over U.S. $1 million in sectors not totally reserved for the weak economic sector or cooperatives, need to have a minimum of 10 percent of the investment in the hands of the weak economic sector or cooperatives. A company is considered to be part of the weak economic sector if at least 50 percent of the capital is owned by *pribumi*, the majority of the Board of Directors are *pribumi*, and the size of the investment is less than Rp. 100 million in the industrial and construction sector.[38]

Some activities, such as yarn spinning and canvas weaving, are reserved for domestic national indigenous operations. Domestic national indigenous operations are those with more than 50 percent of the equity in the hands of *pribumi* or a 50:50 share of *pribumi*: non *pribumi* in the equity, with most of the management in *pribumi* hands.

The third investment incentive involves the approval of domestic and foreign investment by type and location of industry. Approval of investment now carries requirements of the percentage of domestic components to be used. For example, in 1982/83, investment approval in diesel engine and machinery and equipment manufacture depends on 15–30 percent and 10–30 percent use of domestic components respectively. In the case of motor vehicle spare parts manufacture (except for engines), 100 percent use of domestic components must be achieved by 1985. Other types of industries encouraged are export-oriented, labor intensive and basic necessity industries. To satisfy the regional distribution objective, locations outside Java are preferred.

The *Daftar Skala Prioritas (DSP)* or Investment Priority List (DSP),[39] for both domestic and foreign investment, which has been published since 1977, considers the above factors of the group involved, the types of investment, and the location of the investment, in preparing the priority scale. The DSP has protective effects by promoting the development of certain priority industries and groups.

The use of domestic components

The present five year plan (*Repelita* IV) calls for the 'deepening' of the industrial base by developing upstream industries. Towards this end, protection of domestically produced components and other intermediate products has been increased by increasing tariffs and import taxes, either directly or through placing the good on the negative list, as well as by import restrictions under the TNI system. A more direct approach has been through the Deletion Programs which specify a timetable for the use of certain domestic components by a certain deadline. Some examples of industries subject to the Deletion Program are the assembling of motor vehicles, motorcycles, scooters, diesel motors, and tractors. Most of the Deletion Programs were introduced in the last two years, except that for noncommercial motor vehicles, which was introduced in 1977.

Other discretionary measures exist, usually by specifying the suggested percentage of domestic components to be used. For instance, to protect the domestic dairy cooperatives, the Department of Trade rules that milk processing plants can import powdered milk only in certain proportions. The stipulated ratio of domestic fresh milk to imported powdered milk products was 1:7 in 1982, 1:5 in 1983 and 1:4 in 1984, with plans of increasing the ratio in favor of more fresh milk. There is also an informal rule of thumb that one should use domestic components if the price differential with respect to imports is not more than 15 percent. Furthermore, according to the Presidential Decree no. 29/1984, it is compulsory to use domestic products whenever possible in government projects.

Preferential credit policies

To reduce inflationary pressures from the monetisation of oil revenues, Bank Indonesia resorted to using special credit schemes and credit ceilings since April 1974 to control the expansion of total credit given by the state banks. As with investment incentives, credit has been allocated by priority sector, social and economic category of the recipient, and size of the company. The size of the credit ceilings obtained by lending banks can be affected by to whom and to what sector they intend to lend. However, credit ceilings were abolished in June 1983 along with liberalisation of interest rates of state banks and a reduction in liquidity credits/rediscounting facilities.

There are also several special credit schemes with preferential lending and rediscount rates for small scale enterprises, medium and large scale enterprises, and the economically weak sector and weak exports.[40] State banks also have to lend to government enterprises, such as Krakatau Steel, and provide guaranteed credit to large government projects.

Policies regarding the type of letter of credit (L/C) allowed for different types of imports and exports, and the amount of deposit required on opening the L/C, can influence the liquidity of importers and exporters, and in turn result in different protective effects. For example, in 1980 the deposit requirement for imports of specified raw materials, intermediate goods, and spare parts using the sight L/C and for imports under the usance L/C, was determined by the foreign exchange bank, whereas previously it was fixed at 40 percent.[41]

The exchange rate

The exchange rate has been used to protect the import competing and non oil export sectors, which experienced pressure from the oil boom. The rupiah was devalued by 50 percent in November, 1978 to Rp. 625/US dollar and allowed to float.[42] In March, 1983 the rupiah was devalued again to Rp. 970/US dollar. The 1978 devaluation was in part protectionist, although concern about dwindling reserves and expectations of a fall in the price of oil in 1979 also played a role. The second devaluation was more in reaction to falling oil prices.

The pressure from the oil boom can be explained briefly as follows. In the case of Indonesia, an increase in the world price of oil translates itself into increased foreign exchange reserves or the foreign component of government revenue, owing to an increase in revenue from oil company taxes. The increased revenue enters the domestic economy when the government uses the increased revenue to buy domestic or non traded goods. With fixed exchange rates and given world prices, the price of traded goods is exogeneous. The price of non traded goods, on the other hand, is determined by domestic supply and demand conditions. Given that both types of goods are normal goods, the excess demand for non traded goods induces an increase in the price of non traded goods relative to the price of traded goods. This will tend to draw resources out of the traded goods sector towards the non traded goods sector.[43]

The change in relative prices appears to have occurred in Indonesia, prior to the 1978 devaluation, as indicated by the rise in various measures of the relative price of non traded goods to traded goods.[44] This price change played an important part in the policy discussions leading to the 1978 devaluation, which raised the relative price of traded goods. The main justification given for protecting the non oil traded goods sector by this policy is that the oil resources are depletable and, thus, the non oil traded sector should not be allowed to deteriorate in the short run, since the sector will be required to expand when the oil resources run out.

Other policies and practices

Various other policies and practices have protective effects. The more important ones are: first, the increase in the cost of imports due to delays caused by customs clearance and other administration procedures, as well as to invisible payments. In the case of imports from Japan, Hong Kong, Singapore, and Penang there is an additional transportation charge, since the goods have to be stored in specified warehouses for special inspection and the warehouses are located away from the ports. Since 1980 there have been some efforts to reduce the regulations and bureaucracy involved at customs so that, as a result, the number of days a ship remains in port for clearance purposes at Tanjung Priok has supposedly fallen to two days compared with five days in 1979.[45]

Second, illegal as well as administrative smuggling[46] of goods subject to duties or quantitative restrictions implies a fall in effective protection. Another policy is the requirement for inspection of goods by the government corporation, P.T. Sucofindo, whereby the importers have to pay the inspection fee.

A recent policy is to save foreign exchange by prohibiting imports through a third country. Thus, imports from Japan can no longer come through Singapore and Malaysia and it appears that the cost of imports is, in fact, higher.

Government subsidies to particular sectors play an important role in increasing the effective protection of subsidised sectors. Examples of such sectors are fertilisers, sugar refining, rice milling and polishing and other petroleum and coal products. The subsidy on domestic oil consumption benefits all sectors with some variation, depending on which type of fuel is used by the sector, since the subsidies vary according to the type of fuel. In recent years due to the fall in revenues from oil, government subsidies have been falling, especially for domestic fuel consumption. The averge price of domestic fuel has been continually increased since 1981.

Finally, government procurement policies provide protection to the industries affected and the groups favoured. Under Presidential decrees 14/1979 and 14A/1979, *pribumi* entrepreneurs received preferential treatment in bidding for government purchases or projects. For contracts less than Rp. 20 million, entrepreneurs from the weak economic sector can be appointed directly without a tender. In April 1984, Presidential decrees 29 and 30 overrode the above decrees and the term *pribumi* was replaced by *Indonesia asli* or Indonesian national.

The trend in protection

The aim of this section is to look at the trends in the level and structure of protection in the manufacturing sector in Indonesia in the New Order period. Throughout the section, an attempt will be made to relate any changes or trends in the protection level or structure with any changes in the system of protection discussed in the previous section.

The level of protection from tariffs and import sales taxes

Two major forms of protection are tariffs and import sales taxes. Table 1.4 indicates the *average* burden that these levies had on imports during 1969–1981. On average, tariffs are more than twice as important as import sales taxes in producing the wedge between domestic and international prices. The two levies together represent one sixth of the c.i.f. value of imports.

The table also indicates that over the period, the average collection rate had in fact declined from about 25 percent to 10 percent, except between 1975/76 and 1978/79. This fact, however, should not be taken to mean that protection from the two levies to the domestic

Table1.4 Revenue from import duties and import sales tax and average collection rate 1968–1981: Indonesia (Rp billion)

Year[a] (1)	Import duties (2)	Import sales tax (3)	(2)+(3) (4)	Import value[b] (5)	Average collection rate (%) (6)
1969/70	57.7	15.9	73.6	291.0	25.3
1970/71	70.7	22.1	92.6	373.0	24.8
1971/72	69.4	22.4	91.8	449.2	20.4
1972/73	73.2	27.8	101.0	635.5	15.9
1973/74	128.2	50.7	178.9	1,114.4	16.1
1974/75	160.6	68.9	229.5	1,518.4	15.1
1975/76	174.0	72.5	246.5	1,874.3	13.2
1976/77	257.4	102.2	359.6	2,172.7	16.6
1977/78	286.9	114.6	401.5	2,281.8	17.6
1978/79	295.3	125.5	420.8	2,254.8	18.7
1979/80	316.7	137.2	543.9	4,005.6	11.3
1980/81	448.0	195.1	643.1	5,681.5	11.3
1981/82	536.2	223.3	759.5	7,219.3	10.5

Notes:
a Fiscal year April 1 19(t) to March 31 19(t + 1).
b Calendar year non oil imports c.i.f. converted to rupiah value by using prevailing exchange rates.
Source: Biro Pusat Statistik.

manufacturing sector had declined. Further examination indicates that most of the drop is attributable to the increasing share of low tariff imports, which do not directly compete with domestic activities. The share of capital goods in total imports has risen from 33 percent between 1969–1972 to 41.2 percent between 1979–1981, while that of consumption goods has fallen from 29.6 percent to 21.7 percent.[47] The result reflects the policy of import substitution of final goods. Examination of tariff collection rates for manufactured goods listed in the input-output tables also shows that collection rates for a variety of imports directly competing with domestic activities have risen between 1971–1978.[48] Furthermore, since most of the low tariff imports of intermediate and capital goods were in fact used by those very activities, one is tempted to conclude that the average level of effective protection for such activities has risen.

The structure of protection

In order to analyse the structure of protection over the past decade, estimates of nominal rates of protection (NRP) and effective rates of protection (ERP) by Pitt (1981), the World Bank (1981), and ourselves will be used.

Measurement of protection is no easy task since the effects of many of the protection instruments, except for tariffs and import sales taxes, discussed in the previous section, are not directly measurable.

The ideal estimate of nominal protection in the small country case is to compare domestic wholesale prices with the c.i.f. import prices of the goods and then aggregate accordingly with the input-output classification to enable estimation of effective rates of protection. An even more accurate analysis of the structure of protection would necessitate protection measures on a detailed product by product basis for each industrial subsector and would involve collecting primary data from production outlets. A two year study of such a nature, aided by the Harvard International Institute for Development (HIID), appears to be in progress under the direction of the Department of Finance.

Given the limitations of the data and time that would be required to conduct necessary price estimates and meaningful aggregation exercises, we were only able to calculate effective protection rates based on tariffs and import sales taxes for 1975 and 1980. The Pitt and World Bank studies attempt to carry out estimates for some products by using direct price comparisons, although it is not clear what prices were used or how the various products under each input-output classification were aggregated. Nevertheless, we feel it is still possible to draw a broad and approximate picture of the structure of protection using the available estimates.

The World Bank and Pitt Studies: 1971, 1975 and 1978

The World Bank and Pitt studies both used input-output tables to estimate effective rates of protection. The main factors included in the nominal protection estimates were tariffs, import sales taxes, MPO taxes, export taxes, direct price comparisons for goods subject to quantity restrictions, subsidy to petroleum production, and discriminatory price fixing. Some adjustments were made for imported inputs from the same sector, such as assembling operations, since imports of completely knocked down (CKD) packages are discounted.[49] The Corden formula was used to calculate effective rates of protection.

Owing to the different classifications used in the 1971 and 1975 input-output tables, estimates of NRPs and ERPs are not directly comparable. Comparisons subject to this limitation indicate that the overall protection level provided to manufacturing tradeable goods in 1975 was 29.7 percent, compared with 33 percent in 1971 (see Table 1.5). Exportables in both periods received negative average protection, although it was a lower rate in absolute terms in 1975, owing to the increase in ERP experienced by most exportable sectors. However, in both periods, 11 out of the 15 exportable sectors had negative ERP. Thus, the bias against exports inherent in the trade system still existed in 1975.

Table 1.5 Distribution of effective rates of protection in Indonesian manufacturing, 1971 and 1975

	Effective rates of protection (%)		
	1971	1975	
		(1)	(2)
All tradeables	33	29.7	39.2
Exportables	−11	−2.4	−6.4
Importables	65	57.6	97.6
Import competing	69	61.0	108.6
Noncompeting	15	9.2	9.2

Notes:
(1) Includes rice milling and sugar refining.
(2) Excludes rice milling and sugar refining.
Source: Table 5.11 in Pitt (1981) and Table 3 in *World Bank* (1981), Annex 2. (See Appendix 1 for trade classification of goods)

The trade system favoured the importables sector, especially the import competing sectors, in both periods, with the overall average ERP of 65 percent in 1971 and 58 percent in 1975. If the major agricultural processing sectors, rice milling and sugar refining, were excluded, the average ERP for importables in 1975 was even higher at 98 percent and for import competing sectors 109 percent. In both periods the protection of import competing sectors was much higher on average than that for noncompeting import sectors.

Furthermore, in Indonesia, as with many developing countries, the structure of protection is cascading, with durable consumer goods receiving the highest ERP followed by intermediate, capital, and nondurable consumer goods.[50] The cascading structure is indicative of the different incentives given to each major industry group as part of the industrialisation strategy of the government. The overall average ERP has also declined between 1971 and 1975. This trend is consistent with our earlier observation that the average collection rates were falling over the same period.

The ranges of ERP in both periods were also very wide. In 1971 the range was from 38 percent for Batik industries to 5400 percent for canning and preserving of fruits and vegetables. In 1975 the range was from 35 percent for Batik industries to 4315 percent for tyre and tube industries.

The World Bank study also updated ERP for 1978 for eleven manufacturing sectors using the production structure from 1975 input-output tables. They note that such a comparison between 1975 and 1978 would reflect the changes in the trade system only, since the production structure is taken to be constant. The estimates show that there was an increase in ERP of seven sectors between 1975 and pre devaluation 1978. The ERP of the four sectors which experienced a fall in ERP between the two periods still remained above 100 percent. Based on this small sample, there appears to be a reduction in the variance of ERP especially, owing to the significant fall in the ERP of the tyres and tubes industry. The World Bank study concluded, based on this small sample and on evidence from increased collection rates of custom duty and import sales tax (as discussed in the above section) on the level of protection, that protection was greater in pre devaluation 1978 than in 1975. This conclusion is strengthened by the fact that since 1975, as pointed out in the previous section, credit, import prepayment, and MPO regulations have also become more protective. The data on import duty and sales tax collection also indicated that the collection rates on simple, labour complementing capital goods imports, such as agricultural tools (87 percent), were relatively higher than on more sophisticated capital goods, such as industrial electrical machinery (17 percent). This pattern could be attributed to the tendency for exemptions on the import duties and sales taxes on sophisticated capital goods used under the BKPM investment incentives.

Nominal and effective rates of protection: 1975 and 1980

Our estimates for 1975 and 1980 are based on the production structure of the 1975 input-output table and import duty and sales tax collection data for 1975 and 1980.[51] Thus, the estimations isolate the changes in protection due to changes in import duties and sales taxes between 1975 and 1980, since the production structure is taken to be constant between the two periods. The import sales tax figures from the input-output tables net out domestic sales taxes so that the figures reflect the discriminatory nature of the import sales tax. A detailed table of the results is presented in Table 1.A1 in Appendix 1. Tables 1.6 and 1.7, presented here, show the average rates of protection for major industrial groups by trade classification.

There appear to be significant[52] differences between our estimates and the World Bank (1981) estimates for 1975. On average, our estimates of the NRP and ERP of exportables are higher since we did not include the effect of export taxes. Our NRP estimates for the noncompeting import and import competing sectors were also higher on average, except for assembly industries, such as electrical appliances, motor vehicles, and motorcycles, where the World Bank estimates were higher since exemptions for the inputs or CKD packages used in those industries were included. The sectors which had noticeably higher NRPs were the agriculture processing industries such as rice milling and wheat flour, macaroni, cigarettes, and garments. The main reason for the higher NRPs in those sectors is probably due to price comparisons being used in the World Bank case

Table 1.6 Average effective rates of protection by major industrial groups, 1975 and 1980: Indonesia

	1975	1980
Overall (1)	*72*	*41*
(2)	*137*	*58*
Consumer goods (1)	*72*	*41*
(2)	*137*	*58*
Tyres & tubes	185	3
Motor vehicles	28	33
Dairy products	41	15
Drugs & medicines	85	0
Cigarettes	246	105
Rice milling, cleaning & polishing	354	−1
Sugar refinery	−8	0
Intermediate goods (1)	*43*	*22*
(2)	*81*	*61*
Plastics	117	70
Weaving industries	135	127
Spinning industries	30	16
Capital goods (1)	*4*	*2*
(2)	*8*	*2*
Non electrical machinery	16	15
Structural metal products	3	−1

Notes:
(1) simple average
(2) weighted average based on production
Source: Table A.1 in Appendix 1. Categorisation of major industrial groups follows the *World Bank* (1981), Annex 2, Table 4 and *BPS*, Statistik Industri, 1975, for production figures.

because those goods are subject to quantitative restrictions or government fixing of prices. The percentage of imports in domestic demand for those sectors is in fact very low, ranging from 0 to 8 percent.

As for the ERP estimates, our estimates are on average lower than the World Bank estimates. Our ERP estimates are probably lower because, on balance, the effect of the higher NRP of the good in question is more than offset by the higher NRP of the inputs used in the production of that good.

Bearing in mind that our estimates of protection are based on import duties and sales taxes only, the structure of protection can still be analysed as follows. The overall protection level did fall significantly between 1975 and 1980, as would be expected, owing to the reduction in tariffs and changes in import sales taxes that occurred in 1979. This fall is reflected in all the three subgroups of exportables, noncompeting import and competing import sectors, which is indicative of the widespread nature of the tariff reductions and import sales tax changes (See Table 1.6).

The ERP for the exportable sector was not negative on average since export taxes were not included. If export taxes and other export

Table 1.7 Average nominal and effective rates of protection, 1975 and 1980: Indonesia

		Nominal rates of protection		Effective rates of protection	
		1975	1980	1975	1980
Overall	(a)	48	27	67	33
	(b)	70	33	115	56
Exportables[c]	(a)	41	23	61	29
	(b)	56	27	91	32
Non competing import sectors[c]	(a)	20	8	17	7
	(b)	20	9	21	9
Import competing sectors[c]	(a)	56	32	80	39
	(b)	74	37	121	60

Notes:
a simple average.
b weighted average based on production figures.
c trade classifications defined in Appendix 1.
Source: Table 1.A1 in Appendix 1 and *BPS*, Statistik Industri, 1975, for production figures.

promotion policies were included, we feel that the ERP for this sector would have registered an increase between the two periods. The important changes that occurred between 1975 and 1980 that would potentially lead to such an increase were the reduction in export taxes in 1976 and the introduction of export certificates in 1978. Some partial offset to the increase could be expected from the export restrictions on logs and palm oil.

However, even based on import duties and sales taxes, the favouring of import competing sectors compared with the exportable sector is evident in both years (Table 1.7). The average rates of the import competing sector were above the overall average rate, while for the exportable sector they were below the overall average rate for both years. The lower ERP of the noncompeting import sector is as expected, since the goods in that sector are still used as inputs into other sectors.

In both years the import competing sector enjoyed the highest levels of NRP and ERP on average, reflecting a continuation of the import substitution industrialisation policy, although shifting from final goods to intermediate goods. As noted earlier, the level of protection for the import competing sector has, on average, fallen between the two periods. On closer observation (see Table 1.A1), it was found that while most industries experienced a significant fall in both NRP and ERP, some industries experienced little or no change or even an increase in NRP and ERP. The industries are: textile bleaching, printing and dyeing; paints and varnishes; nonmetallic mineral products; structural metal products and the assembly industries (which include components manufacture at this level of aggregation); electrical appliances, accumulators and dry batteries; motor vehicles and motorcycles. The promotion of intermediate goods is evident, although the cascading structure with both average and weighted average rates, ranging from consumer goods to intermediate goods to capital goods, still prevails , as shown in Table 1.6. In addition, the fall in the weighted average ERP of intermediate goods is less than for the overall weighted average, as well as for the weighted average of the other industrial groups.

There are many deficiencies to analysing the structure of protection based on just import duties and sales taxes, as we are not including the effects of import restrictions, government imports, subsidies and adjustments for imported inputs from the same sector, such as assembling operations. At best we can hypothesise whether policies from 1975 to the present can be expected to increase or decrease ERP for major subsectors. We will deal with this exercise in the following section.

The trend and structure of protection: qualitative assessment

Besides import duties, import sales taxes and export taxes, most of the government policies which have protective effects are difficult to assess quantitatively. Without detailed data and information, we can at best make qualitative assessments as to whether certain policies adopted would have changed our basic result of a fall in ERP between 1975 and 1980. We will also speculate on the trend of ERP from 1980 to the present time.

The two oil booms and the 1978 devaluation occurred in the 1975–1980 period, which is the ideal background under which to reduce trade barriers, as indicative of the 1979 tariff reductions. There are indications that the government intended to reduce excessive protection by undertaking broad tariff reductions and unifying the structure, and the 1979 action could have been the first step. Besides discretionary changes and the export promotion policies in 1976, which have already been discussed, no other major policy change would lead us to expect a substantial reversal of the downward trend in ERP during the period. The only notable development is the increased protection of *pribumi* by giving them preferential treatment under the presidential decree controlling government procurement in 1978 and 1980. However, these measures appear to have had a limited impact on the ERP. On the whole, the general downward trend of ERP during this period would probably still be true even if the effect of all other protective policies were included.

However, since 1980, several developments lead us to expect a visible increase in the ERP, particularly of intermediate and capital goods industries, relative to the ERP of final goods industries. The most important factor is the TNI system which has been more restrictive towards imports of intermediate and capital goods. The increased price of imports under the TNI system would result from the restrictions on imports as well as from the increased administration procedure. In addition, the domestic content policy, by direct restrictions on the use of imported inputs or through the BKPM negative list or by moral suasion, will provide substantial protection to the domestic components industries, especially in the case of motor vehicle components and industrial machinery.

The increased protection and thus increased prices of intermediate and, eventually, capital goods will also imply a fall in the effective protection of finished goods production or assembly using the now protected inputs. This would mean an increase in the cost of production of these goods, some of which have export potential. The continuation of the SE and other export promotion policies will offset some of the reduced protection towards potential final goods exportables. However, it is difficult to judge whether such moves are sufficient to enable Indonesian manufacturers to compete in the world market.

Several other developments need also to be considered. The 1983 devaluation has, of course, increased prices received by producers of tradeable goods. Also, in the past, complicated procedures underlying government regulations partially offset the protection given to domestic producers. As such, the planned simplifications of government regulations will remove this offset on protection. Finally the substitution of the *pribumi* term by *Indonesia asli*, or Indonesian national, in regulations governing government procurement will eliminate part of the preferential treatment received by this group.

Causes of protection

A host of factors can be attributed to the evolution and formation of the Indonesian trade system. The prevailing broad structure of protection and instruments used can be explained by economic and politico-institutional factors. We also examine the role of international agencies. However, for an explanation of the variation between industries we turn to the political economy explanation of protection.

General explanations

In Indonesia, the general structure of protection is supposed to reflect the government objectives under the various five year plans or

Repelita. The main emphasis of *Repelita I* (1969/70–1973/74) was to provide basic necessities (*sandang dan pangań* or clothing and food), with the industrial sector seen as complementing the agricultural sector. Thus, in the beginning of the New Order period protection was provided to consumption goods industries, such as the textile and food processing industries. Beginning in *Repelita II* (1973/74–1978/79) besides development towards processing of raw materials, intermediate, and finally capital goods production, there was also a shift of emphasis to the more social objectives. In *Repelita II* among the most important social objectives were those of creating employment, achieving an equitable distribution of income and increased participation of the week economic sector. In *Repelita III* (1978/79–1983/84) the emphasis on social objectives was even more apparent. Aside from those emphasised in *Repelita II*, the notable changes were the Presidential Decrees no. 14/1979 and no. 14A/1980, which give preferential treatment to *pribumi* entrepreneurs in bidding for government purchases or projects introduced in the beginning of *Repelita III*. The promotion of nonoil exports and the objectives of speeding up development of industries producing intermediate products and 'upstream industries', were more highlighted during *Repelita III*.

While in *Repelita I* the protection structure apparently reflected the relatively clear cut objective of increasing the production of basic necessities, it seems that beginning in *Repelita II*, the correspondence between the prevailing protection structure and the stated objectives has become less defined. The reason for this development is that the objectives themselves have become much more complex, with one objective often conflicting with other objectives. For instance, the objective of developing the upstream industries needs to be reconciled with the employment objective and the promotion of nonoil exports. The objective to speed up the growth of the manufacturing sector in general needs to take account of the objective to promote the economically weak groups and the equity goals.

The result is the wide and rather unsystematic variations in ERP across industrial activities. The government is well aware of these problems and in an effort to streamline industrial structure towards the stated aims there seems to be an increasing reliance on nontariff instruments. The system that existed up to two years ago afforded many loopholes and did lead to evasions. The government seems to have responded by using more direct measures such as domestic component requirements and the TNI system.

The objectives of the TNI system were to increase policing over dumping, to foster specialisation in importing and thus prevent speculation and 'follow the leader' type of imports; to control the flow of imports by mandatory reporting and, at times, specification of the quantity allowed to be imported. Such direct measures to control imports are easier to achieve than basic tariff rate changes, which have to be passed by Parliament. Furthermore, quotas implicit in the TNI system are not as 'visible' as explicit quotas.

The greater reliance on more direct measures appears to be partly related to the worsening balance of payments situation. By 1983 the balance of payments situation was precarious. The need to resort to extensive direct controls was avoided by successful implementation of a series of measures to reduce imports, most pertinent of which were the postponement of the implementation of large import intensive government projects and a currency devaluation in March 1983. While the TNI system was also directed at foreign exchange conservation, its main objectives seem to be to promote an orderly trading system and industrial structure.

Protection of state owned enterprises and private sector pressure have also been important factors in forming protection policy. The formal process of obtaining increased protection, whether increased tariff protection, bans or other restrictions on imports, is to petition the interdepartmental team. Petitioners must show that their firms can satisfy domestic demand at a reasonable price and that their increased production would satisfy one or more of the objectives of government economic strategy: saving foreign exchange and increasing employment. It appears that in the cases of the recent bans, such as of canvas fabrics and corrugated boxes, the restrictions were given after the Department of Industry was satisfied that the domestic plants could meet domestic demand at reasonable prices. However, as occurs everywhere to a certain degree, there appears to be 'indirect' lobbying through channels close to the people in charge of policy making by those who have a vested interest in receiving protection. It is, of course, difficult to measure the degree to which this factor influences the structure of protection. We feel that this factor probably had some sporadic influence, but the broad structure of protection results from the general objectives of the government's development strategy. Variation in protection can be attributed to lobbying, as discussed in the next section.

As noted earlier, state owned enterprises comprise a sizeable portion of the modern industrial sector, especially in rubber and palm oil estates, yarn, steel, and other natural resource based industries. It is often stated that such state owned enterprises have a dual function as a business concern and as agents of government policy. In practice, these two functions often conflict and pose difficult management and operational problems. However, because of their dual function, the state enterprises enjoy a unique position in economic policy. The hypothesis is that such enterprises are in a better position to obtain protection whenever it is deemed necessary.

The revenue raising motive of the trade system which was important in the Old Order period and in the beginning of the New Order period, is not an important motive any longer owing to the reliance on corporate tax from the oil sector. Table 1.2 showed that 17 percent of domestic revenue came from international trade taxes in 1981/82, out of which 4 percent was from tariffs. In 1968 tariffs accounted for 25 percent of domestic revenue.

The income distribution motive is also apparent in the design of some of the policies. For instance, imports on luxury goods are subject to higher levies, and the regulations, as well as levies on sedans (CKD) are stricter and higher than for commercial vehicles. These policies are aimed at taxing luxury goods consumption by the upper income groups. Of course, the income distribution motive also underlies policies to increase the participation of the weak economic sector.

In many countries, international agencies such as the World Bank and the (International Monetary Fund) IMF have been influential in the liberalisation of their trade regimes. In Indonesia, 'since the early 1970s the (World) bank has been accused of pressing Indonesia into following its advice ... to nudge Indonesia away from increased state intervention towards a free market economy.'[53] It appears that the technocrats in charge of economic policy are partly sympathetic to the liberalisation line; however, in the actual formulation of policies, such advice is always modified to be more in line with Indonesian political and social realities. Indonesian nationalism, which leads to policies to increase Indonesian participation at all levels of economic activity and which means sensitivity to any control by international institutions over its economic policies, is a strong force with which to contend. Glassburner notes that 'the assertion that they have allowed themselves to be dictated to by foreign advisors, most notably the IMF, AID, IBRD and the Harvard Advisory Group, is quite at variance with the facts. Such advice has been sought but, more frequently than not, ignored.'[54] As such, it appears that the international institutions did not play a major role in designing the trade regime in Indonesia, although the various studies undertaken by the Harvard Advisory Group (now HIID) in conjunction with several government departments, have had some influence on policy making.

For instance, there were some moves in the beginning of 1981 to reduce tariffs and import sales taxes on a substantial number of commodites produced domestically.[55] The reason given by the Department of Trade for the reduction was that it was in accordance with the GATT and Tokyo round negotiations. Tariffs and import sales taxes were to be reduced by 50 percent with, for example, such results: cotton yarn was to be reduced from a 60 percent tariff and 10 percent import sales tax to a 35 percent tariff; batik from a 100 percent tariff to a 50 percent tariff, and detergent soap from a 50 percent tariff and 20 percent import sales tax to a 25 percent tariff and 10 percent sales tax. It is not clear whether the reductions were a result of the World Bank Report[56] that came out around the same time and recommended a reduction as well as simplification of the protection system (based on the already discussed estimates of protection levels), or whether it was already planned prior to the report. Given the fortuitous economic conditions at the time, with no threat of falling oil prices, the reductions could plausibly have been planned prior to the World Bank report. It appears that the substantial reductions were thwarted by the subsequent oil glut and falling oil prices.

The political economy explanation

In recent years there have been many studies, theoretical and empirical, on the political economy of protection.[57] The main idea is that while the public or national interest in protection can explain the broad structure and average level of protection and its change over time, one has to look at the motives and actions of private interest groups to seek for an explanation of interindustry variation of protection. The government is the 'supplier' of protection and the import competing producers are 'demanders' of protection and will 'invest' in lobbying activities to obtain protection. There is likely to be opposition, though normally weak, from consumer groups, exporters, and firms who use the protected product as an intermediate input.

We attempt to explain interindustry variation of protection by estimating the reduced form of the political economy of protection model. Both our own and the World Bank NRP and ERP estimates by industry were regressed against several independent variables postulated by the political economy of protection literature (adjusted for the less developed country case) and using Ordinary Least Square (OLS). The variables, expected effects, and measures used are given in Table 1.8. In 1975 estimates were used, as the completed 1980 input-output (I-O) tables were not yet available and some of the independent variables used required data from the I-O tables. It should be stressed that our estimates represent simply the preliminary probe to a necessarily more comprehensive and detailed study of the political economy explanation of the determinants of protection.

A few points on lobbying in the case of a less developed country should be made at this stage. In developed countries, the number of workers in an industry can be thought a sign of strong lobbying power and also as a voting bloc to be wooed by politicians. In most less

Table 1.8 Explanation of variables used in the regressions[a]

1.	PTC (+)[b]	— Pitt's trade classification; importables are positive and exportables are negative (see Appendix 1).
2.	WVA (−)	— Wages as share of value added in output.
3.	VAO (?)	— Share of value added in output; negative because capital owners will be induced to lobby (i.e., demand) for protection to increase their incomes, but positive because the government will tend to want to protect (i.e., supply protection) sectors with a high value added.
4.	NF (−)	— Number of firms in the sector. Negative because a few firms in the industry can lobby (i.e., demand) for protection more effectively.
5.	NEMO (+)	— Number of employees in the sector. Positive because the government feels it should provide (i.e., supply) protection on equity grounds.
6.	AVW (−)	— Average wage per worker. Negative for the same reason as NEMO.
7.	EMPO (+)	— Number employed per Rp.lm output. Positive, as EMPO is a measure of labour intensity.
8.	OPF (+)	— Value of output (Rp.lm) per firm in the sector. Positive as it is a measure of concentration and is an indication of lobbying power (i.e., demand for protection).
9.	GR (−)	— Growth rate of output. Negative as producers can always justify a lower rate of growth as being caused by import competition and use that ground to obtain (i.e., demand) higher protection.
10.	IMP71 (+)	— Import share out of total demand (i.e. 401/310 in the 1971 I-O tables). Higher levels of past import penetration imply higher levels of present protection (i.e., supply) under an import substitution policy.
11.	IMP75 (−)	— Import share out of total demand for 1975. The lower the present import penetration ratios, the higher the level of protection expected (i.e., supplied) for these import substituting industries.

Note: a All of the data are for 1975 except for IMP71. All the data were taken from *BPS*, Statistik Industri, 1975 except for IMP71 and IMP75 which were taken from the I-O tables. Only 62 observations (before omission of one or two outliers) were used since many sectors in the I-O tables could not be corresponded with the SITC sectors in the industrial statistics.

b The expected sign of the coefficient is in brackets.

developed countries such as Indonesia, there is no organised trade union system under which to lobby for protection, and the voting aspect is also not as important. Lobbying in Indonesia takes the form of the special relationships and informal connections that exist between the political power centres and the various industries. This relationship is, of course, difficult to quantify as it necessitates a study of how the structure of ownership of particular industries is related to key people in power and, as such, was not included in our estimates. The only variable included as a rough proxy for lobbying is gross output per establishment, since the larger the gross output per establishment, the more likely it is that it has the 'resources' to foster connections with power centres and lobby for protection.

When all the explanatory variables were included, the t-statistics and F statistics were insignificant for all four dependent variables. Various regressions were then tried to include the more significant independent variables, and by looking at the residual plots, some outlier observations[58] were eliminated with the result of improving the R^2 of the regressions. It was found that using the World Bank-Pitt

NRP and ERP estimates as dependent variables yielded more significant results.

The following NRP equation gives the regression with the highest R^2:

$$\begin{aligned} NRP = {} & \underset{(0.52)}{6.7} - \underset{(-0.04)}{0.007\,WVA} + \underset{(1.69)^*}{0.341\,VAO} + \underset{(3.18)^*}{0.0739\,AVW} \\ & + \underset{(7.25)^*}{245\,PTC} - \underset{(-7.09)^*}{3.20\,IMP75} - \underset{(0.94)}{0.0062\,OPF} + \underset{(0.80)}{0.0340\,GR} \end{aligned}$$

$R^2 = 67\%$ and * indicates significance at the 5% level

Any explained variance of NRP appears to result from the trade classification, value added share of output, average wage levels, and the share of imports in demand in the same period, 1975. The size of the coefficient of the trade classification variable is large, but the range of that variable is small (−0.65 – 0.98), so the net effect should not be dominating. The positive coefficient implies that importables receive a higher level of protection compared wtih exportables, once again indicating the bias against export oriented industries.

Average wage levels were significant, but the sign of the coefficient was positive, contrary to expectations. The higher average wages could be interpreted as higher levels of capital intensity or more labour skill intensive industries (i.e. modern industries). However, the industrial statistics used are based on surveys of medium and large industries and exclude the labour intensive small industries.

The value added share in output was barely significant and positive, indicating a tendency to protect industries which contribute to national output. Variables which are measures of lobbying power, such as output her firm, growth rate of output, and profits per value added did not appear to be significant. However, this result does not necessarily mean that lobbying was not important since, as mentioned before, it is difficult to quantify lobbying.

Finally, import share of demand was significant and negatively related to NRP, indicating that established import substitution industries which have succeeded in providing for the home market are likely to get protection.

The estimated ERP equation with the highest R^2 is:

$$\begin{aligned} ERP = {} & \underset{(4.88)^*}{301} - \underset{(-3.79)^*}{5.53\,VAO} + \underset{(1.65)^*}{0.285\,AVW} + \underset{(5.51)^*}{1371\,PTC} \\ & - \underset{(-5.66)}{18.71\,IMP75} + \underset{(0.11)}{0.005\,OPF} + \underset{(0.92)}{0.1348\,GR} \end{aligned}$$

$R^2 = 43\%$ and * indicates significance at the 5% level.

As with the NRP regression, the significant variables are the trade classification, value added share of output, average wage levels, and the share of imports in demand in the same period. The signs associated with the independent variables are also the same except for value added share of output. The sign is negative, indicating that the lower the value added share of output, the higher the expected level of protection, since protection will increase the income of the capital owners of the industry and this will induce them to lobby for protection.

The relatively poor fit of the regressions, especially in estimating the determinants of ERP, is probably due to several factors. First, owing to limitations in the data, the variables used do not accurately reflect the attribute to be included and thus there are variables that are excluded and which may be important explanators. This is especially true for lobbying power and, in fact, most empirical studies on the political economy of protection suffer from this problem. Furthermore, in our case the industrial statistics used exclude state corporations and small scale industries. State corporations usually have a high concentration in output amongst a few firms such as steel, cement, and fertilisers, and no doubt receive protection, as explained in our earlier discussion. If we could account for the state corporations, value of output per firm may have been significant. The omission of small scale industries excludes the labour intensive and low average wage sector, as mentioned before. Total employment figures could also be misleading, since employment figures differ among sectors owing to disaggregation. A more detailed sector classification would be more desirable. Finally, the extent of *pribumi* or non *pribumi* ownership in the protected industries has also not been included as it is difficult to quantify, just as with the special relationship between the power centres and various industries.

Second, our study also suffers from the limitations of other empirical studies on the political economy of protection. We have estimated the reduced form equation of protection, but the variables used act as proxies for both demand and supply of protection. There is clearly a problem of identification with respect to whether the resulting estimates can be interpreted as coming from either the demand or the supply side. There are also problems with regard to elements that constitute 'causes' and those that comprise 'effects'. The low import share of demand could be caused by the high protection rather than by the established import substituting industries being protected. Similarly, it may be that protection is inducing and maintaining high growth in an industry and not that high growth industries attract protection in the national interest to promote growth. These problems call for estimation methods beyond OLS. One should consider simultaneous equation estimation methods and causality tests to overcome the identification and cause-effect problems in future research.

Conclusions

While Indonesia has come a long way in liberalising her foreign trade regime compared with the system during the Old Order period, the existing system still has unnecessary complications and inconsistencies. As in the Old Order period, although of course to a much lesser extent, the prevailing system of protection is burdened with too many objectives which tend to result in conflicts.

In the early 1970s, under the import substitution policy, the focus of protection was on final consumer goods. Empirical estimates tend to show a rise in protection up to 1978 and a persistently cascading structure. The lobbying pressure of vested interest groups and the political difficulty of letting inefficient firms close down once they are set up can partially explain this result. Unlike the rapid growth of final goods industries behind high protection, the development of intermediate and capital goods industries was slow. Such industries require large capital to establish and such funds have not been easy to come by in the present international and domestic situation. However, over the long run, an appropriate protection structure is a necessary condition for achieving the objective of 'deepening' the industrial structure. To this end, an overhaul of the existing system of protection to make it more rational and systematic is necessary. The inherent bias against exports has also hampered the growth of nonoil exports and unless this bias is reversed, it will continue to be an obstacle in any policy to promote nonoil exports. Given the fall in oil prices and production quotas, it has become more crucial for Indonesia to develop its nonoil exports.

A rational and systematic protection system implies keeping at a minimum the unplanned and unintended effects of the system. In our view, a simpler structure of ERP with smaller variations among industries would help reduce the incidence of such unintended effects. Such a structure would be easier to monitor and its effects more readily evaluated. The overhaul should take a step in this direction.

We have noted that nontariff protection is important in Indonesia. It is, therefore, imperative to take steps toward a better synchronisation of the protective effects of all existing policies, ranging from credit policy and direct industrial subsidies to investment incentives, the export certificate system, the approved traders scheme, and import quotas. At present, with the many principal decision makers for these policies, coordination of policies from the viewpoint of their protective effects is weak. There seems to be a case for the formation of an institution or, at least, a more permanent forum which can serve as a vehicle for reviewing and coordinating those policies with special attention to their protective effects on domestic industries. The institution (or forum) undertakes the overhauling of the tariff system, makes decisions on tariff and any other protection policy changes, and determines the timetables for protection afforded to individual industries or groups of industries based on the country's industrialisation strategy. It seems that only in this way can the multiple objectives of reducing import content, 'deepening' the industrial structure, promoting nonoil exports, promoting investments, and improving income distribution be balanced and reconciled with one another.

Furthermore, it is not sufficient to just design a protection system with all the discussed qualities. In the final analysis, it is the implementation of policy that matters. While the simplification of the protection system will make policy implementation easier, it will still require consistency and toughness in the actual application of the policy. The policies need also to be designed to give the correct incentives to encourage efficient production by domestic producers, and not just to protect a sector because it is part of the industrialisation strategy. Finally, putting aside arguments of whether a protection policy is desirable or not, given that a protection policy is part of the country's development policy, the protection policy needs to be dynamic. It needs to have a long term outlook built into it and be able to foresee and plan for industrial restructuring.

Appendix 1

The World Bank and Pitt Study

The Pitt (1981) and World Bank (1981) studies essentially used the same methods of estimation to estimate effective rates of protection (ERP) based on the 1971 and 1975 input output tables.

Nominal protection estimation

The factors included in nominal protection estimation were:

i Actual collected customs duties and import sales taxes (i.e. code 403 in the Input-Output tables).

ii The discriminatory effect of MPO taxes, which was higher on average for imports, was included by adding 1 percent in 1971 and 0.5 percent in 1975 to nominal protection for most importable sectors. The downward adjustment in 1975 was to reflect the lower collection rates of MPO taxes compared with scheduled rates.

iii General export taxes of 10 percent were incorporated for all exportables, except for finished goods and handicrafts and certain 'weak' agricultural products and specific export taxes for certain products.

iv Direct price comparisons were used for goods subject to quantitative controls, whose imports are performed by or for government agencies (e.g. wheat flour, fertilisers and pesticides) and where check prices on imported items have been used. However, neither study explains what price comparisons were used nor what products were included in each sector.

v Adjustments for imported inputs from the same sector such as assembling operations where the CKD packages receive tariff discounts.

vi Subsidies to petroleum products, especially fuel oils.

vii Discriminatory price fixing.

Effective rates of protection

The Corden formula was used to calculate ERP based on the estimated NRP. The version used was:

$$ERP_i = ((DVA_i/IVA_i) - 1) \times 100$$

where ERP_i = effective rate of protection of activity i

DVA_i = Domestic value added of ith activity
IVA_i = International value added of ith activity

The sectoral aggregation used are:

i In 1971 there were 171 sectors in the input-output table out of which 83 were used to represent the manufacturing sector. The sectors used were sectors 45 to 126 and sectors 172–173.

ii In 1975 there were 179 sectors in the input-output table out of which 95 sectors are included as the manufacturing sector. The sectors used were sectors 58 to 141 and the agricultural processing activities 2, 5, 13, 15, 18, 19, 22, 24, 26, 34, and 43.

Comparability between 1971 and 1975

There have been some changes in classifications although in the main the sectors are quite comparable. Some of the changes are:

i Meat and fish canning and preserving sectors in 1971 became the two sectors of slaughtering in 1975.

ii Farm coconut oil and processed tea were not separated out from the coconut oil and tea processing sectors in 1971.

iii Fertilisers and pesticides in 1971 became pesticides and agricultural chemicals in 1975.

iv Palm oil was not included separately in 1971.

Trade classification

The trade classification, t, is defined as

$$t = \frac{\text{imports} - \text{exports}}{\text{production} + \text{imports} - \text{exports}}$$

And, exportable sectors are those for which $t < 0$; importables for which $t \geq 0$ where import competing sectors lie in the $0 \leq t \leq 0.8$ range and $t \geq 0.8$ for noncompeting sectors.

1975 and 1980 Estimates

The data used to estimate NRP in 1975 and 1980 were import duty and sales tax collection figures (i.e. I-O code 403) from the 1975 input-output tables and preliminary

1980 input-output tables; that is, NRP is estimated as 403/401 where 401 is the I-O code for imports valued c.i.f. The formula used to calculate ERP was:

$$ERP_i = \frac{t_i - \sum_j a_{ji} t_j}{1 - \sum_j a_{ji}}$$

where $t_{i,j}$ = NRP of activity i or j
a_{ji} = the share of input j in the production of i.

Since we did not estimate domestic and international value added figures, the average broad sector ERP and NRP figures were obtained by using simple averages and weighted averages based on production.

Table 1.A1 Effective and nominal rates of protection in the Indonesian manufacturing sector, 1975 and 1980

1975 I-O Code	Sector	Nominal rates of protection		Effective rates of protection	
		1975	1980	1975	1980
	Exportable sectors				
5	Cassava & tapioca flour	56.77	0.00	173.11	−0.98
13	Smoking & remilling of rubber	14.10	0.00	10.39	−10.89
18	Small scale coconut oil production	0.00	0.00	−30.40	−65.09
19	Palm oil	0.00	11.76	−0.98	13.33
22	Processed tobacco	90.43	40.44	140.44	65.89
24	Small scale coffee processing	n.a.	n.a.	38.16	22.08
26	Small scale tea processing	0.00	0.00	−38.25	−0.18
34	Slaughtering	26.78	35.28	27.26	58.74
40	Dried & salted fish	49.92	32.75	76.68	10.22
62	Coconut & cooking oil	14.25	18.60	33.99	44.39
69	Cocoa, chocolate & sugar confectionery	74.02	61.59	122.46	120.33
73	Other food products not elsewhere classified	55.60	15.08	13.37	17.35
80	Batik industry	96.38	49.56	191.99	83.39
85	Tanneries & leather finishing	54.15	34.33	103.77	74.27
88	Sawmills & other wood processing	56.19	18.58	68.66	−20.02
89	Wood & cork products	29.63	13.79	22.90	44.58
	Non competing imports				
118	Iron & steel basic industries	17.93	7.54	18.38	7.61
124	Nonelectrical machinery	17.20	12.81	16.39	14.52
125	Electrical industrial machinery & apparatus	18.96	7.40	15.05	5.40
129	Other electrical apparatus	20.05	7.12	−0.30	0.91
135	Aircraft industry	0.17	0.23	−26.42	−6.52
136	Professional & scientific instruments & repair	13.42	10.55	4.26	9.81
137	Photographic & optical goods	29.47	26.84	29.49	27.66
94	Basic industrial chemicals, excluding fertiliser	18.81	3.28	20.32	2.71
101	Pesticides & agricultural chemicals	16.76	0.00	11.52	−5.43
102	Other chemical industries	52.05	9.48	79.18	12.67
	Import competing sectors				
58	Canning & preserving meat	66.41	66.47	133.10	128.78
59	Dairy products	29.94	11.96	40.68	14.54
60	Fruits & vegetables	71.71	41.15	81.58	32.96
61	Canning & preserving fish & other sea food	63.44	21.04	103.58	−9.44

Table 1.A1 Continued

1975 I-O Code	Sector	Nominal rates of protection 1975	Nominal rates of protection 1980	Effective rates of protection 1975	Effective rates of protection 1980
63	Other vegetable & animal oils & fats	30.52	6.87	42.17	−0.07
64	Rice milling, cleaning & polishing	75.78	0.00	354.49	−0.59
65	Wheat flour & other grain mill products	11.28	2.02	24.22	1.60
66	Sugar refinery	−2.94	1.86	−7.66	−0.22
67	Bread & bakery products	20.62	94.49	25.96	212.40
68	Macaroni, noodles & similar products	64.51	21.15	124.09	49.99
74	Alcoholic beverages	103.33	69.01	107.58	80.74
75	Soft drinks & carbonated water	76.10	63.91	94.09	83.39
76	Cigarettes	155.09	67.55	246.17	105.16
77	Spinning	15.67	8.59	30.11	15.85
78	Weaving	59.05	51.88	134.89	127.08
79	Textile bleaching, printing, dyeing (excluding batik)	0.00	6.49	−16.10	8.23
81	Knitting	88.60	71.91	199.05	169.62
82	Made up textile goods (excluding garments)	63.19	22.14	52.25	−11.86
83	Garments	84.65	19.61	116.71	−20.02
84	Carpets, rugs, ropes & other	37.03	25.33	65.39	45.58
86	Leather products, excluding footwear	67.12	40.40	76.47	46.21
87	Leather footwear	78.90	49.17	94.53	59.06
90	Furniture & fixtures, excluding those from metal	38.25	15.13	36.82	12.26
91	Pulp, paper & cardboard	28.00	11.79	31.41	12.31
92	Paper products	55.54	54.79	65.18	73.44
93	Printing, publishing & allied industries	27.62	19.65	26.15	23.83
96	Paints, varnishes, lacquers	22.63	18.92	19.26	25.43
97	Drugs & medicines	73.38	2.70	84.62	−0.05
98	Soap & cleaning preparations	54.17	23.13	107.80	50.53
99	Cosmetics	87.05	54.93	101.01	97.97
100	Matches	38.72	0.00	43.36	−3.24
109	Other petroleum & coal products	60.19	4.98	135.97	11.56
110	Tyres & tubes	112.37	11.83	184.92	2.75
111	Other rubber products	52.32	11.73	90.78	21.01
112	Plastics	56.07	29.15	116.69	70.33
113	Ceramics	95.40	12.06	117.37	13.30
114	Glass & glass products	416.65	27.14	698.87	42.14
115	Structural clay products	20.13	3.89	20.31	3.77
116	Cement	22.50	5.40	23.08	4.04
117	Other nonmetallic mineral products	50.27	91.37	69.29	139.94
120	Cutlery, hand tools & general hardware	26.24	17.67	27.19	27.59
121	Furniture & fixtures, metal	41.31	29.26	59.18	47.82
122	Structural metal products	19.65	19.94	3.05	−1.24
123	Other fabricated metal products	36.60	45.44	10.43	11.98
126	Radio, T.V., communications equip't	27.07	25.25	22.98	23.73
127	Electrical appliances & houseware	29.15	27.12	34.53	57.72
128	Accumulators & dry batteries	43.57	63.66	15.69	22.37
130	Ship & boat building & repair	14.81	−5.70	0.30	−12.59
131	Railroad equipment	3.07	−1.28	0.18	−2.35
132	Motor vehicles	43.95	46.10	27.74	33.40

133	Motorcycles, bicycles & other vehicles	36.75	33.90	13.36	13.12
138	Jewellery & related articles	50.23	79.32	34.32	57.39
139	Musical instruments	58.84	73.41	10.98	10.92
140	Sports & athletic goods	34.34	42.18	36.35	61.80
141	Other manufacturing industries	69.64	87.41	30.32	35.60

For details on calculations, see Appendix 1.
Source: *Biro Pusat Statistic*, input-output table, 1975 and (preliminary) input-output table, 1980.

Endnotes

1 Furnival, I.S. (1944), *Netherlands India*, New York: Macmillan, p. 439.
2 Translation of statement by Department of Industry to Chamber of Commerce, February 1984.
3 For a more detailed discussion of trade policy in this period, see Boeke, J.H. (1946), *The Evolution of the Netherlands Indies Economy*, New York: Institute of Pacific Relations, chapters 3–5; Broek, J.O.M. (1942), *Economic Development of the Netherland Indies*, New York: Institute of Pacific Relations, chapters 5 and 7; Furnival, I.S. (1944), chapter 12; Barber, A. (1939), 'Six Years of Economic Planning in Netherlands India', *Far Eastern Survey*, 16 August; Van Eeglren, G.M. (1937), 'The Beginnings of Industrialization in Netherlands India', *Far Eastern Survey*, 9 June; Sitsen, P.H.W. (1942), 'The Industrial Development of the Netherlands Indies', Paper submitted to 8th Conference of Institute of Pacific Relations, Mont Tremblant, Quebec, Canada; and Van Gelderen, J. (1939), *The Recent Development of Economic Foreign Policy in the Netherland Indies*, London and New York: Longman, Green and Co.
4 Booth, A. (1983), 'Foreign Trade and Economic Growth in the Colonial Economy', Paper given at Conference on Indonesian Economic History in the Dutch Colonial Period, Table 3.
5 Goods affected under the free quota were cement, rice, tyres, garments and frying pans. The national quota was aimed at protecting imports from the Netherlands such as bleached and unbleached cotton goods, cotton blankets, and rubber tyres.
6 Booth, A. (1983), Table 3.
7 For a detailed discussion of foreign trade policies during this period, see Boediono (1972), 'Foreign Exchange Policies for Stabilization: The Indonesian Experience 1950–1965', MEc. thesis, Department of Economics, Monash University; Corden, W.M. and Mackie, J.A.C. (1962), 'The Development of the Indonesian Exchange Rate System', *Malayan Economic Review*, vol. 7, no. 1; Glassburner, R. (ed.) (1971), *The Economy of Indonesia: Selected Readings*, Ithaca: Cornell University Press, chapters 1, 2 and 12; Higgins, B. (1957), *Indonesia's Economic Stabilization and Development*, Institute of Pacific Relations, New York: American Book-stratford Press Inc.; Kanesa-Thasan, S. (1966), 'Multiple Exchange Rates: The Indonesian Experience', IMF Staff Papers; Paauw, D.S. (1963), 'From Colonial to Guided Economy', in R. McVey, *Indonesia*, New Haven: Hraf Press; Thomas, K.D. and Panglaykim, J. (1973), 'The Effect of Past Policies and President Suharto's Plans for the Future', Committee for Economic Development of Australia, P. series, no. 11, November; and Pitt, M. (1977), 'Economic Policy and Agricultural Development in Indonesia', PhD. dissertation, University of California, Berkeley.
8 The inducement schemes differed in the percentage of the value of exports received in negotiable certificates (the remainder was received at the official exchange rate) and whether the scheme applied to the majority of exports. The percentage ranged from 5–10 percent to 100 percent. Only two schemes, the Foreign Exchange Certificate and *Bukti Pendorong Ekspor* (Export Promotion

Certificates) schemes applied to the majority of exports. The requirement to use the export inducement certificates applied to only certain non essential imports and only to all imports during 1957–1959. Most essential imports such as rice and fertilisers were exempted from exchange taxes and essential imports such as raw materials were subject to low exchange taxes.

9 The majority of exports originated from the outer islands of Sumatra and Sulawesi. Thus, the regions controlled a significant proportion of the foreign exchange supply and were able to use barter trade and refusal to use the BE system to wield bargaining power over the central government.

10 The pressure to nationalise came from the increasingly socialist factions of the government and also as a protest against Dutch refusal to yield on the West Irian issue; Paauw (1963), p. 209.

11 The foreign exchange shortage was due to low production levels caused by nationalisation, as well as due to the low prices of Indonesia's main export commodity, rubber and the practice of illegal barter trade in the outer islands.

12 Paauw, (1963), p. 212.

13 The export inducement and import surcharge schemes of the earlier years prevailed during this period. There were some variations in the schemes such as the requirement for exporters to surrender foreign exchange receipts up to the check price and the difference between the check price and the f.o.b. price could be used to import or could be sold on the black market.

14 Thomas K.D. and Panglaykim, J. (1973).

15 Arndt, H.W. in Glassburner, R. (1971).

16 ADO replaced the BE-Daerah or Regional-BEs, where one half of the BEs for imports was tied to the province of the export's origin. A large proportion of the proceeds from the ADO scheme went to the outer islands through the provincial government subsidy scheme. The policy was aimed at reducing smuggling and ensuring the outer regions a share of imports.

17 The B category was not subject to the foreign exchange tax. The A list was comprised of rubber, copra, coffee, tobacco, palm oil, palm kernels, pepper, tin, tea, sisal and sugar before July 1967. The B list: all other exports except petroleum which was subject to its own regime. The A list comprised approximately 75 percent of non-oil exports.

18 Up to July, 1967, exporters received a certain percentage of the check price in BEs and the remainder in Rupiah payments at the official exchange rate.

19 Restrictions on access to credit was an important limitation on imports because suppliers' credit was often not available and importers had to purchase foreign exchange before a Letter of Credit could be opened.

20 Glassburner, (1971), p. 405.

21 The Domestic and Foreign Investment Laws of 1968 and 1967 were aimed at encouraging the flow of investment by providing several incentives such as tax holidays, exemptions from tariffs on inputs, accelerated depreciation and so on.

22 Various government actions in intervening in the BE market, such as reducing the validity period of the BE certificates, requiring all BEs to be sold to a foreign exchange bank and abolition of the special Aid-BE rate, were undertaken in an attempt to prevent exchange rate increase and speculation.

23 Pitt, M. (1971), p. 91.

24 World Bank (1981), *World Development Report, 1981*, Washington, D.C.: World Bank, Annex 2.

25 The rate on domestic sales is 2 percent with the seller acting as the tax collector. The rates for imports are stated as Rps/US$1 and are higher for imports using a Merchant's Letter of Credit (L/C), rather than an ordinary L/C. For instance, in 1980 the rates were Rp. 24/US$1 or 3.8 percent for imports with an ordinary L/C

and Rp. 36/US$1 or 5.7 percent for imports with a Merchant's L/C (using the exchange rate Rp. 633.7/US$1).

26 Pitt M. (1981), 'Alternative Trade Strategies and Employment in Indonesia', in A.O. Krueger et al. (eds), *Trade and Employment in Developing Countries*, vol. 1, Chicago: NBER, p. 205. Profits are presumed to be determined as a margin on sales which in turn are a mark up on costs. Thus, by under reporting the MPO tax credit it will reduce its presumed profits since the total MPO credit indicates the firm's cost and thus increases its margin and profits.

27 The VAT was devised to overcome the multiple taxation outcome where the sales tax was levied on the full value of the product at each stage of the production process. The VAT and excise tax was originally supposed to become effective in July, 1984. The deadline is now January 1, 1986.

28 World Bank (1981), estimates that the undervaluation could be around 5–10 percent.

29 The demand for higher check prices could also be because the existing check price is unrealistically low due to slowness on the part of the Department of Finance.

30 From interviews with domestic producers.

31 At present there are nine categories out of which seven are industrial goods. The seven categories are electrical and electronic goods, chemical products, metal industries, machinery and spare parts, heavy equipment and spare parts, motor vehicle components and textiles.

32 The petition will go to the relevant technical department first. A rough guide to evaluate whether the domestic price is competitive with imports is whether the domestic price is less than or equal to 15 percent more than the import price.

33 Based on reported changes from various issues of *Business News* during the period.

34 Asher, M. and Booth, A. (1983), *Indirect Taxation in ASEAN*, Singapore: Singapore University Press, p. 51.

35 *Business News*, 22/2/1984. On this date, tenders under the counterpurchase agreements involved suppliers from 19 countries.

36 Most of the cases were in the export of garments to nearby destinations, such as Singapore.

37 Hill, H. (1984), 'Survey of Recent Developments', *Bulletin of Indonesian Economic Studies (BIES)*, vol. 20(2), April.

38 As defined by KEPPRES (Presidential Decree) /14/1979.

39 The objectives underlying the preparation of the DSP are: increase employment, increase foreign exchange, promote exports, increased use and processing of domestic raw materials and resources, increase indigenous Indonesian participation in all economic activities, protect existing domestic industries and maximizing investment.

40 Kredit Investasi Kecil or Small Investments Credit and Kredit Modal Kerja or Working Capital Credit for small scale enterprises, Kredit Investasi Biasa or Odinary Investment Credit for large scale enterprises and Kredit Kelayakan or Feasible Credit Scheme for the economically weak sector.

41 The aim of this policy was to facilitate the flow of imports to be used in production and to reduce the cost of financing imports by import-producers.

42 Prior to this date the Rupiah was pegged to the US dollar. The float was according to a basket of currencies of Indonesia's trading partners.

43 For a fuller theoretical discussion of this effect see for instance Corden, W.M. and Warr, P.G. (1981), 'The Petroleum Boom and Exchange Rate Policy in Indonesia, *EKI*, 29(3); and Corden, W.M. and Neary, J.P. (1982), 'Booming Sector and Deindustrialisation in a Small Open Economy', *Economic Journal*, 92, p. 368.

44 Warr used various indices to represent the relative price of traded goods to non

traded goods such as import prices over the consumer price index, import prices over the housing price ratio, the nominal exchange rates of the major trading partners multiplied by the country's consumer price index over Indonesia's consumer price index, and wholesale price for imported textiles or transport equipment over the housing component of the CPI (see Warr, P.G. (1984), 'Exchange Rate Protection in Indonesia', *BIES*, August, p. 66.

45 *Business News*, June 1, 1984.

46 That is by underinvoicing, false declarations with regard to type of good or quality of good and underassessment of duties with or without coperation of customs officials as defined by Cooper, R.N. (1974), 'Tariffs and Smuggling in Indonesia', in J. Bhagwati (ed.), *Illegal Transactions in International Trade*, Amsterdam: North Holland.

47 See Indonesia, Department of Finance, *Nota Keuangan*, 1983/84. Regression of the shares of consumption good imports (C), intermediate good imports (I) and capital good imports (K) on the average tariff collection rates (T) for 1969–81 produces the following result: T = 0.68408 + 0.48C + 0.071 + 0.02K R2 = 0.352. This indicates that the *average* collection rate for consumption good imports was 48 percent, for intermediate good imports was 7 percent and for capital good imports was 2 percent.

48 See World Bank (1981), Annex 2, Table 6.

49 A fuller explanation of the method of estimation and problems of comparison across years is given in Appendix 1.

50 Pitt, M. (1981), Table 5.A.1. The average ERP for each subsector was calculated for 1975 from data in World Bank (1981), Annex 2, Table 4.

51 See Appendix 1 for a fuller explanation of the data and estimation methods. At the time of writing the full 1980 input-output tables were out available yet so that it was not possible to use the input-output coefficients for 1980.

52 Let d_i be the difference between the World Bank (1981) and our results. Then if $d_1, d_2, \ldots, d_n$ $N(u, o^2)$, whether there is no significant difference between the two estimates reduces to the hypothesis test, H_o: $u = 0$ versus H_A: $U \neq 0$. The hypothesis was rejected for all estimates except for the ERP of non competing imports. Thus, except for that sector there is a significant difference between the two estimations (at the 5% level of significance).

53 *Far Eastern Economic Review*, May 29, 1981, p. 44.

54 Glassburner, R. (1978), 'Political Economy and the Soeharto Regime', *BIES*, November, p. 28.

55 Sinar Harapan, May 1, 1981.

56 See *Far Eastern Economic Review*, 29 May, 1981 for a background of the report.

57 See Anderson K. and Garnaut, R. (1983), 'Australian Protection and Trade with Developing Countries: Lessons from the 1970s and Prospects for the 1980s', monograph to be published by World Bank (forthcoming); Anderson K. and Baldwin R. (1981), 'The Political Market for Protection in Industrial Countries: New Empirical Evidence', World Bank staff Working Paper, no. 492; Bhagwati, J.N. (1982), *Import Competition and Response*, Chicago: University of Chicago Press, and various World Bank Staff Working Papers on specific industrial country studies. See Alikhani, I. and Hanrylyshyn, O. (1982), 'The Political Economy of Protection in Developing Countries: A Case Study of Colombia and South Korea', World Bank, Division Working Paper, no. 1982–4, July, for a study on developing countries.

58 Examination of the studentised residuals of the dependent variables (i.e. the residual is the difference between the actual and fitted value of the variable) and the plot of residuals against fitted values of the independent variable, suggested that some observations were outliners (i.e. these were observations which failed to

conform to the hypothesised model while the bulk of the data appeared to correspond quite well to the model). The observations omitted were paper products (cardboard boxes, paper bags etc.) and musical instruments for the NRP regression, and cutlery, agricultural handtools and general hardward, as well as the musical instruments sectors for the ERP regressions.

It could be that the omitted variables, such as the supply of protection for small and cottage industries by the government, were more important in explaining protection in these two sectors compared with other sectors. Examination of the range of values of the independent variables for these two observations did not reveal any abnormal pattern that could also explain for the 'outlier' characteristic of the dependent variables. (This section has benefited from discussion with Yudianto Pawitan; see also Weisberg, S. (1980), *Applied Linear Regression*, New York: John Wiley and Sons, chapters 5 and 6 for a discussion of outliers and residual plot analysis).

2 The structure and causes of manufacturing sector protection in the Philippines

NORMA A. TAN*
December 1984

THE importance of a system of protection in determining the direction of resource allocation in a country is increasingly recognised. It appears especially among developing countries, however, that the incentives system often does not consistently support national objectives and priorities. In studies of economic incentives arising mainly from protection it has been shown that in a number of developing countries, including the Philippines, the structure of protection has actually created biases within a range of industries, which generally leads to inefficient allocation of resources and inhibits overall economic growth. It is for this reason that increased attention is being given to the estimation and study of industrial protection to provide some basis for developing appropriate policy.

The determination of the causes of protection, in particular, is useful in understanding how systems of protection have evolved and have been influenced or changed to serve various purposes.

This chapter aims to describe the system of protection in the Philippines as it has evolved, to estimate the effective rates of protection on the basis of the existing system, and to present the changes in both the system of protection as well as in the effective rates since 1970. Then, these estimates are examined in relation to a number of variables to determine the causes of protection in the Philippine manufacturing sector.

* At the time of writing, Norma Tan was a consultant to the ASEAN—Australia Joint Research Project.

Table 2.1 Gross value added in agriculture, manufacturing, and other nonagricultural activities: Philippines

Sector	Percentage share		Annual growth rates (compounded)
	1948	1961	1948–1961
Agriculture	49.1	33.6	3.8
Manufacturing	17.5	28.0	10.9
Other nonagriculture	33.4	38.4	8.1

Source: R. Baldwin, (1975), *Foreign Trade Regimes and Economic Development: The Philippines* (New York: NBER, Inc.), p. 122.

Table 2.2 Measurement of import substitution in manufacturing 1948, 1956, 1960 and 1968: Philippines

ISIC Code	Industry	Ratio of value of imports to value of production			
		1948–1949	1956	1960	1968
	TOTAL MANUFACTURING	1.3	0.55	0.42	0.42
20	Food manufactured)				
21	Beverages)	0.47	0.24	0.16	0.15
22	Tobacco products)				
23	Textiles	6.32	1.11	0.27	0.22
24	Footwear and other wearing apparel	0.46	0.02	0.02	0.02
25	Wood and cork products)	0.03	0.01	0.01	0.01
26	Furniture and fixtures)				
27	Paper and paper products	43.78	0.98	0.35	0.35
28	Printing and printed products	0.53	n.a.[1]	n.a.[1]	n.a.[1]
29	Leather and leather products	11.60	1.22	0.35	0.07
30	Rubber products	n.a.	1.65	0.07	0.11
31	Chemicals and petroleum)				
32	products)	8.76	0.34	0.31	0.28
33	Nonmetallic mineral products	1.73	0.22	0.15	0.12
34	Basic metal products	n.a.	10.26	2.12	0.61
35	Fabricated metal products)		0.41	0.17	0.32
36	Machinery except electrical)	6.98	10.26	7.65	6.76
37	Electrical machinery)		2.65	0.56	0.71
38	Transportation equipment)		1.12	2.66	1.01
39	Miscellaneous manufactures	4.38	0.22	0.19	0.20

Note:
[1] n.a. means not available.
Source: R. Baldwin, *Foreign Trade Regimes and Economic Development: The Philippines* (New York: NBER, Inc., 1975), p. 126.

History of protection policy in the Philippines since 1950

Growth and structural change: an overview

There is some evidence that industrial protection in the country has had important effects on the structure of production. High protection to manufacturing has meant that growth was substantially higher than that of the agriculture and other nonagriculture sectors, resulting in the shift toward manufacturing in the composition of output, as shown in Table 2.1. This suggests that substantial import substitution which fueled manufacturing growth must have taken place in response to the favoured position of manufacturing in terms of protection. This appears to be confirmed by Baldwin's estimates of imports to gross value of output ratios in various Philippine manufacturing industries shown in Table 2.2. Between 1948–1956, the ratio of imports of manufacturing to gross value of manufacturing output declined from 1.13 to 0.55. Between 1956 and 1960, little further import substitution seems to have occurred. By 1960, import substitution appears to have levelled off. This observation is consistent with Power and Sicat's estimates of the growth rates for manufacturing and agriculture between 1948 and 1968, shown in Table 2.3. The explanation they offered for the sharp decrease in the growth of manufacturing and of national product after 1956 is that this might be attributed to declining opportunities for easy import substitution in the processing of consumption goods.[1]

The manufacturing sector, while showing some improvement after 1965 in response to the government's expansionary policies in the 1960s, shows no sign of picking up to the rates of growth exhibited in the early years of exuberant import substitution in the 1950s. Its average annual growth rate between 1965 and the 1970s is no higher than 7.5 percent as compared to the 14 and 11 percent recorded in the 1950s.[2]

In the light of the relative smallness of the manufactured exports sector, manufacturing output continues to rely on the expansion of the domestic market from import substitution. Opportunities for this appear to be limited for manufacturing as a whole judging from the growth of manufacturing output. However, there appears to be evidence that the structure of protection has led to difference in the performance of industries within manufacturing. The result of a study[3] implies the predominance of the effective rate of protection (ERP) in influencing sectoral difference in import share in domestic supply. This seems to confirm the effectiveness of the ERP as a policy instrument in changing the structure of resource flows and import substitution in the economy. This also supports the importance of protection in the Philippines where policy has sought for more than two decades to bring about the process of import substitution as an industrialisation strategy.

Major policy changes

Use of foreign exchange controls and multiple exchange rates up to 1960

Protection policy in the Philippines may be said to have started

Table 2.3 Growth rates of national products, manufacturing, and agriculture, 1948–68: Philippines
(Average annual percentage changes at 1955 prices)

Period	Gross national product	National income	Value added in manufacturing	Value added in agriculture[1]
1948–52	9.2	9.1	10.5	6.6
1950–60	7.9	7.8	12.3	6.2
1952–60	7.7	7.6	12.9	6.3
1956–60	4.4	4.6	6.3	3.3
1960–64	5.6	5.1	4.8	3.5
1964–68	6.1	5.3	4.7	6.9

Note:
[1] Includes fishing and forestry.
Source: J. Power and G. Sicat, *The Philippines: Industrialization and Trade Policies* (London: Oxford University Press, 1971), p. 39.

by the end of 1949 when, in response to a balance of payments crisis, import and exchange controls were imposed. The deficit in the country's trade balance mounted as a result of the huge expenditure on imports for reconstruction and rehabilitation after the war as well as from large imports of food and consumer goods. At the same time, export earnings were slow to recover and provided little offset to the continuing trade deficits. The policy response to this crisis was to ration foreign exchange. This was implemented in a manner that allocated more foreign exchange to the importation of essential goods, such as capital and intermediate goods, than to the less essential finished consumer goods. Controls, then, were intended to solve a balance of payments problem but provided at the same time protection for the domestic manufacture of consumer goods.

Import control was instituted through the creation of an import control board which had the authority to restrict the importation of luxury and inessential items. Initially, import control was intended to run for only a year, but eventually it took on a more permanent nature with the passage in 1950 of a new law which established the Import Control Administration. This time the intention to protect local industries was evident as the law provided greater percentage reduction in imports if there appeared to be sufficient domestic supply of a commodity.

The Import Control Administration was later replaced in 1951 by the Import Control Commission which continued the function of import control until 1953 when the Central Bank took over this function through its control of foreign exchange. The Central Bank administered exchange controls such that imports of nonessentials were restricted and preference was given to the importation of producer goods. This meant the elimination of any significant foreign competition in domestic nonessential industries, which made these industries very much protected in the domestic market. Moreover, a system of multiple exchange rates provided additional protection to industries considered 'new and necessary'. This resulted from the exemption of these industries from the 17 percent tax on the sale of foreign exchange in 1951. This tax was enforced until 1954, when it was replaced by a special tax on imports. This tax on imports was waived for the machinery and raw materials importation by the new and necessary industries and was periodically reduced until its complete elimination in 1965. In 1959 too, the Central Bank was given the authority to charge a 25 percent fee on the sale of foreign exchange. Again, the new and necessary industries were exempted from this fee, which further enhanced their status as compared to other industries. This rate was reduced gradually and phased out by 1962.

Exchange controls and the multiple exchange rate system were abolished with a change in administration in 1962 in response to their growing notoriety as a source of graft and corruption. Moreover, it was increasingly apparent that after a decade of exchange controls there was little room for further favouring the producer and the capital goods considered as essential imports relative to the inessential consumer items. At least 85 percent of total imports by 1962 consisted of producer goods and the rest were also essential in the maintenance and expansion of output and employment in the existing industries. The exchange rate, however, was not completely unified until November 1965, when the requirements to surrender 20 percent of export earnings at the lower than the free market exchange rate were removed and the peso was devalued officially.

Thus, the system of import licensing and exchange control was liberalised and full decontrol was achieved by the mid 1960s. By 1965 the main source of protection to domestic industries was the imposition of high tariffs, which were strengthened by discriminatory sales taxes as well as tax exemptions to so-called 'new and necessary industries'. The 1967–1970 period saw the reestablishment of moderate exchange controls as a result of the rapidly worsening balance of payments situation brought about by substantial expansion of domestic credit and government investment expenditures. These were in the form of tighter controls of credit to importers, particularly of less essential goods. In February 1970, with the deficit in the balance of payments again reaching crisis proportions, the government decided to put the peso under a floating exchange rate system in which exchange controls were gradually removed. Initially, a system of multiple exchange rates providing for lower peso conversion rates for exporters was adopted but was later replaced by a special export tax to exporters. Even as the peso has been allowed to depreciate since 1972, import licensing has been reinstated for some items. This results from a provision of the Tariff and Customs Code of 1978 which grants the President authority, among others, to impose quota or surcharges and ban imports for specific purposes. Also, the Central Bank has been charged with the responsibility of regulating the importation of certain commodities. In particular, import ceilings are set for items classified as nonessential and unclassified commodities. By early 1983, however, more stringent import and exchange controls once again appeared necessary in the face of the rapid deterioration of the balance of payments resulting from a combination of economic and political factors.

Increasing use of tariffs

Tariffs as a protection instrument became established only in 1957 with the passage of Republic Act No. 1937, known as the Tariff and Customs Code of the Philippines. Before this, tariffs, as provided for in the Philippines Tariff Act of 1909 which was enacted by the United States Congress before Philippine independence, were intended to serve primarily the purpose of raising government revenue. However, Philippine government revenues from tariffs were eroded by the preferential free trade with the United States, which was the country's predominant trading partner. The clamour for the removal of this preferential trade culminated in the Laurel-Langley Agreement which provided for the gradual phasing out of tariff exemptions for US goods starting in 1956. This led to the enactment in 1957 of the Tariff and Customs Code by the Philippine Congress which clearly added the protection objective to that of mainly collecting revenue in the prewar years.

As noted earlier, protection of the manufacturing sector was carried out through a system of foreign exchange controls and multiple exchange rates between 1949 and the mid 1960s. With the virtual dismantling of this system by 1966, however, the tariff became the main instrument of protection. This tariff structure exhibited a protective bias in favour of consumer goods relative to intermediate and capital goods similar to that provided by the system of exchange controls. It appears that the same pattern of industrial structure and resource allocation that the system of exchange controls encouraged was perpetuated by the tariff system that replaced it. This structure became even more pronounced by 1965 when, in response to pressures to offset the elimination of controls, some upward adjustments in the tariff rates were made. Tariff rates were relatively high by 1965, estimated to average 70 percent for consumer goods, 55 percent for inputs into construction, 21 percent for intermediate goods and 16 percent for capital goods. Thus, the structure of protection remained virtually unchanged between the 1950s and 1960s although the instruments of protection changed from import and exchange controls to tariffs.

Modifications to the Tariff Code in 1973 and 1981

Between 1965–1973 there have been minor changes in the tariff system. A major revision of the tariff law was implemented on 1 January 1973, with the primary objective of simplifying the tariff rates to a six level schedule in order to improve the administration of customs. The amended Tariff Code of 1973 also provided for special duties to protect Philippine industries against unfair competition such as dumping, subsidy, and subvention. Furthermore, it added a provision on the imposition of an additional duty of 10 percent on the products of any country which discriminated against Philippine products.

An important modification was the imposition of export duties, which became a permanent feature of the Tariff Code. The present schedule of rates for export duties as amended by subsequent legislation consists of a basic rate and a flexible rate known as the premium duty. The export tariffs fall on the traditional agricultural and mining export sectors of the country. An export tariff of 4 percent is applied on most of these commodities with the exception of logs, with an export tariff of 20 percent, copra with 7.5 percent, and centrifugal sugar with 6 percent. The flexible rate or the premium duty is applied on a lesser number of industries with the maximum of 30 percent imposed on copra and 20 percent on all the rest. The objective of export taxation is to induce the processing of these traditional export products and give an advantage to and encourage the export of manufactured and semi manufactured goods as a whole. The higher tariffs for logs, copra and centrifugal sugar reflect the policy of phasing out the export of these commodities in their unprocessed forms and promoting their export in more processed forms.

Since 1973 and prior to 1981, there have been some tariff changes embodied in the Tariff and Customs Code of 1978 and subsequent amendments. The new Code provided for the same six levels of *ad valorem* rates, namely, 10, 20, 30, 50, 70 and 100. While there are no duty free rates, some articles may be imported duty free under certain conditions as specified in the Code. This was later amended in 1979 to include other tariff rates not exceeding 100 percent *ad valorem*. In the main, the Tariff changes between 1973–1981 consisted of downward revisions in the tariff rates as well as tariff reclassification of specified commodities aimed primarily at easing the supply and bringing down the prices of essential food and consumer items. The collective effect of these, however, did not substantially change the structure that consistently characterised the Philippine Tariff and Customs Code since its inception in 1957.

A major revision of the Tariff Code was implemented from 1

January 1981. This resulted from a policy decision to effect a major reform of the protection system in the country aimed at improving the competitiveness of domestic industries and promoting better allocation of resources in relatively efficient export and import substituting industries. The removal of the long standing biases in the protection structure against exports, against backward integration of industries, and against the more efficient users of resources was recognised as an important prerequisite to the improvement of the country's balance of payments and the overall efficiency in the manufacturing sector.

A maximum duty of 50 percent and a minimum rate of 10 percent, with a few exceptions, are set. This narrows down the range of tariffs substantially from the previous maximum rate of 100 percent while maintaining the base rate of 10 percent. The maximum 50 percent tariff rate provides some allowance for adjustment to industries which for more than two decades have existed under a regime of protection, but still represents a sharp reduction of protection for these industries.

Tariff rates are reduced as well as evened out across the board within specified industrial or product categories. The classification of industrial or product categories is based on the degree of processing involved, i.e. raw materials, intermediate goods, and finished goods. For raw materials, the outputs of which are either insufficient to meet local demand or not domestically produced at all, the tariff rate ranges from 5–10 percent. The lower rate of 5 percent is given as a temporary concession in the first or second year of the program implementation period. For raw materials produced domestically in sufficient quantities to meet local demand, the range of 20–30 percent is applied.

Intermediate goods receive a 20–30 percent tariff. Finished goods consist of two categories: capital equipment and producer goods, and consumer goods. On the first is applied a range of 20–30 percent which is the same as that on intermediate goods. Consumer goods get the maximum range of 40–50 percent. A movement toward greater uniformity in tariffs is apparent as the 20–30 percent rate is applied on certain categories of raw materials, on intermediate goods, and finished goods in the form of capital equipment and producer goods. The unifying effect of the application of the 20–30 percent range on broad economic categories irrespective of the degree of processing is one of the more significant features of the tariff reform. This revised schedule of tariffs minimises but does not remove the cascading structure of tariffs, so that inputs, in general, continue to receive lower protection than final products.

Thus, the tariff reform program is staged over a five year period beginning on 1 January 1981, and reduces the previous maximum rates of 70–100 percent to 50 percent covering largely nonessential items. In addition, a generally reduced tariff is applied on the other categories, ranging from a minimum of 5–10 percent to 50 percent. The range of tariff rates being limited to 10 the percentage point within each category leads towards a closer uniformity of tariff rates. All these have the overall effect of lowering tariff protection for the nonessentials relative to other items and reducing substantially the dispersion of rates across industries. With the tariff reform, the average tariff rate falls from 43 percent to 28 percent. The possibility that infant industry reasons exist for some of the industries is allowed for by staging or gradually lowering the tariff through a period of five years in two to five stages. For the peak rate industries, the staging period spanned a two year period, from 1 January 1981 to bring the 100 percent duty to 50 percent. The 70 percent duty was lowered to 50 percent on 1 January 1981 and maintained thereafter. For the lower rates, further reduction is staged over a period from 1 January 1981 to 1 January 1985 or less.

Other policies

Indirect taxes

There are two basic types of indirect taxes in the country. One is the sales tax and the other is the excise tax or specific tax. The sales tax is imposed on most products while the specific tax is imposed on a few products, namely tobacco, alcoholic beverages, gasoline, and oil. Both the sales tax and the specific tax are imposed on products whether domestically produced or imported. Products are subject to either sales tax or specific tax but not both.

The sales rate varies according to essentiality. Nonessential items are taxed at 25 percent. The sales tax rate on essential articles is 5 percent and that on agricultural products is 1 percent. Moreover, graduated tax rates ranging from 10–25 percent are levied on certain locally produced articles, such as watches and clocks, fountain pens and ballpens, electric fans, stoves and ranges, phonographs, radio, television sets, refrigerators, and the like. The imported versions of these articles are subject to higher graduated sales tax rates. Manufactured cars likewise are subject to graduated sales tax rates with lower rates on locally manufactured cars ranging from 10–200 percent depending on the price, and a higher range of 100–200 percent on imported cars. Imports for direct personal use are charged a compensating tax equal to the sales tax rate. The specific tax is in terms of a specific amount collected on goods per unit quantity or amount sold whether produced locally or imported. The specific taxes on imported goods are higher than the specific taxes on domestically produced products. The product coverage of specific taxes is very limited and indirect taxation is mainly in the form of *ad valorem* sales taxes.

Indirect taxes, in the form of sales taxes and specific taxes, are a source of protection to domestic manufacturing because of the discriminatory treatment of imports as compared to domestic produce. For products subject to sales taxes, the discrimination is in terms of adding a mark up on the transaction value of the products, which is then used as the base of the tax while domestic produce is taxed without the mark up. The rate of mark up progresses with the degree of inessentiality. For some products, this discrimination is increased by the higher rates of the sales tax on imports relative to the domestic produce. A further advantage to the latter is provided in terms of a lower base for domestic produce as a result of allowing the application of the sales tax rate not on the total transaction value but only on the remaining value after deducting the material inputs on which percentage tax payments were previously paid. This was modified in 1978 by granting tax credits equal to indirect taxes paid on inputs. For products subject to specific taxes, the discrimination is in terms of the higher specific taxes on imports than on domestic produce. The structure of indirect taxes have changed very little since the post war period. Some reduction in the rates and mark ups was implemented since 1978 but this has not modified the structure significantly.

Tax and fiscal incentives

There are various tax and fiscal incentives provided to manufacturing firms in the country. To the extent that these incentives increase producer receipts from domestic output as compared to the free trade value of imports and exports, these incentives can be considered a subsidy. These constitute a source of protection and must be included in any measure of protection to industries.

The tax and fiscal incentives are granted to firms mainly by virtue of the provisions of two major incentives laws. These are the Investment Incentives Act 1967 and the Export Incentives Act of 1970. These two laws have been amended in 1973 and 1974 to further liberalise certain incentives provisions. In addition to these two basic laws, there are other laws which grant tax incentives to a number of products in the form of tax exemption from the payment of duties on imported raw materials and capital goods and from compensating sales taxes.

The Investment Incentives Act of 1967 provides for the creation of the Board of Investments (BOI), the agency in charge of administering the Act. This Act was primarily intended to encourage import substituting industries but also encourages expansion of exports. The Act reiterates the constitutional guarantees of freedom from expropriation and requisition except in the interest of national welfare or defence, and only upon payment of just compensation. All foreign investors are allowed to remit earnings on investments and to repatriate their investments in the currency in which they were originally made. Remittance of foreign exchange to meet payments of interest and principal on foreign loans as well as obligations arising from technological assistance contract is also assured.

The Act provides for the yearly preparation of an Investment Priorities Plan (IPP) wherein are listed the industrial sectors which merit priority consideration for investment and incentives pruposes. It grants fiscal and other incentives to firms which register with the BOI under priority classifications, e.g. 'pioneer' and 'preferred'. Investments which introduce new products or new processes into the country are classified as pioneer projects, while investments which are in areas where existing capacity is less than what would meet domestic demand and export demand are classified as preferred projects. Enterprises can register for preferred status if at least 60 percent of the owners of voting capital stock as well as of the membership of the Board of Directors are Philippine citizens. Enterprises can register for pioneer status with no citizenship constraint.

The incentives that enterprises may avail themselves of under this Act include: 1. accelerated depreciation; 2. net operating loss carry-over; 3. tax deduction for expansion reinvestment; 4. tax exemption on imported capital equipment; 5. tax credit on domestic capital equipment; 6. exemption from all internal revenue taxes except the income tax; 7. post-operative tariff protection; 8. double deduction from taxable income of promotional expenses; 9. double deduction from taxable income of shipping costs; and 10. tax credit equivalent to the sales, compensating the specific taxes on raw materials used in export production for ten years. Preferred enterprises may avail themselves of incentives 1 to 5. Pioneer enterprises may avail themselves of incentives 1 to 7. BOI registered enterprises, preferred or pioneer, which export or propose to export, are allowed incentives 8 to 10.

The Export Incentives Act of 1970 was aimed at intensifying export activities by increasing the extent and coverage of incentives to export enterprises already provided for in the Investment Incentives Act of 1970. This Act provides for the yearly drawing of an Export Priorities Plan (EPP), which is a list of export products which are to be given priority in terms of incentives depending on their ability to increase foreign exchange earnings, encourage the utilisation of excess manufacturing capacities of exports, develop new markets for Philippine products, utilise indigenous raw materials, create employment opportunities, and encourage regional dispersal in industries.

Availment of incentives under the Export Incentives Act is granted to Philippine citizens or to corporations and other entities which are at least 60 percent owned and controlled by Philippine citizens who are pioneer export producers of products which are classified as pioneer under the Investment Incentives Act of 1967 and which are listed as export products in the EPP. If such products are not listed in the EPP, at least 50 percent of such products must be exported.

The incentives under the Export Incentives Act are: tax and duty free importation of capital equipment and spare parts; tax credit for taxes paid on domestic capital equipment, raw material and semi-manufactured products used in export products; antidumping protection; reduced income tax; exemption from export tax; post operative tariff protection; expansion reinvestment deduction; infrastructure subsidy for regional development; market development loss carryover; and market development deduction.

Subsequent legislation has amended the two basic investment laws discussed above. These amendments further liberalised certain incentives provisions, notably: deduction of expansion reinvestment regardless of whether or not the profit reinvested is realized from registered operations; availment of special tax credit on semimanufactured products used in export production for an indefinite period instead of the earlier 10 year provisions; additional deduction from gross income of 50 percent of the value of labour training expenses incurred for upgrading the efficiency of installed labour; and additional deduction from taxable income of an amount equivalent to labour and local raw material costs up to a maximum of 25 percent of export revenue.

There are tax incentives laws, besides the two incentives laws discussed so far, which are sources of subsidies to domestic industries. These incentives are mainly in the form of exemption from import duties on capital goods and raw materials and from compensating sales taxes. Among the industries which are allowed such tax incentives are cottage industries, chemical fertilisers, mining, textiles, and ship building industries. The product coverage of all these exemptions is quite extensive, with the special laws providing exemptions numbering about 75 in all.

A major avenue for granting of fiscal incentives is through the operation of the Export Processing Zone Authority (EPZA), which is charged with the development and operation of the Export Processing Zone in Battan and also with the establishment and development of other export processing zones in the country. The EPZA is authorised to give incentives to manufacturing firms which locate their plants inside the export processing zones.

The structure of protection before 1981

Variation in sectoral protection rates

Estimates for the Philippines in 1974 presented in Appendix 1 show that the nominal rates and the effective protection rates on ERPs differ substantially among sectors. The calculations refer to the effective protection of value added by both primary factors and nontradeable inputs and the nominal rates reflect tariffs, export taxes and the differential impact of indirect taxes.[4] The ERPs, however, are more dispersed around their mean than are the nominal rates, as shown by the coefficients of variation of 203.1 percent for the effective rates compared with the 99.7 percent for the nominal rates. This suggests the magnitude of costs involved or the benefits forgone by the economy, which allows the levels of efficiency as indicated by the incentives required by its industries to operate and survive, to vary widely from industry to industry.

The nominal rates of protection, which were derived from the tariff and indirect tax rates, yielded a supply weighted mean of 44.8 percent in 1974. On the whole, the nominal rates of protection exhibit characteristics widely ascribed to most tariff structures such as the 'cascading' of tariff rates, meaning zero or low rates are identified with raw materials and fuels, higher rates with semimanufactures and highest rates with finished products. In general, intermediate products and semimanufactures including semiprocessed food and industrial materials and equipment fall within 20–70 percent range while finished products and manufactures recorded much higher rates. Nominal rates on the basis of tariff rates alone are generally lower than those based on both tariffs and indirect taxes. Protection of industries, while mainly resulting from tariff protection, appears to be markedly strengthened by domestic indirect taxes.

The country's structure of ERP seems to be penalising a number of domestic industries relative to their foreign competitors as indicated by the negative ERPs of a number of sectors. Despite the penalty, they continue to survive and therefore can be considered economically efficient. On the other hand, there are also many industries with very high ERPs indicating that either these industries are so relatively inefficient that they require very high ERPs in order to continue operating, or enjoy very high excess profits due to the very high ERPs offered them by the protection system.

ERPs of major industry groups In general, the primary and agricultural sectors have much lower ERPs than the manufacturing sectors, which indicate, the protective bias for the domestic processing industries. Export oriented industries[5] have mostly negative ERPs implying a bias against exports in the protection structure. This can be traced to the imposition of the export tax on a number of

Table 2.4 Average effective protection rates according to major industry group, 1974: Philippines (In percent)

Industry Group	ERP
Agriculture and Primary	9
Manufacturing	44
Exports	4
Nonexportable	61
Import, competing	37
Import, noncompeting	148
Overall Average	36

Source: Norma A. Tan, 'The Structure of Protection and Resource Flows in the Philippines', Unpublished Ph.D. dissertation, School of Economics U.P., March 1979.

agricultural and mineral exports, as well as to the tariffs and taxes on the inputs of these commodites in combination with zero tariffs and taxes on exports.[6] These results are summarised in Table 2.4.

Under the drawback system being implemented, all tariff and tax payments on inputs into exports are returned to the exporter in the form of tax credits. This tends to give exports zero protection, except for those export commodities subject to export taxes, in which case drawbacks serve to reduce the penalty from input taxation. Board of Investments (BOI) registered firms, however, get zero protection for their exports since tariff and tax payments are given back in the form of tax credits and are also exempted from the payment of export taxes.

Relatively higher effective protection would seem to characterise the import noncompeting[7] industries for which a weighted average of 148 percent is estimated. For the import competing industries, the average ERP is 37 percent, much lower than the 148 percent average derived for the import noncompeting group. This would seem to suggest a correlation between an industry's level of protection and the proportion of the domestic market locally supplied. Where effective protection is less adverse to imports, it would seem plausible for imports to gain a bigger share of the domestic market while they encourage cost reduction in the home industries through competition.

On average, the ERP estimates are significantly higher than the nominal rates. A close similarity is observed in the ranking of industries by the nominal rates and by the ERPs. In general, the effect of differential tariffs on inputs and on output is to exaggerate the range of protection given to industries.

In terms of discrimination according to the level of processing, however, effective protection in 1974 hardly deviates from the pattern observed in 1965. Consumption goods receive very high ERPs. Intermediate goods receive substantially lower protection.

Table 2.5 Average effective protection rates according to end use, 1965, 1974: Philippines (in percent)

Industry Group	1965[a]	1974[b]
Consumption goods	70	77
Intermediate goods	27	23
Inputs into construction	55	16
Capital goods	16	18
Total manufacturing	51	44

Sources: a Estimated in J. Power and G. Sicat, *The Philippines: Industrialization and Trade Policies*, (London, New York and Kuala Lumpur: Oxford University Press, 1971).
b Norma A. Tan, 'The Structure of Protection and Resource Flows in the Philippines', Unpublished Ph.D. dissertation, School of Economics U.P., March 1979.

Capital goods and inputs into construction receive the lowest rates. Estimates for these industry groups are shown in Table 2.5.

Three elements of bias are, therefore, apparent from the structure of effective protection. First, the bias in favour of manufacturing over the other sectors; second, the penalty imposed on exports, both within manufacturing as well as in nonmanufacturing industries; and third, the bias in favour of the finishing stages of producing consumption goods over intermediate goods and, especially, over capital goods.

Subsidy effects of BOI incentives on ERPs

In addition to tariffs and indirect taxes, the incentives available to firms registered with the Board of Investment (BOI) constitute another source of protection, in view of the number and magnitude of these incentives availed of by firms in recent years.[8] Incorporating the BOI subsidies into the ERP measure provides a second set of ERPs covering a smaller number of industries for which subsidy rates could be estimated. These are shown in Table 2.6.

Thus, compared to tariffs and indirect taxes, BOI subsidies prove to be a very minor source of protection to domestic industries. Benefits to export in terms of BOI subsidies in the national investment and priorities plans, which are heavily directed toward export oriented industries, are too small to offset the general penalty on exports that the system of protection imposes. The overall relative incentive effect of the combined tariff indirect tax BOI subsidies to industries largely reflect the incentive effects of tariffs which dominate the effects of the indirect taxes and tax subsidies. The estimated effect of BOI subsidies on industries slightly reduces the inefficiency—promoting variations in the ERP.

The structure of protection after 1981

Tariff reform for industrial development

The above findings confirm the misgivings on the appropriateness of

Table 2.6 Implicit tariffs and effective rates, as affected by BOI subsidies, classified by products of BOI registered firms, 1974: Philippines (In percent)

I-O Sector	Products	Effective rates: Without subsidies	Effective rates: With subsidies
	Agro-based		
3	Fruit production	−6.0	−6.0
39	Livestock production	123.0	128.0
25	Marine products	116.0	116.0
59	Cassava starch, cornstarch	650.0	670.0
103	Coconut oil	−5.0	−2.0
57	Processed coconut products and byproducts	−10.0	−10.0
40, 41	Processed food and)		
45, 46	beverage products)	445.0	449.0
48, 52	)		
54, 56	)		
16	Ramie	8.0	154.0
74, 75	Fibre products	14.0	19.0
83, 82	Wood and wood products)	10.0	14.0
84, 86	)		
87	Pulp and paper; pulp of straw and abaca	38.0	43.0
85	Handicrafts	0.0	0.0
	Mining and mineral processing		
32	Iron ore	−13.0	−10.0
33	Copper	−4.0	0.8
35	Nickel, silicate ore	2.0	2.0
121	Primary steel	0.0	11.0
119	Rock aggregate	26.0	27.0
117	Dinnerware)		
	Ceramics, ceramics products)	31.0	36.0
116	Glass products, laminated safety glass	45.0	49.0
	Metal based		
129–134	Machinery, equipment and parts	91.0	92.0
120–128	Metal products	28.0	29.0
136–138	Electrical equipment and electrical products	50.0	62.0
	Chemical based		
101, 111	Industrial chemicals, chemicals and chemical products	0.0	6.0
72	Synthetic fibres)		
	Textiles, textile products)	78.0	84.0
150	Plastic products	194.0	194.0
107	Medicinal and pharmaceutical products	9.0	10.0
79	Garments	−26.0	−22.0

Source: Norma A. Tan.

this, the structure of economic incentives which has prevailed for well over two decades in this country. The study[9] of the system of protection in 1965 yields the implication that the economy's resources have not been efficiently allocated among industries, as the more socially profitable and efficient industries appeared to have been put to a greater disadvantage by protection than the less socially profitable and inefficient ones. It also indicates the overwhelming bias of the protection system in favour of manufacturing over agriculture, the penalty against exports, and the bias in favour of the finishing stages of producing consumption goods over intermediate goods and, especially, over capital goods. This problem, however, is not peculiar to the Philippines. In a study[10] of economic incentives arising mainly from protection, it was shown that in developing countries, the Philippines included, the structure of protection has often actually created biases within the range of industries, which generally are against exports and primary sectors of the economy, and in favour of manufacturing for domestic consumption.

More recent studies[11] indicate the inhibiting effect on productivity that arises from a structure of market incentives heavily influenced by protection. Particularly important is the finding that the relatively more inefficient and high cost industries are also the more highly protected industries. The implication is that resources may have been misallocated from the more efficient sectors to the less efficient ones through the attraction of profits artificially raised by tariffs.

In view of the above consideration, some changes in the tariff system were made in 1981. These revisions mainly consisted of overall reduction and evening out of tariffs and effective protection rates across industries. Details of the tariff reform program have been discussed earlier in the policy section of this paper.

Greater uniformity

The structure of protection to industries implied by the tariff reform can be examined on the basis of nominal tariffs and effective rates of protection (ERPs). By aggregating tariff headings or lines and the revised schedule of tariff rates according to the industrial classification of the Input-Output (I-O) Table of the Philippines, the impact of the tariff reform on industrial sectors can be assessed.

Rates of protection were estimated for the 127 tradeable sectors of the I-O Table and are presented in Appendix 1. It appears that there remains a substantial variation in both the nominal rates and ERPs across sectors. This, however, represents a more even schedule of rates compared to the schedule estimated for 1974. The coefficient of variation for the ERPs is 176.4 percent while that for the nominal rates is 83.4 percent. Both coefficients are lower than the earlier estimates before tariff reform.

Nominal protection

A lowering of nominal rates for most sectors is apparent as a result of reform. Nominal rates derived from the revised tariff rates in combination with the prevailing indirect tax rates yield a weighted average of 21 percent. Protection to industries can be expected to continue to result mainly from tariff protection, although it is markedly strengthened by the effects of indirect taxes. The protection to industries from indirect taxes arises from the fact that

Table 2.7 Average effective protection rates according to end use, 1974 and post tariff reform: Philippines (In percent)

Industry group	Nominal	ERP	
	Post tariff reform	1974	Post tariff reform
Consumption goods	40	77	42
Intermediate goods	22	23	33
Inputs into construction	27	16	31
Capital goods	28	18	25
Total manufacturing	29	44	36

Source: As for Table 2.5 and Norma Tan, 'A Program for Tariff Reform in the Philippines', Tariff Commission of the Philippines, 1980.

the country's tax structure taxes imports more than it does the that the country's tax structure taxes imports more than it does the domestically produced output of an industry. The tax rates applied on sectoral imports are greater than the tax rates applied on their domestic counterparts by some percentage mark up of the basic tax rate. This is also true for the *ad valorem* equivalents of specific taxes applied on a number of sectors.

In general, the tariff reform does not significantly change the cascading structure of tariff rates. Low rates continue to be associated with raw materials, higher rates with semiprocessed products, and highest rates with finished products. This is shown in Table 2.7, which gives the average nominal rates for industrial categories according to the degree of processing and end use.

The nominal rates, determined largely by the tariff rates, continue to be highest for consumer luxury and inessential items. These are substantially lower than the prevailing rates because of the reduction of the peak rates from 100–70 percent to the maximum 50 percent. However, the effect of indirect taxes brings the nominal rates for luxury and inessentials to higher than 50 percent, notably distilled, rectified and blended liquors, cigarettes, wines, motor vehicles and jewellery.

Effective rates of protection

The combined effects of tariffs and indirect taxes on both the final product and inputs into the final product yield the effective rates of protection on these sectors. The ERP estimates on the basis of the revised tariff rates and existing rates of indirect taxes give an average of 28 percent. This is higher than the average of 21 percent for the nominal rates. The implication remains that, in general, the weighted average of the tariffs on the inputs must be less than the weighted average of the tariffs on the products. This appears to be consistent with the cascading structure of nominal tariffs. In general, the effect of differential tariffs on input sectors and on output sectors is to

Table 2.8 Average protection rates according to major industry group: Philippines (In percent)

Industry group	Nominal	ERP	
	Post tariff reform	1974	Post tariff reform
Agriculture and Primary	4	9	3
Manufacturing	29	44	36
Exports	4	4	12
Nonexportables	36	61	45
Import, competing	35	37	35
Import, noncompeting	37	148	67
Overall Average	21	36	28

Source: As for Tables 2.4 and 2.7.

strengthen the degree of protection given to industries. The general restructuring of the nominal rates due to the tariff reform reduces the level of overall protection to 36 percent for all industries. The ERPs of individual industries, however, still vary across sectors, although a marked narrowing of dispersion across sectors is apparent.

ERP structure of industry groups

The primary and agricultural industry group continues to receive very low protection (Table 2.8). In contrast, manufacturing as a group continues to be more protected. Its ERP, though lower than it was for 1974, still demonstrates the same overwhelming protective bias for the domestic processing industries. The very high ERPs of manufacturing industries compared to those of agriculture indicate the continuance of the policy to encourage the domestic processing in manufacturing sectors of raw materials and agricultural commodities.

Export oriented sectors, on the whole, continue to receive much lower protection than nonexportables. The tariff reform, however, reduces the relative bias against exports. The low ERP of exports can be traced to the continued imposition of the export tax on a number of agricultural and mineral exports, as well as to the tariffs and taxes on the inputs of these commodities in combination with zero tariffs and taxes on exports.

With the same drawback system being implemented, all tariff and tax payments on inputs into exports are returned to the exporter in the form of tax credits. This tends to give export zero protection except for those export commodities subject to export taxes. The drawbacks serve to reduce the penalty resulting from taxes on inputs. Board of Investments registered firms, however, get zero protection for their exports since tariff and tax payments are given back in the form of tax credits and are also exempted from the payment of export taxes.

The nonexport oriented industry group continues to get an overwhelming, though lower, advantage in the protection system, with this group getting about four times the level of protection of export oriented sectors. However, the tariff reform means a significantly lower ERP of 45 percent compared to the 89 percent it received. This drop in the ERP is chiefly due to the sharp reduction in the ERP of the import noncompeting industries for which an average of 67 percent is derived compared to the previous ERP of 185 percent. This is expected to encourage efficiency among this previously highly protected group of industries through competition from imports.

For the import competing industries, the overall level of protection goes down from 49 percent to 35 percent after the tariff reform. Compared with the ERP of 185 percent for the import noncompeting industries, this represents a narrowing of the edge of import noncompeting industries over import competing industries from 4 to 1 to 1.9 to 1 ratio. This is expected to realign the overly protected import noncompeting industries with the import competing industries by making it easier for imports in the former group to gain a bigger share of the domestic market while they encourage cost reductions in the home industries through competition.

The reduction in industrial ERPs is seen to be borne mainly by the consumer goods industries. This industry group receives a much lower level of protection of 42 percent than the 102 percent it enjoyed before. The structure of protection remains biased against the capital goods industry as it continues to receive the least protection.

Thus, the protection structure continues to favour the relatively finished and more processed products as compared to the semi-finished and intermediate goods. It also continues to be particularly biased against agricultural and raw materials and capital goods. However, the spread of protection levels is more even than before. The levelling effect is attributable to the sharp reduction of protection to consumer goods combined with the significant improvements in the protection to intermediate goods and inputs into construction.

Causes of protection in the Philippines

Hypotheses

Information on the causes of protection can give useful insights into the continuing importance given to it in the development processes of many nations. The economics literature provides certain arguments by proponents of the protection of specific industries and of protectionism in general. In addition, there may be certain circumstances in the socio-political and economic environment which are conducive to maintaining or increasing the level of protection. These arguments and propositions form the threads in the tapestry of the causes of protection under study.

Several of these arguments and propositions may be articulated to form the basis for the hypotheses on the causes of manufacturing protection in particular.[12] First, protection is perceived to be desirable for improving the income distribution within the country; protection is desired if it would benefit the poorer sectors and would maintain or expand employment and output. Second, protection is desired for nurturing 'infant industries', which would otherwise not grow without protection. Third, protection is desired for providing a 'balance' in the development of both manufacturing and agriculture sectors to insure the country's self sufficiency in strategically essential commodities and to avoid the risk of specialisation in agricultural production.

A fourth proposition points to the readiness to seek and grant protection to industries which are seen to be subject to injurious import competition. This view particularly gains strength if injury to the domestic industry is believed to be made through such unfair methods of import competition such as patent infringements. A fifth proposition would support protection in retaliation for 'unjustifiable or unreasonable tariff or import restrictions or other acts of policies which burden, restrict, or discriminate against the country's trade.' A sixth proposition would explain the readiness to grant protection to an industry by the severity of its overall impact on the national economy. A number of factors that may be considered include: the effects on the price to consumers of the imported article and of its domestic substitute; the effect on other domestic industries of any retaliation; the economic and social costs to be incurred by tax payers, communities and workers if protection is not granted; the probable effectiveness of not granting protection to the industry as a means of promoting adjustment; and the flexibility or mobility of factors of production to be displaced if protection is not granted.

All these arguments stress the role of protection in serving the national interest. The arguments for less protection to serve the national interest would be the income benefits from greater specialisation in production, the improvement in international relations with other countries, and a more equitable distribution of income among nations through better access by developing countries to the markets of the industrialised countries.

Other propositions would stress the importance of political factors in influencing protection policies. This would explain protection policy decisions in terms of the net effect of opposing electoral pressures representing various political and economic interests in the country. The decision to protect a certain industry would then depend on whether it had an edge in terms of overall support from the electorate. In this context, petitions for protection would tend to get favourable action if the industry employment was large or the potential injury to the industry would greatly affect employment in that industry, if the request and decision for protection occurred before an election, if these petitions were made at about the time the country was participating in multilateral trade negotiations. On the other hand, favourable decisions for protection might be unlikely if these appeared to deviate from previous pronouncements of major policy.

Another proposition would focus on the ability of groups in organising and lobbying for the protection of their vested interests. This involves the ability to lobby, use press releases, influence the politicians, threaten layoffs, or use production and capacity utilisation rates to achieve their objectives. The strength of this ability and the likelihood of protection would depend on the number of firms in the industry, the size of the labour force, and the financial capacity and voting power of the industry.

In the context of the historical perspective in which Philippines industrial protection has taken place in a span of three decades, some other explanations of protection may be advanced.[13] These take off from perceptions of the avowed policy goals which have evolved over the years of planned development in the country. To the extent that protection is considered an important policy instrument for attaining national development goals, then the nature of protection may be expected to have evolved to reflect the influences of certain biases towards these goals.

The Philippine tariff structure was started primarily as a revenue raising instrument. Then, it was made to restrict imports to reduce the balance of payments deficits which had historically characterised the Philippine trade position. The tariff was set to be higher on goods from which revenues were potentially greatest, i.e. the high value luxury and inessential items. The import volume and value of these items had historically been highest, so the tariff was applied to them in order to achieve the greatest import reduction effect on the balance of payments deficits. Thus, tariff levels may have been determined by the degree of inessentiality of commodities as a result of the country's pursuit of revenue expansion and balance of payments deficit management.

Also, an important concern in policy pronouncements in the country is the efficiency with which resources are allocated among various sectors of the economy. It may be expected that greater encouragement to more efficient relative to inefficient industries has been granted in terms of protection. In view also of the national concern to promote industries which utilise resources in proportions suited to the domestic factor price ratio, it may be expected that protection had been granted to favour the country's abundant resource—labour—by tilting protection towards labour intensive industries. In the light, too, of efforts to conserve the country's supply of scarce foreign exchange, it may be expected that protection has favoured the industries which use relatively less foreign exchange. Finally, the priority given to exports, particularly of manufactures, in the national development strategy would yield the expectation that protection had been influenced by the export potential of industries.

Regression results

The causes of protection in the manufacturing sector can then be specified in terms of the variables suggested by the various hypotheses. The number of possible explanatory variables mentioned in the above discussion is quite large. These variables represent a combination of economic, social, and political relationships that underlie the particular setting for the determination of industrial protection in a particular country. There is, however, a lack of sufficient accurate and consistent data on most of these variables which corresponds with the effective protection rates estimates at I-O classification level. Therefore, the explanatory variables in the model are limited to a few for which some confidence with the data can be had.[14] The data for these variables were obtained or derived mainly from the 1974 I-O tables to insure consistency with the ERP estimates. Data were directly derived from I-O data and other sources.[15] The preferred equation was:

$$\begin{aligned} ERP_j = 176.42 &- \underset{(1.84)^{*}}{425.66\,WSP} - \underset{(3.07)^{***}}{0.61WGR} \\ &+ \underset{(5.98)^{***}}{266.10\,MIP} - \underset{(9.31)^{***}}{162.66\,MSS} - \underset{(1.96)^{*}}{135.83\,XSP} \end{aligned}$$

$(R^2 = 0.17, F = 4.50, n = 119)$

where ERP_j = effective rate of protection in industry j in 1974;
WSP = wages share in value of output;
WGR = wage rate;
MIP = imported inputs to output ratio;
MSS = import share in supply;
XSP = exports to output ratio.

Some implications may be drawn from the above regression results. The negative and very significant coefficient for MSS indicates that industries with lower import share are associated with higher protection. A likely explanation for this is that high protection has tended to be effective in inducing import substitution and reducing the share of imports in domestic supply. The positive coefficient for MIP indicates that the national concern to conserve scarce foreign exchange, by discouraging import intensive industries, may have given way to the clamour for higher protection by industries to cover higher production costs due to tariff inflated price of imported inputs. It would seem that protection, instead of being biased in favour of the less foreign exchange using industries, has instead tended to favour industries which use greater amounts of foreign exchange for imported inputs.

The nagative and significant coefficients for WGR and WSP suggest that protection tends to be higher for low wage industries as well as for those where the wages to output ratio is low. However, the insignificant coefficients for other factor intensity indicators, such as wages share in value added or capital/labour ratio, suggest that protection has not necessarily favored labour intensive industries nor is it significantly correlated to any factor intensity bias.

The insignificant result for value added share in output suggests that protection is not necessarily correlated with the ability of industries to contribute to national output as measured by the proportion of value added in value of output of industries. The insignificant results for the variables which may be considered as indicators of financial and lobbying power, such as value of profits, profits to output ratio, growth rate of output or capital utilisation, suggest that these factors have not been of significant influence.

A significant negative correlation for XSP is indicated. Greater export orientation appears to be associated with low protection, suggesting that the policy concern to promote comparatively competitive industries has not been pursued through protection; rather a bias against the more export oriented industries is apparent.

Overall, however, the regression results explain only a small proportion of the variation in effective rates. This implies the included variables fail to capture the combined influences of public interest motivation and private interest lobbying.

Conclusion

The Philippine experience in protecting its manufacturing sector started inadvertently as a temporary measure to meet a balance of payments crisis in 1949. This was initially in the form of quantitative import controls which were later to be made more permanent through the institution of Central Bank control of foreign exchange plus the adoption of a system of multiple exchange rates. The protective effect of this spawned the growth in the 1950s and the 1960s of many inessential and luxury goods industries which the system favoured. These very same industries were soon unwilling to give up the protection, which was not intended at the inception of control. This, as well as the vast opportunities for graft and corruption which accompanied the administration of controls, combined to perpetuate the system of controls until the mid 1960s. It is not surprising, therefore, that the same structure of protection emanating from import exchange control was virtually left unchanged by the tariff system of 1957, which took over as the principal instrument of protection when controls were dismantled.

Prior to 1981, protection of Philippine maufacturing appeared to differ substantially among sectors. A number appeared to be penalised but the majority received very high effective protection rates. The protection bias for manufacturing is indicated by the much higher average rate of ERP than that of the primary and agricultural sectors. Export oriented industries have low or negative ERPs, implying a bias against exports in the protection structure. There is also evidence of discrimination among industry groups. Consumption goods received very high protection rates. Intermediate goods, inputs into construction, and capital goods received a comparatively much lower rate.

After the tariff reform of 1981 and particularly by its full implementation in 1985, there will be greater uniformity in the levels of protection within manufacturing. The protection structure will continue to favour the relatively finished and more processed product as compared to the fairly unfinished and intermediate goods. It will also remain to be particularly biased against agriculture and raw materials. However, this spread of protection level will be much more even than at present. The levelling effect is attributable to the sharp reduction in the ERP of consumer goods.

On the basis of observation of looking at results of broad categories of data pertaining to a number of characteristics of manufacturing industries, there appear just a few variables that strongly explain the variation in protection level. Low share of imports in supply appears to be associated with high ERP, but there are two directions of causation at work in that case. The high ERP associated with high ratio of imported inputs to output may indicate the success of domestic producers in getting higher protection so as to offset the tariff inflated costs of imported inputs. The high ERP associated with low wage industries may indicate the sensitivity of policy makers to such a politically important segment of the populace as the low wage workers. Finally, the low ERP associated with the more export oriented industries confirms the bias against exports in the protection system, which is explained by the penalty on export industries arising from tariffs applied on their inputs.

Appendix 1

Table 2.A1 Nominal and effective protection rates: Philippines (In percent)

1975 I-O Code	Sector	Nominal rates of protection		Effective rates of protection	
		1975	1980	1975	1980
3	Bananas	−4	−4	−6	−4
4	Citrus	62	42	70	44
5	Pineapple	0	0	−3	−3
11	Coffee	50	51	50	52
12	Cacao	7	10	6	4
14	Coconut including copra (in farms)	−6	−10	−6	−7
15	Abaca	−4	−3	−6	−4
16	Ramie and other fibre crops	10	23	8	15
17	Tobacco (native and virginia)	5	7	21	16
25	Commercial fishing, ocean, and offshore	100	51	116	51
29	Logging	−10	−19	−10	−21
32	Iron ore mining	−4	−4	−13	−9
33	Copper ore mining	−2	−4	−4	−6
34	Chromite mining	−4	−4	−13	−8
35	Other metallic mining	2	2	2	3
38	Other nonmetallic mining and quarrying	20	25	16	13
39	Slaughtering and poultry dressing	25	11	128	NT
40	Meat products, canned	25	9	5	14
41	Meat products, uncanned	39	26	68	61
42	Evaporated and condensed milk	16	11	5	6
44	Butter, cheese, and other dairy products	33	35	52	50
45	Fruits and vegetables products, canned	19	11	30	15
46	Other preserved fruits and vegetables	13	10	19	7
47	Fish canning	11	8	−24	14
48	Other fish and seafood products	55	24	76	37
49	Rice milling	0	0	−2	−2
50	Corn milling	0	0	−2	−2
51	Flour milling, cereal and flour-blended	41	39	925	804
52	Bakery products	110	51	480	78
53	Sugar milling and refining	−6	−6	−12	8
54	Candy and chewing gum products	86	54	519	119
55	Cocoa and chocolate products	51	29	1,325	153
56	Processed coffee	46	37	36	30
57	Dessiccated coconut products	−4	−3	−10	−3
59	Starch and starch by products	80	34	650	99
60	Macaroni, spaghetti and noodles	63	31	78	41
61	Vegetable lard and margarine	82	45	262	412
62	Prepared feeds for animals and fowls	33	29	35	91
63	Flavouring extracts	9	4	7	9
64	Miscellaneous food manufactures, n.e.c.[1]	59	38	156	83
65	Distilled, rectified and blended liquors	209	159	394	222
66	Wines	148	79	113	120

Table 2.A1 Continued

1975 I-O Code	Sector	Nominal rates of protection 1975	Nominal rates of protection 1980	Effective rates of protection 1975	Effective rates of protection 1980
67	Brewery and malt products	72	51	69	33
79	Cigarettes	149	69	205	120
70	Cigars, chewing, and smoking tobacco	0	0	−21	−13
71	Leaf tobacco processing	115	66	*	*
72	Textile mill products	57	44	78	65
73	Knitting mill products	23	18	14	6
74	Cordage, twine and net industries	0	2	−2	2
75	Carpets, rugs, and linoleum including mats	43	23	43	22
76	Other textile products	48	44	61	42
77	Footwear, except rubber and plastic	19	11	18	11
79	Ready made clothing	18	12	27	4
80	Manufacture of embroidered products	68	36	92	34
81	Other made-up textile goods	19	18	13	17
82	Lumber	−4	0	16	50
83	Plywood and veneer plants	−4	−1	5	28
84	Doors, windows, and other millworks	0	2	−2	7
85	Other wood, cane and cork products	0	0	0	8
86	Furniture and fixtures	8	7	21	17
87	Pulp, paper, and paperboard manufacturing	43	32	38	34
88	Paper products	109	39	195	51
89	Paper and paperboard containers	128	52	181	50
90	Miscellaneous converted paper products, n.e.c.	72	27	478	28
92	Books and pamphlets	25	28	19	36
95	Tanning and leather finishing	68	28	145	51
96	Leather products except footwear apparel	20	21	−27	30
97	Rubber footwear	108	59	454	69
98	Tyres and inner tubes manufacturing and retreading	103	40	323	50
99	Other rubber and related products	27	43	21	38
100	Compressed and liquefied gas	25	28	17	24
101	Basic industrial chemicals	17	22	−7	31
102	Fertiliser and lime	31	34	41	23
103	Coconut oil	−4	−4	−5	−2
104	Other oils and fats	24	32	32	42
105	Paints, varnishes, and related compounds	95	52	221	49
106	Plastic materials	45	33	56	28
107	Medicinal and pharmaceutical preparations	23	28	9	15
108	Cosmetics and toilet preparations	249	138	*	108
109	Soap and other washing and cleaning compounds	61	51	175	155
110	Insecticides, germicides, and agricultural chemicals	19	31	17	41
111	Other chemicals products	42	44	35	41
112	Petroleum refineries	21	24	21	30
113	Other products of petroleum and coal	25	27	16	27
114	Structural clay products	5	11	−11	16
115	Structural concrete products	57	61	110	107
116	Glass and glass products	56	38	45	36
117	Pottery china and earthenware	47	26	31	38
118	Hydraulic cement	−4	0	−36	−8
119	Other nonmetallic mineral products	28	32	26	35

120	Basic ferrous metal products	35	30	27	19
121	Basic nonferrous metal industries	10	24	0	20
122	Metal cans, boxes, and containers	57	36	110	59
123	Cutlery, handtools, and general hardware	39	34	34	35
124	Structural metal products	65	39	95	51
125	Stamped, coated and engraved metal products	37	37	38	87
126	Fabricated wire products	28	34	14	42
127	Heating apparatus, lighting, etc.	59	43	85	47
128	Other fabricated metal products	62	41	79	43
129	Agricultural machinery and equipment	22	33	14	28
130	Other special industry machinery and equipment		13	34	436
131	General industry machinery and equipment	18	22	7	17
132	Office computing and accounting machines	29	26	27	40
133	Electrical distribution and control apparatus	25	22	18	19
134	Other electrical industrial machinery and equipment	33	22	30	20
135	Communication equipment	31	23	31	17
136	Batteries	56	37	73	38
137	Electrical lamps and fixtures	38	26	27	23
138	Electrical wires and wiring devices	40	23	51	18
139	Household radio, TV sets, phonos, and supplies	164	80	204	65
140	Refrigeration and air-conditioning equipment	142	70	195	69
141	Other household electrical appliances and wares	95	61	103	62
142	Shipbuilding and repairing	17	23	26	17
143	Motor vehicles, manufacturing/assembled	93	90	127	20
144	Motor vehicles, engines, bodies and parts	29	40	23	35
146	Motorcycles, bicycles, and parts	50	55	52	53
147	Other transport equipment, n.e.c.[1]	6	24	9	12
148	Jewellery, silverware, and related articles	91	82	133	97
149	Musical instruments	60	73	61	71
150	Fabricated plastic products	92	54	194	70
151	Measuring, controlling, scientific equipment	15	20	12	34
152	Medical, orthopaedic and surgical supplies	14	21	9	35
153	Photographic and optical goods	37	40	30	48
154	Sport equipment and supplies	91	39	93	41
155	Pen, pencil, office and artist's supplies	66	40	68	43
156	Toys, dolls, parlour games excluding plastic	74	22	72	34
157	Miscellaneous manufactures, n.e.c.[1]	61	48	91	72

Note:
1 n.e.c. means not elsewhere classified.
* ERP cannot be calculated for the sector because its derived international value added is negative.

Endnotes

1 Power, J. and Sicat, G. (1971), *The Philippines: Industrialisation and Trade Policies*, London, Oxford University Press.
2 See Table 1 in R. Bautista, J. Power and Associates, (1980), *Industrial Promotion Policies in the Philippines*, Philippine Institute for Development Studies, p. 6.
3 See Tan N. in R. Bautista, J. Power and Associates, p. 163–168.
4 The estimation methodology is presented in more detail in Tan (1979).
5 Industries which export more than 10 percent of their production.
6 Because of the aggregative level of the input-output sectors, some activities classified under a sector considered as export oriented by definition, are import

competing (industries which import at least 10 percent of domestic supply) which have positive tariffs and taxes. This resulted in ERPs which are positive for several export oriented sectors.

7 By definition, less than 10 percent of total domestic supply is imported.

8 Apart from BOI subsidies, there are possibly other important subsidies available to firms such as accessibility to low interest and long term loans. Their inclusion in the ERP measure was not attempted because of the unavailibility of consistent data on a sectoral level of aggregation.

9 See Power, J.H. (1967), 'Import Substitution as Industrialization Strategy', *Philippine Economic Journal*, Vol. V, (spring 1967), 167–204.

10 See Balassa, B., (ed.) (1971), *The Structure of Protection in Developing Countries*, Baltimore: The Johns Hopkins University Press.

11 See Bautista, Power and Associates (1980), *Industrial Promotion Policies*.

12 See Anderson, K. and Baldwin, R. (1981), 'The Political Market for Protection in Industrial Countries: Empirical Evidence', World Bank Staff Working Paper No. 492, Washington, D.C., October.

13 Tan, Norma A. in R. Bautista, J. Power and Associates (1980).

14 Variables tested included wages share in value added, wages share in value of output, value added share in output, growth rate of output, wage rate, capital per worker, value added per worker, value added to capital ratio, value of profits, profits to output ratio, capital utilisation rate, domestic resource cost, imported inputs to output, import share in supply and exports to output ratio.

15 Bautista, R.M., and Teeson, G.R. (1978), 'Plant Characteristics and Export Orientation in Philippine Manufacturing', *Philippine Review of Business and Economics*, XV, September, 1978; and Bautista, R.M. (1978), 'Domestic Resource Costs in Philippine Manufacturing Industries', *Philippine Economic Journal*, SVII, November 3, 1978.

3 Manufacturing industry protection in Thailand: issues and empirical studies

NARONGCHAI AKRASANEE AND JUANJAI AJANANT*

April 1984

The history of trade and tariffs

DURING the 17th century, Siam was brought much closer to the European powers through trade. In the reign of King Narai (1657–1688), foreign trade was at a high level.[1] Descriptions of goods imported and exported are numerous but, unfortunately, statistics of value and volume of trade were not kept. As far as we know, trade was conducted with China, the Dutch East Indies, India, Japan, and the European powers (i.e. France, Britain, Holland and Portugal).

In the early part of the 18th century, Siam had fought wars with Burma, Cambodia, Laos, and Vietnam. Foreign trade was not significant during this period. Later that century and until 1824, the primary concern of the Siamese government was to keep foreigners away from Siam and, hence, trading with other countries did not expand. There was a change in policy in the reign of King Rama III (1824–1851), when the Burney Treaty was signed to foster trade with Britain. Nevertheless, the treaty put a constraint on trade. Under this treaty, British subjects were placed under Siamese jurisdiction. They were forbidden to build, hire or buy shops and houses. The King could, through his provincial governors, prevent British merchants from trading in any locality. Opium was an illegal commodity. Additionally, British ships entering the port of Bangkok were levied 1,700 baht (210 pound sterling) per fathom as duty. The exportation of rice and paddy was prohibited in the treaty. Hence, there were serious restrictions of trade between Siam and foreign powers.

* Narongchai Akrasanee is Managing Director, The Industrial Management Co. (IMC) Ltd. and Senior Vice President, The Industrial Finance Coporation of Thailand. Juanjai Ajanant is Assistant Professor of Economics, Chulalongkorn University, and Senior Consultant to IMC.

The Burney Treaty was enforced until 1855 when Sir John Bowring, the British envoy, went to Bangkok with the single aim of signing a new treaty which removed all restrictions on trade with Britain. King Rama IV (1851–1868) realised that the fate of his kingdom depended on negotiating with the British envoy. Western occupation of China and the opening up of Japan exemplified the changing trend of world affairs, and Rama IV was pragmatic enough to negotiate with Sir John Bowring. The Bowring Treaty, concluded in April 1856, liberalised Siamese trade as never before. It abolished the measurement duties, and import and export duties were fixed.

The import duty was fixed at 3 percent for all articles, with two exceptions: opium was to be free of duty, but it had to be sold to an opium farmer; and bullion was to be free of duty. Articles of export were to be taxed just once, whether the tax was called an inland tax, a transit duty, or an export duty. The Siamese government reserved the right to prohibit the export of salt, rice, and fish whenever these items were deemed to be scarce. The most favoured nation clause was also included.

The Bowring Treaty represented a substantial surrender of control by Siam. The 3 percent ceiling on import and specialised export duties took the fiscal control out of the hands of Rama IV and his successors. Of the 64 commodities which were enumerated in the tariff attached to the treaty 51 were not to be subject to land taxes of any kind. The remaining 13 commodities were not to pay any export duty and were to be subject to certain inland or transit duties (but not both), the rates of which were also specified in the treaty.

The treaty set a pattern which other countries were quick to follow. The following countries concluded treaties with Siam, all of them similar in most respects to the first:[2]

US	1856	Norway	1868
France	1856	Belgium	1868
Denmark	1858	Italy	1868
Portugal	1859	Austria – Hungary	1869
Netherlands	1860	Spain	1870
Germany	1862	Japan	1898
Sweden	1868	Russia	1899

The absence of fiscal autonomy lasted for 70 years. During those years, the Siamese government had to finance expenditure from other taxes. The Siamese government tried to negotiate with foreign nations to amend the treaties. Such attempts were not successful. The foreign powers were adamant to keep the treaties as they were. In 1883, the treaties were amended to raise duty on liquor imports to 10 percent, but the general rate of duty remained at 3 percent. As a percentage of total revenue, import duties amounted to less than 3 percent in 1898/99, 5 percent in 1908/09 and 4 percent in 1918/19. Even in free trade Britain, customs revenue was over 20 percent of the total. Thus, import duties were not significant in Siam even after 1856.

During King Rama VII's reign, Siam's efforts to renegotiate treaties began to show positive results. In 1920, US and Siam signed a treaty which recognised the complete fiscal autonomy of Siam, subject to the conditions of equality of treatment with other nations. By 1925/26, all nations except Britain agreed to amend their treaties with Siam. The most difficult task was to amend the treaty with Britain, and the end result of negotiations was a compromise. The new Britain-Siam Treaty placed a limit of 5 per cent *ad valorem* for 10 years on tariffs on cotton, iron and steel manufactures and machinery. These commodities comprised the major imports from Britain. Notes were also exchanged in which Thailand stated that she did not intend to increase the export duties on rice, tin, or teak. These were the items in which Britain had an interest at the retail stage. By 1927 renegotiations were completed and Siam was again free to levy import duties on incoming commodities.[3]

Siam lost no time in altering its tax structure. A new code of tariffs became effective in March 1927, and the revenue from import duties jumped from 7 million baht to 16 million baht within one year. The general rate was increased to 5 percent. In addition, certain articles were subject to high taxes. Beer and wine were taxed at 12 percent; manufactured tobacco at 25 percent; and motor cars and equipment at 10 percent. Special duties were also placed on kerosene, benzine, matches, and sugar, all of which were widely consumed. The new tariff was designed to increase the revenue rather than protect local industry.[4]

During the following years import duties contributed an ever larger part of total revenue. The tariff was increased frequently from 1927 to 1941. As the tariff was increased, imports of some items declined and local production began to emerge. One of the consequences of tariff was that, in some instances, imports declined to a low volume and the revenue accruing to the government fell accordingly. The government had to switch to excise taxation to balance the budget. Through trial and error, the government learned to adjust tariffs and excise taxes so as to trade off revenue against the size of the local industry. By the 1930s tariffs encouraged local production of sugar, soap, cigarettes, and matches. During the decade, the government learnt how to use taxation to influence the structure of industry. Without tariffs, some industries would not have been promoted. The

Table 3.1 Average height of tariff 1892–1960: Thailand

Year	Tariff (%)
1892	11.08
1905	3.91
1906	3.91
1915	4.63
1917	4.76
1926	3.64
1927	7.97
1938	23.59
1941	20.00
1944	7.50
1950	20.05
1951	12.40
1952	n.a.
1953	17.40
1954	17.10
1955	18.70
1956	19.20
1957	17.90
1958	20.10
1959	20.20
1960	21.70

Source: Calculated from Table XV and Appendix B of Ingram

tariff on matches decreased imports and led to the establishment of match factories. In order to replenish the revenue loss, an excise tax on matches was levied on locally produced matches. The excise tax and tariff rate were set to achieve a local production target. Similar experience was gained with sugar, tobacco, and other products.[5]

The purpose of import duties in Thailand was mainly that of revenue collection, not protection. Quantitative restrictions, concessions on revenue duties on materials, and tax holidays were the main protective devices. In general, after some experience, duties were not raised to the levels where they might have a serious effect on the volume of particular imports. The overall ratio of customs duties collected to the total value of imports gives a reasonable indication of the average tariff. This ratio was about 20 percent in 1958–1959, and a little lower in the previous years. Table 3.1 illustrates the average height of tariffs from 1892 to 1960.

At the end of World War II, Pridi Pranomyong headed the civilian government. He was ousted by Field Marshall Pibulsongkram in 1947. Pibulsongkram's government lasted for 10 years before Field Marshall Sarit Thanarat took over. It was recently revealed that Pibulsongkram was reluctant to propose forward looking economic activities.[6] Two reasons underlay his deliberate reluctance: first, the fear of Chinese domination of industry and trade; second, the belief that state capitalism could foster the growth of economy without loosening controls on foreigners.

Table 3.2 Remittances[a] 1953–1962: Thailand (Million baht)

Year	Remittance to Thailand	Remittance from Thailand	Balance
1953	2	6.4	−4.4
1954	0.7	7.8	−7.1
1955	3.8	11.9	−8.1
1956	2.3	11.2	−8.9
1957	1.8	11.1	−9.3
1958	2.2	3.5	−1.3
1959	5.0	1.9	3.1
1960	4.9	2.4	2.5
1961	6.9	2.2	4.7
1962	7.1	2.0	5.1

Source: IMF, Balance of Payments Yearbook, various issues, Washington: IMF
Note:
a Other nationalities (e.g. Indian) are included but they account for a mere 5% of the foreign nationalities in Thailand.

The fear of Chinese domination of the private sector was pervasive long before Pibulsongkram's time.[7] Before the end of Rama IV's reign, retail and wholesale businesses were already in the hands of the Chinese. Rice milling, wood sawing, tin mining, and rubber plantations were controlled by the Chinese. The Chinese community in Thailand was also involved in banking. Had the government remained passive, Chinese business acumen could have dominated other products and services.

Landon, who investigated the Chinese involvement in the Thai economy, estimated that overseas Chinese in Thailand remitted income to China, but that remittances were less than 2 percent of the wealth created by the Chinese. This estimate may not be accurate since it is difficult to define who is Chinese and what items are wealth. Landon also suggested that, 'there would be no progress without the Chinese'.[8] Another view was based on the absolute size of remittances to China. These were to the value of about 1,250 million baht between 1890–1941, which exceeded the amount of fixed investment of railways and irrigation during the same period of only 250 million baht.[9] The figures, however, were guesswork since remittances were sent by remittance agents not through the banks.

After World War II, the amount of remittances did escalate for some years (see Table 3.2) before they tapered off. The decline of remittances was due to the Communist government's control of China in 1949, when the Chinese in Thailand felt the money they sent never reached their relatives. In 1960, remittances under 500 baht formed 90 percent of the total.

It is hypothesised that the size of Chinese remittances and the fear of Chinese domination of business encouraged Pibulsongkram to discriminate against the Chinese. Certain professions such as rice farming, fishing, hairdressing, and public service were preserved for indigenous Thais. Also Pibulsongkram's nationalistic viewpoint complemented his views on the role of state capitalism and, consequently, he expanded the role of state enterprises.[10]

At present there are 72 state enterprises in Thailand and the majority were created during Pibulsongkram's time. In the early 1950s, the government established several industries—textiles, paper, glass, gunny-bags and others. Muscat echoed the view that 'the real driving force behind the government's industrial program was the desire to prevent the Chinese community in Thailand from dominating industry'.[11] The state enterprises also had another function. They were the means of rewarding the loyal friends of the Pibulsongkram clan. This pattern has become the tradition in Thailand.

After World War II, the tariff retained its status as the major revenue earner for the government. However, Pibulsongkram was probably constrained in his use of the tariff for protection.[12] The use of tariffs even for revenue purposes has an unavoidable protective effect on the industrial structure. The constraint may have stemmed from the fear that to further promote the industries in Thailand through the use of tariffs would have encouraged Chinese participation in industries. The structure of the tariff, as indicated by import duties as a percentage of the total imports, remained fairly stable after the war (see Table 3.1). By 1960 import duties were about 30 percent of government revenue, compared to 27 percent in 1950.[13]

Revenue from foreign trade could also be earned by taxes on export items such as rice, tin, and teak. Table 3.3 illustrates the extent to which export duties contributed to government revenue during 1947–1957. These data indicate that export taxes raised a significant and increasing proportion of government revenue over the period. These data are consistent with the hypothesis that Pibulsongkram was deliberately avoiding the protective effects of tariffs and keeping the budget balanced by taxing exports, especially rice.

Another of Pibulsongkram's strategies was to foster the growth of the economy using state capitalism. The significance of government sector investment is illustrated in Table 3.4. While the ratio of government sector investment to total investment was high in the early 1950s, part of the government investment was for social infrastructure. The state enterprises were concentrated in the manufacturing sector of the economy. State enterprises were inefficient and did not become a dynamic sector. The government did hope for revenue from the enterprises, but the end result was a drain on the treasury.

Table 3.3 Significance of export taxes in government revenue 1947–1957: Thailand

Year	(Ratio of rice premium + export tax)/G revenue
1947	3.21
1948	9.81
1949	17.62
1950	15.85
1951	6.72
1952	13.12
1953	23.97
1954	19.34
1955	13.49
1956	19.32
1957	19.99

Source: Mousney, A. (1964), *The Economy of Thailand: An Appraisal of a Liberal Exchange Policy*, Social Science Association of Thailand, p. 37.

Table 3.4 Public and private investment 1952–1957: Thailand (Billion baht)

Year	Total (1)	Private (2)	Public (3)	(3)/(1) (4) (%)
1952	3.7	2.5	1.2	32
1953	4.6	2.8	1.8	39
1954	4.7	2.9	1.8	38
1955	4.7	2.9	1.8	38
1956	5.1	3.5	1.6	31
1957	6.4	5.2	1.2	19

Source: National Income Statistics of Thailand, 1964.

Later, in 1959, a World Bank Mission recommended the government alter the industrial strategy for development. The mission suggested that the government encourage the private sector and concentrate its investment on power, roads, water, and other infrastructural facilities. It was believed by the mission that the presence of state enterprises inhibited foreign investment and that the prospect of receiving aid and loans would be enhanced if the government restrained the role of state enterprises.

Thailand's foreign trade policies and tariff structure can be thus summarised. From 1850 to 1926, the Thai economy was bound by the Bowring Treaty and other treaties with European powers. The consequence of such treaties was the absence of fiscal autonomy for 70 years. The tariff was fixed at 3 percent during the period. The Thai economy remained agrarian in production and exports, with the major exports consisting of rice, tin, rubber, and teak—the rice exports contributing more than half of the total export value. Tariffs played the role of a revenue earner in the fiscal budget with few protective effects. After World War II, tariffs continued to play the same role, with additional revenue from export duties. The Pibulsongkram administration deliberately avoided protecting industries for fear of promoting the Chinese community. Instead of providing incentives in the form of tariffs, it chose to promote state enterprises. In the final analysis, tariffs were never to become a significant protective device for the Thai manufacturing sector between 1850–1960.

The history of trade and tariffs: 1960–1980

Field Marshall Sarit Thanarat ousted Field Marshall Pibulsongkram in 1957 and the change of government signified the beginning of the government's promotion of the private sector. Unlike the Pibulsongkram administration, Sarit's economic policy was to stop constraining the private sector. It was believed the private sector had a dynamism which could be utilised for the welfare of the population. Sarit expected that he could manipulate the Chinese factor, about which Pibulsongkram had been so worried. Tax incentives were used to encourage the investment by private enterprises. During Sarit's time and during subsequent periods, until 1980, there were significant changes in tariff policy. Tariff and other policies from 1955 to 1981 are summarised in Table 3.5.

When Field Marshall Sarit came into power in 1957 he went along with the practices of Pibulsongkram's administration. But this was merely because its trade and economic development policies were already in place. The World Bank Mission of 1959 appears to have convinced Sarit that the private sector needed encouragement.[14] A period of modernisation and import substitution followed in which the economy expanded very rapidly.

The first National Economic and Social Development Plan was drawn up in 1961. It was a six year plan, and was implemented by the newly formed National Economic Development Board. This was followed by the Second Plan (1967–1971).[15] Broad economic policies could be assessed from the plans. Essentially, both plans dealt with sectoral programming, the organisation of government revenue, government expenditures on infrastructure, and project evaluation over a longer period of time. The plans contained no explicit restrictions or guidelines with respect to foreign trade. However, the plans had an indirect effect on trade through their guidelines on industrial development in the private sector, the effect of which was import substitution.

The policy of import substitution during the 1960s can be seen in the system of taxation of foreign trade and of domestic production, and the promotion of industrial investment. Initially, the structure of

Table 3.5 Summary of industry policy 1955–1981: Thailand

Period	Description of activities
1955–1960	Promotion of public enterprises in various economic activities besides the public utilities; lack of economic planning; the role of private sector in the economy ignored by the government; taxation of rice through rice export premium; rubber and tin taxes introduced in 1955; a multiple exchange rate replaced by a fixed exchange rate and the official exchange rate varied between US$1:20.838 baht and US$1:21.117 baht during this period.
1961–1967	Application of mild tariff protection; import substitution by application of strong investment incentives in industrial production; continuation of heavy taxation on rice; fixed exchange rate with US (US$1:20.45 baht).
1967–1971	Reduced tax incentives in the industries promotion scheme; reduction of taxation on rice export.
1971–1976	Further reduction of tax incentives in the general promotion scheme; stronger tax incentives for industries in promoted areas; import substitution through stronger and more escalated tariff protection; introduction of an option of an import surcharge for short term protection (but rarely applied); more frequent applications of controls on trade, industry, and the price level; stronger promotion of manufactured exports; increase in export tax on rice and imposition of an export tax on sugar; continued fixed exchange rate with US dollar but at a generally lower value of Thai currency than previously.
1977–1981	Beginning active investment promotion drive for heavy investment projects; resumption of potentially strong tax incentives in general and even stronger incentives for promoted areas (remote locations) and for exports; more application of import surcharge for short term protection; beginning of structural adjustment in manufacturing industries by changing business (sales) tax, excise taxes and tariffs; strong trend towards export promotion which replaced import substitution; depreciation of Thai baht vis a vis US dollar (US$1 = 23.00 baht) after a period of over valued currency in 1980.

tax concessions provided most of the protection. However, towards the end of the period, while investment incentives were reduced, the degree of tariff protection was increased.[16] Some primary exports were taxed, and there was no effective promotion scheme for manufactured exports throughout the period.

Tariffs

In the early 1960s tariff rates were not very high—except for a few luxury consumer goods which were taxed at more than 40 percent—but they did not provide protection for domestic production.[17] The highest rates were levied on commodities in the groups of processed food, alcoholic beverages, and tobacco. These are not usually competing with domestic production. For the other groups, in 1964, the rates were scattered around 25–30 percent for durable and nondurable consumer goods; 20–30 percent for intermediate products; and about 20 percent for machinery and equipment. In 1971, the rates for durable and nondurable consumer goods were scattered around 30–55 percent, and there was little change in the rates for intermediate products and machinery and equipment.

The divergence of tariff rates for intermediate products and finished products indicates the presence of tariff escalation and an incentive for a greater degree of local processing than might otherwise have been the case.

Tax concessions

Another measure which was used as part of the import substitution strategy was the promotion of industrial investment. The Board of Investment (BOI) was set up in 1959.[18] The main purpose of the investment promotion scheme was to facilitate the setting up of private enterprises.[19] The provisions included tax concessions on imported machinery, equipment, raw materials, and intermediate inputs needed directly for the production. It was this set of provisions which could be considered as policy to encourage importation of capital goods and intermediate products including components for local assembly. Tax concessions on these imports depended upon the classification of promoted industries. Industries were classed A, B, or C according to the importance to the national economy. Group A received full exemption from tariffs and business taxes, group B and C received one half and one third exemption respectively. Towards the mid 1960s, most industries under group A were reclassified into groups B and C and there was only group C left in 1967. This indicates a tendency to use fewer tax concessions on imports. The structure of tariffs became more escalated towards the end of the 1960s and the early 1970s. This shift, therefore, could be viewed as a substitute for tax concessions on imported intermediate goods, which were being phased out.

Other measures

There were also many other measures to assist import substituting industries. These included industrial controls (which regulated entry and expansion and assisted existing firms in the industry), import and export quotas, and credit assistance. Towards the end of the period, attention was focused on the promotion of manufactured exports. But import competing industries, which had been sustained by various forms of assistance, still demanded other measures. Thus, there was a proliferation of measures to assist import substituting industries while, simultaneously, efforts were being made to introduce measures to promote manufactured exports in 1972.

Exchange rate policy

During the 1960s the exchange rate was fixed to the US dollar. The US dollar, in turn, was over valued with other major currencies such

Table 3.6 Exports as proportion of GNP, Thailand, 1970–1983 (M. baht)

	Exports	GNP	%
1970	14,772	136,439	10.83
1971	17,281	144,637	11.95
1972	22,491	164,299	13.69
1973	32,226	216,119	14.91
1974	49,799	270,774	18.39
1975	45,007	296,409	15.18
1976	60,797	331,329	18.35
1977	71,198	368,925	19.30
1978	83,065	464,550	17.88
1979	108,179	546,449	16.90
1980	133,197	672,400	19.80
1981	153,001	786,166	19.46
1982	159,728	846,136	18.88
1983	146,472	928,548*	15.77

* Estimated
Source: National Accounts Division, NESDB and Customs Department.

as the Japanese yen and the Deutsch mark. The Thai baht was also over valued. Imports from Japan and Germany, two of Thailand's major trading partners, were therefore cheaper in Thai currency than they would have been with a market-related exchange rate. The price of Thai manufactured exports was too low for producers in Thailand. Consequently, there was a reduced incentive to export. Also, the incentive effects for import competing goods created by other policies were offset to some extent, while the effect was to the advantage of users of imported inputs.

Policy in the 1970s

Promotion of exports was carried out actively from 1972 (see Table 3.6). This began with the revision of the Investment Promotion and Export Promotion Acts in 1972, to provide full tax exemption on imported inputs plus other usual promotional privileges.[20] In addition, there were tax rebates of all taxes in the production process at the Fiscal Policy Office of the Ministry of Finance, a discount facility on short term loans at the Bank of Thailand, and an exemption of business (sales) tax on the products. An Export Promotion Committee was set up in 1972 to coordinate these promotion activities.

It is extremely difficult to pinpoint when and why the trade regime changed simply from observing the change from the import substitution policies towards export promotion. There is no document which gives us details of this deliberate attempt to shift from the restrictive regime to a more open one. Our guess is that the World Bank convinced the Thai government that it was feasible, and politically attractive, to move towards that direction. The worldwide commodity boom in 1974 would have helped this change. This appears to be borne out by the fact that in that year the export performance of Thailand in all sectors was spectacular (Table 3.6).

The export boom in certain products, such as animal feeds, thread and yarn, was so great that the government banned the exports temporarily. In early 1974 several other commodities were added to the temporary list of products subject to export control (meaning that approval was needed).

During the year 1974 there were a number of cuts in tariff rates, but these resulted in a more escalated structure of tariff. The change involved a reduction in the rates on consumer goods regarded as 'necessities' which were not competing with local products, some to one half and others to one tenth of the old rates. A large number of raw materials and intermediate products had their rates reduced to one tenth of the old rates. The rates for machinery and equipment both for agricultural and industrial uses were reduced from 15–20 percent to 10 percent. The rates for other consumer goods remained the same.

In 1975 and 1976, changes were made in the tariff rates of only a few items. These included the reduction in the rates of some paper products (30% to 10%), equipment for movie production (30–40% to 10%), motors, and orange peel.

A major change also took place in the system of business (sales) tax in 1973. Previously the same rates were applied to imports as to domestic products. In December 1973, the rates for many domestically purchased inputs were reduced in an effort to lower the inflation rate. The rates for several items of domestic raw materials and foods were reduced to zero, and for others the rates were reduced if they used purchased inputs. Thus the new practice of collecting business tax gave rise to a protective effect.

The significance of the changes in tariff and business tax rates was that the effort to assist some import substituting industries continued into the period of export promotion. Another measure used was that of import control on some products.

During the early 1970s the exchange rate continued to be pegged to the US dollar. The value of the dollar had been declining with respect to other major currencies, with the exception of the pound sterling. Thus the value of the Thai currency with respect to these currencies had also been declining.[21] The effect of the implicit devaluation of the Thai baht is beyond the scope of this study, but it is believed that the devaluation during 1970–1973 must have contributed substantially to an increase in manufactured exports.

In 1977, a year after the launching of the Fourth National Plan, industrial policies underwent another major change. The government was reluctant to develop the industrial sector via heavy protection.

The Fourth Plan indicated: continuation of the promotion of manufactured exports; government participation in large scale industrial projects; conditional promotion of foreign investment, reorganisation of state enterprises; redefinition of the roles of government agencies in industrial development; and the promotion of employment and regionalisation of industrial production.

On 29 April 1977, the new Investment Promotion Law was enacted. The new act was different from the 1972 act in two respects. First, tax exemption on imported or local machinery became subject to the discretion of the BOI. Second, tax exemption on imported or local materials could be up to 90 percent on an annual basis, at the discretion of the Board, compared to no exemption in the 1972 act. Another major change was the use of import bans and the import surcharge. The import surcharge had been in existence prior to 1977, but it was rarely used.

In order to reduce the heavy concentration of industries and processing units within the Bangkok metropolis and urban centres, the government created promotion zones. Firms locating in the promotion zones could produce for local consumption or for exports. All firms in those zones would be 'promoted' firms. However, those which produced for exports received such additional benefits as lower taxes on machinery and raw materials.

For industries in promoted regions, the new law generally provided the same incentives as the 1972 act, except that tax incentives may have been less for machinery and more for raw materials. A more important change was the location of the promotion zones. The promotion zones were those areas lying outside the Bangkok metropolis and four neighbouring provinces. As of December 1979, 122 firms producing for exports had been promoted out of the total of 897 promoted firms. For industries in rural areas, there were 257 firms which were located in the central region outside Bangkok and Samutprakarn. In all other regions of the country there were altogether 141 promoted firms.

From 1977 to 1981, there were several changes apart from the 1977 investment promotion law. The rates of protection during the period have been on an upward trend. The import tax system still provided protection to durable and nondurable consumer goods more than to intermediate products and capital goods. Additionally, the frequent use of import surcharge to provide protection and/or to prevent dumping confused the structure of protection, to the extent that it was difficult to understand which industries had been given priority in the protection policies. In 1980, there were about 20 products and product groups which were subject to import surcharges ranging from 10 to 40 percent of c.i.f. prices.

Controls included a total import ban on 12 new items, adding to the total list of 23 items. Permission to import was required for about 22 items. There was also a cxport ban on 16 categories of goods and permission to export was required for 22 categories which exported under quota.

The Ministry of Commerce exercised price control through the Price Setting and Anti-Monopoly Act of 1979. The implication of price setting (price control) is that it retards the inflows of commodities from abroad. Price setting implies that the profit margin is regulated for the benefit of consumers. Importers were not receptive to this kind of control and avoided the regulated products. In 1980 there were 34 categories of goods under price control. Until 1980, 19 categories of products remained on the price control list.

The second round of oil price rises was not as severe as the 1973/74 crisis, but it contributed to large deficits in Thailand's balance of trade account. This effect exacerbated any deficit present because of the import substitution policy. Though the Thai baht was pegged to an unidentified bag of currencies by March 1977, it was believed that the exchange rate remained pegged to the US dollar. The balance of payments deteriorated owing to weak export prices of commodities and limited access to international markets. The Thai baht was over valued with respect to major currencies outside the US dollar area. The pressure on the exchange rate to devalue was evident given the trade deficit of 47 billion baht in 1980. Furthermore, the depreciation of the Thai baht value was, according to the Bank of Thailand's (BOT) calculation, on average −4.2% in 1980 (compared to the rates on March 1978). By May 1981, the devaluation of 1.07 percent was brought on, and in July 1981, a further 8.7 percent devaluation was announced. The over valuation of the Thai baht in 1980, a year after the 1979 oil crisis, blunted the competitiveness of Thai exports.

Summary

The last twenty years have seen the deliberate use of tariff, tax, and other control measures to protect Thai industries. With the onset of the First (1961–1967) and Second Plans (1967–1971), the mild protective tariff was introduced to encourage local production. However, the protection umbrella was only for durable and nondurable consumer goods. By the Third Plan (1971–1972), there were attempts to discourage import substitution by relaxing tax incentives, but an escalation of tariff rates was evident. The pressure to assist import competing industries remained and the attempt to change to export promotion was not very successful. High prices in the international markets in 1973 led to an export boom and the government began to control directly selected items of exports. The Fourth Plan

(1977–1981) further relaxed tax incentives and the tariff structure remained the same. That is, industries which were established with the help of tariff continued to receive some assistance. The new direction was to promote industries in rural areas. Intermediate inputs were not encouraged while consumer goods were protected. Tax rebates and import surcharges began to become significant judging from the frequency of their use during 1977–1981. Other forms of controls were also on the rise toward the end of 1970s. Thus, the trade regime of Thailand has progressed during the last two decades from a policy of implementing mild tariffs to implementing high tariffs and then to high tariffs plus rebates and controls of imports.

Liberalisation effort and revenue from trade taxes

A trade policy which goes against the free trade regime has been criticised by trade theorists on the grounds of economic inefficiency. Thai economists, in general, have accepted this line of thought. However, they have also been willing to accept the infant industry argument, so long as the protected industry in question was prepared to grow up. After two decades of protection, some industries have yet to grow up and they could not survive without the government's assistance. The economic inefficiency which these industries have generated plus the effort by some economists to restructure industries have prompted the government to shift the policy from a restricted regime to a more liberal regime.

It is difficult to pinpoint the liberalisation course since it is interwoven with some domestic inertia. The group which favours liberalisation comprises some academics and government officials. Its close allies are the international agencies such as the World Bank, International Finance Corporation (IFC), United Nations Industrial Development Organisation (UNIDO) and so on. People have misunderstood the alliance and seen it as one in which the international agencies impose their will on the Thai system. One should view the relationship as Thai economists' attempt to liberalise trade with the support of the international agencies. The restrictive trade regime is supported by the private sector on which the international agencies have no influence at all. Had the intellectuals not done their job to convince the government, the international agencies' recommendations of liberalisation measures would not have been effective.

As far as can be ascertained, there has been no study on the lobbying activities by the private sector but the following account of the activities should suggest that there is a real force that prevents trade liberalisation. In 1982, a World Bank study on electronics and

Table 3.7 Tax revenue of Thailand, 1972–1981

	Unit: Billion baht year ending September 30									
	1972	1973	1974	1975	1976	1977	1978	1979	1980	1981
Tax revenue	18.8	22.6	35.2	33.9	37.4	46.9	57.0	69.3	84.8	98.8
Income tax	1.5	1.7	2.0	2.6	3.0	3.6	5.0	6.0	7.2	8.5
Corp. tax	0.9	1.5	2.6	3.5	3.8	4.5	6.4	7.6	9.6	13.0
Taxes on goods & services	9.8	11.7	16.8	16.8	19.5	24.3	28.3	34.8	43.8	50.2
Int. trade taxes	6.0	7.2	12.9	10.4	10.4	13.5	15.9	19.2	22.4	25.1
Share	(31.9)	(31.9)	(36.6)	(30.7)	(27.8)	(28.8)	(27.8)	(27.7)	(26.4)	(25.4)

Source: IMF, *Government Finance Statistics Yearbook*, Volume VI, 1982.

electrical industries recommended a tariff reduction from 80 percent to 40 percent on finished products. Firms in these industries did not take the issue for granted. They sought some support from the Ministry of Finance, which was worried about the drop of revenue from the tariff cut. The managers of the industries attempted to discredit the study and deployed a delay tactic so the measures were not decided at cabinet level. It is our belief that they sought help from officials at various ministries to delay the measures. The end result was that the Ministry of Finance did not support the tariff reduction on the ground of loss of revenue.

The Thai government has regarded the tariff as an important revenue device. This situation is not unusual for a developing country which has limited revenue sources. Table 3.7 indicates that about 10 percent of total tax revenue came from income tax over the period 1972–1981. Other sources of revenue consist of taxes on goods and services, international trade taxes. In particular, taxes on international trade contribute about 25–31 percent of the total tax revenue.

When the international trade taxes are broken down into import duties and export duties, export duties contribute roughly between 6.7–33.4 percent of the international trade taxes. Rice and sugar are the two products subject to export duties on occasion. Trade tax revenue comes mainly from the import duties. The trend of nominal value of import duties is rising, as indicated in Table 3.8. the revenue from import duties is in the range of 80 percent of the total international trade tax revenue. These data suggest that the government has not considered resource allocation problems seriously. Instead, they suggest that revenue was a major objective of implementation of tariffs.

Review of the empirical studies of protection in Thailand

There have been several empirical attempts to measure the protective system in Thailand and this common result is a bias against the export

Table 3.8 International trade taxes, 1972–1981: Thailand

	Unit: billion baht									
	1972	1973	1974	1975	1976	1977	1978	1979	1980	1981
Total trade taxes	6.0	7.2	12.9	10.4	10.4	13.5	15.9	19.2	22.4	25.1
Import duties	5.6	6.4	8.6	8.3	9.1	11.8	14.2	16.5	18.9	21.9
(% of total)	(93.3)	(88.8)	(66.6)	(79.8)	(87.5)	(87.4)	(89.3)	(85.9)	(84.4)	(87.2)
Export duties	0.4	0.8	4.3	2.1	1.3	1.7	1.7	2.7	3.5	3.2
(% of total)	(6.7)	(12.2)	(33.4)	(20.2)	(12.5)	(12.6)	(10.6)	(14.1)	(15.6)	(12.8)

Source: IMF, *Government Finance Statistics Yearbook*, vol. VI, 1982.

industries. We shall outline some of the important works completed during the 1970–1980 period. The first attempt was made by Trairong Suwankiri (1970).[22] His calculations were based on the coefficients derived from Balassa's standardised input-output table, which in turn was based on the Belgian and Dutch data. The results of his work showed that the structure of protection was in favour of the production of consumption goods and intermediate goods but discouraged the production of capital goods.

Narongchai Akrasanee (1973) in his doctoral thesis calculated the ERPs by using the coefficients from the industrial survey for the year 1969[23] and covered 58 industries in his work. The results showed that the ERPs for the export industries were all negative, under both the Balassa and Corden methods. Most import competing industries received higher protection. In the non import competing industries, the potential ERPs (based on the tariff) were high with the exception of some industries having negative value added measured at world prices. According to the level of fabrication, the potential effective rates ranged from −50 percent for processed food to 488.2 percent for beverages and tobacco. When realised effective protection (based on price comparisons) was considered, the lowest rates, according to the Balassa method, went to processed food and petroleum products. The rates for other classes were higher. It was found that in the 1960s the protective system was biased in favour of import competing industries and strongly discriminated against export industries.

Pirote Wongwuttiwat (1975) made ERP calculations for the years 1964, 1971, and 1974.[24] The results can be summarised as follows. In 1964, the nominal rates ranged from high negative rates for some exports to extremely high rates for consumer goods. The rates were generally high for processed food, beverages and tobacco, and consumer durables and nondurables. The realised nominal rates for the import competing group were similar to the potential rates. For the non import competing group, the realised rates were much lower. The effective rates of protection for the import competing group were not much different from the nominal rates. For the non import competing group, the potential ERPs were very much higher than the

Table 3.9 Average[a] nominal and effective rates of protection, manufacturing sector, Thailand, 1980

	Nominal[b]	Effective[c]	
		C	B
Importables	32.3	49.5	70.9
Exportables	9.4	2.6	−0.6
Other[d]	39.0	62.4	81.7
Manufacturing sector	32.4	49.3	65.1

Notes:
a Value added weights.
b Realised nominal rates.
c C = Corden, B = Balassa.
d Neither exports nor imports more than 20 percent of total apparent consumption.
Source: Appendix 1.

potential nominal rates. The effective rates for export industries were shown to have large negative values. Thus in 1964 the protective structure was biased against the export industries. In 1971, the potential nominal rates were generally higher than those in 1964, exceptions being beverages and tobacco. The realised nominal rates were similar to those in 1964, but they were much lower than the potential rates for the non import competing industries. The ERPs were very high, especially among the non import competing industries. Pirote's study concluded that the protective system was in favour of consumer goods and that it also favoured the production of goods for local sales.

Following the construction of Thailand's Input-Output Table in 1975, ERP calculations took a major stride in terms of readily available technical coefficients. Narongchai Akrasanee made more ERP calculations for 1980. Updated estimates are given in the Appendix. The 1980 ERPs were based on the technical coefficients of the year 1975 and covered 93 sectors (I-O 042 to I-O 134).

The estimates are summarised in Table 3.9. The results are effective rates of close to zero for the exportable industries on average but a positive rate overall for the manufacturing sector. This reflects high rates for importables and other industries.

Summary

Trade taxes, both on imports and exports had initially been imposed in Thailand to raise revenue. This use of trade taxes was especially important after Thailand had regained its fiscal autonomy and the revenue motive for the tariff has continued to be important, even as recently as 1982 in the electronic industries case.[25] Trade taxes inevitably had the effect of promoting local production. Furthermore, the government learnt how to use trade taxes to promote actively the manufacturing sector. These resource allocation effects were to create a protected group in the import competing sector, the presence of this group making it difficult to change policy. In the early 1970s an attempt was made to shift away from import substitution.

Reasons for this change are difficult to pinpoint but it appears to reflect a combination of factors. One was the influence of the World Bank. Another was the work done by Thai economists and the third was the export commodity boom in 1974. But the policy change actually resulted in an escalation of the tariff structure, rather than an across the board liberalisation. This reflected the power of the interest groups at the final goods level. Also, the reform led to the use of greater discretion by the bureaucracy in the application of bans and surcharges. Despite the reform, there was not a significant reduction in the bias against the exporting sector. The attempt at an export promotion policy was therefore not successful.

Appendix 1

Table 3.A1 Effective rate of protection (ERP) by single industry, 1980: Thailand

I-O Code		NRPO	ERP'C	ERP'B
042	Slaughtering	55.22	*	*
043	Canning & preservation of meat	60.00	71.17	81.81
044	Diary products	24.51	28.99	31.51
045	Canning of fruit	0.00[e]	−36.12	−39.53
046	Canning of fish	0.00[e]	−22.86	−23.58
047	Coconut and palm oil	20.70	9.91	10.18
048	Animal oil, vegetable oil	23.69	84.79	92.07
049	Rice milling	5.00[e]	−7.12	−7.80
050	Tapioca milling	0.00[e]	−1.31	−1.38
051	Grinding of maize	0.00[e]	−4.29	−4.57
052	Flour and other grain	34.37	124.78	148.30
053	Bakery products	60.00	297.66	371.35
054	Noodle and similar products	60.00	*	*
055	Sugar	1.50[e]	−32.04	−33.76
056	Confectionery	45.04[e]	87.31	96.81
057	Ice	0.00	−1.76	−2.19
058	Monosodium glutamate	60.00	125.60	162.51
059	Coffee and tea	47.65	59.25	74.07
060	Other food products	58.57	152.04	191.92
061	Animal feed	10.00	−11.06	−11.64
062	Distilling & spirit blending	60.00	88.31	89.29
063	Breweries	60.00	71.35	106.38
064	Soft drinks	44.69	70.07	95.43
065	Tobacco processing	60.00	*	*
066	Tobacco products	60.00	63.14	66.44
067	Spinning	23.87	20.17	25.17

Table 3.A1 Continued

I-O Code		NRPO	ERP'C	ERP'B
068	Weaving	60.84	349.83	532.30
069	Textile bleaching & finishing	0.00	−16.59	−20.32
070	Made-up textile goods	42.48	43.60	54.60
071	Knitting	60.00	29.44	166.12
072	Wearing apparel	63.57	75.10	90.44
073	Carpets and rugs	44.00	58.46	110.46
074	Jute mill products	33.25	36.26	48.06
075	Tanneries & leather finishing	28.83	6.12	7.21
076	Leather products	73.88	172.92	231.74
077	Footwear, except of rubber	59.59	84.36	107.16
078	Sawmills	12.73	20.08	23.17
079	Wood and cork products	30.35	51.66	61.04
080	Wooden furniture & fixtures	51.67	124.38	182.50
081	Pulp, paper & paperboard	31.11	36.50	49.36
082	Paper & paperboard products	29.05	26.60	30.72
083	Printing and publishing	11.79	−1.79	−2.38
084	Basic industrial chemicals	21.99	23.72	39.56
085	Fertiliser and pesticides	4.24	−11.64	−14.31
086	Synthetic resin, plastic & artificial fibre materials	35.70	34.40	81.42
087	Paints, varnishes & lacquers	26.89	26.04	31.50
088	Drugs and medicine	18.46	12.38	16.34
089	Soap & cleaning products	45.58	73.64	112.65
090	Cosmetics	60.00	83.17	245.14
091	Matches	60.00	127.88	300.94
092	Other chemical products	30.87	44.26	55.81
093	Petroleum refineries	36.69	*	*
094	Other petroleum products	20.60	15.72	18.59
095	Rubber sheet and block rubber	5.00[e]	−34.72	−36.56
096	Tyres and tubes	24.48	38.89	41.28
097	Other rubber products	57.70	*	*
098	Plastic ware	45.91	79.74	97.69
099	Ceramic and glass products	61.82	96.45	156.09
100	Glass and glass products	44.77	64.97	94.52
101	Structural clay products	35.94	66.94	79.40
102	Cement	10.85	0.35	0.47
103	Concrete & cement products	28.64	57.46	73.96
104	Other non-metallic products	28.67	138.90	202.62
105	Iron and steel	1.64	−2.21	−2.66
106	Secondary steel products	13.53	20.65	26.05
107	Nonferrous metal	12.51	4.09	4.81
108	Cutlery and hand tools	29.38	68.37	81.10
109	Metal furniture & fixture	58.23	210.81	284.82
110	Structural metal products	25.49	45.18	51.48
111	Other fabricated metal products	35.92	77.23	94.29
112	Engines and turbines	15.00	11.59	12.29
113	Agricultural machinery & equipment	15.16	10.81	13.88
114	Wood & metal work machinery	15.41	6.90	8.43
115	Special industrial machinery	17.23	16.03	17.75
116	Office & household machinery and appliances	50.60	97.33	157.78
117	Electrical industrial machinery and appliances	21.03	19.46	22.93

118	Radio, T.V. & communication equipment & apparatus	35.59	49.19	59.64
119	Household electrical appliances	35.75	44.86	62.10
120	Insulated wire & cable	30.00	44.86	59.38
121	Electric accumulators & batteries	35.62	45.99	65.74
122	Other electrical apparatus & supplies	40.87	49.73	69.08
123	Ship building & repairing	29.96	52.44	61.11
124	Railroad equipment	5.00	−2.91	−3.72
125	Motor vehicles	72.99	181.73	268.09
126	Motor cycles & bicycles	48.51	77.30	103.09
128	Aircraft	0.00	−17.43	−24.68
129	Scientific equipment	21.05	19.44	21.86
130	Photographic & optical goods	28.95	21.21	28.94
131	Watches and clocks	30.96	41.07	68.89
132	Jewellery and related articles	15.96	16.62	17.20
133	Recreational & athletic equipment	27.25	23.79	30.14
134	Other manufactured goods	40.50	60.43	74.16

Notes:
e = Export tax
* = Negative value added at world price
NRPO = Nominal rate of protection for output
NRPI = Nominal rate of protection for input
ERP'C = Corden's ERP; ERP'B = Balassa's ERP

Endnotes

1 Phraya Suriyanuvat, *Political Economy*, Book 3, P. 4. This book is probably the first textbook on the Thai economy. Phraya served during the reigns of King Rama V and King Rama VI as Ambassador to the Court of Saint James and later as a high ranking officer in the Ministry of Finance.
2 See Ingram, James C. (1971), *Economic Change in Thailand, 1850–1970*, London, Oxford University Press, p. 35.
3 ibid., p. 177–182.
4 ibid., p. 138.
5 ibid., p. 132–136.
6 Speech given by H.E. Sunthorn Hongladarom, Deputy Prime Minister of Thailand, Faculty of Economics, Chulalongkorn University, 26 July 1982.
7 Ingram, J.C. (1971), p. 147.
8 Landon, K.P. (1941), *The Chinese in Thailand*, New York, Russell and Russell, 1973 reprint, pp. 43–45.
9 Ingram, J.C. (1971), p. 204.
10 ibid., pp. 229–231.
11 Muscat, Robert (1966), *Development Strategy of Thailand*, New York: Praeger, p. 193.
12 Akrasanee, N. (1973), 'The Manufacturing Sector in Thailand: A Study of Growth Import Substitution and Effective Protection, 1960–1969', PhD. dissertation Baltimore: Johns Hopkins University, p. 21.
13 Ingram, J.C. (1971), p. 283.
14 See IBRD (1959), *A Public Development Program for Thailand*, Baltimore: Johns Hopkins University Press, pp. 90–94.
15 *First National Economic and Social Development Plan 1961–1966*, Bangkok:

NEDB, 1963; and *Second National Economic and Social Development Plan, 1967–1971*, Bangkok: NEDB, 1968.

16 The increases in tariff rates were due to protection, revenue and balance of payments motives.

17 See Akrasanee, N. (March 1975), *The Structure of Effective Protection in Thailand: A Study of Industry and Trade in the Early 1970s*, Thammasat University, Bangkok.

18 The BOI was set up to implement the '1954 Act on Promotion of Industries', but it became active after the 1960 Act (Promotion of Industrial Investment Act. B.E. 2503).

19 Thailand, Board of Investment (1971), *Investors Manual*, Bangkok, p. 15.

20 National Executive Council Decree, No. 227 and 329. Bangkok, 1972.

21 The trade weighted average devaluation rates of the Thai currency were 1971, 3.4%; 1972, 6.23%; 1973, 6.15%. The Bank of Thailand said the value increased 2.8% in 1974; 2.6% in 1975; 4.2% in 1976.

22 Suwankiri, T. (April 1970), 'The Structure of Protection and Import Substitution in Thailand', M.A. thesis, University of the Philippines.

23 Akrasanee, N. (1973).

24 Wongwuttiwat, P. (June 1975), 'The Structure of Differential Incentives in the Manufacturing Sector: A Study of Thailand's Experience During 1945–1974', MA thesis, Thammasat University.

25 See also Akrasanee, N. et al., (1982), 'The Electrical and Electronics Industry in Thailand', mimeograph, National Economic and Social Development Board, 1982, for ERP 1982 calculations for the electrical and electronics industry.

4 The Structure and causes of Malaysian manufacturing sector protection

LEE KIONG HOCK*
May 1985

FOLLOWING Independence in 1957 Malaysia[1] embarked on a policy of industrialisation based on protection. Concerned with problems of high unemployment, diversification of its highly lopsided economy and with overall growth and development, the consequences of protection on efficient resource allocation appear to have been downplayed. Now with the old import substitution industrialisation policy coming to an end but also with the recent emphasis on heavy industries, it is an appropriate moment to re-examine the issue of manufacturing sector protection in Malaysia. This involves an examination of the structure, trends, causes and effects of protection. Knowledge gained here provides insight into possible courses of action which might be adopted to minimise resource misallocation. However, a proper appreciation of these requires, as a prerequisite, an understanding of the underlying ideas about protection.

The evolution of ideas on protection in Malaysia

The early history of Malaysia is the history of each of the separate sultanates and settlements. Nevertheless, the economic growth and development of each is linked, in one way or another, to the development of the two industries which held sway over Malaysian economic activity and life for the greater part of the twentieth century—natural rubber and tin. Both began around the turn of the century and before long, owing to its obvious comparative advantage, Malaysia became the largest single producer of natural rubber and tin

* Lee Kiong Hock is Associate Professor, Faculty of Economics and Administration, University of Malaya, Kuala Lumpur.

in the world.[2] Against this backdrop, protection in Malaysia may be viewed as developing in three stages: the era of free trade, the era of industrial promotion, and the era of protection.[3]

First stage: free trade era

Given the rapid growth and development of the natural rubber and tin industries, there were no grounds for any protectionist policy during the first half of the century, let alone for the manufacturing sector, which was then of minor importance and interest to the British colonial administration. The British colonial administration were satisfied that Malaysia should remain a supplier of raw materials, on the basis of comparative advantage, to the industries of the western world and, in particular, to the United Kingdom and its dominion territories. The policy of free trade reigned supreme. Any form of restriction or hindrance to free trade was incidental. In the initial period in the development of customs and excise duty taxation, all duties were imposed primarily for the purpose of earning revenue though, in some instances, these were imposed in order to control aggregate domestic supplies of harmful products, such as intoxicating liquor and opium.[4]

With the advent of the World War I, and particularly with the desire to contribute toward Imperial defence expenditure, import duties became an important source of government revenue. The British colonial administration not only increased but also extended the coverage of customs duties. By 1929 import duties alone accounted for over two thirds of total inland revenue in the Federated Malay States, with the remaining one third coming almost wholly from excise duties.[5] The role of import duties as a source of government revenue was highlighted by the Great Depression. The slump in the prices of natural rubber and tin greatly reduced export earnings, which in turn led to a decline in imports and the level of customs duty revenues. The catastrophic decline in the prices of natural rubber and tin focused attention on the vulnerability of the economy in its dependence on rubber and tin for the import of food and other consumer and capital goods, and the severe limitations of financing government expenditure from a narrowly based structure of customs and excise duties. Most of the customs duties imposed at the start of the World War I, and removed at the end of the war, were reimposed. From this period until World War II the British colonial administration extended the coverage of customs duty taxation.

Although the extended system of customs duty taxation hampered free trade to some extent, a new protectionist policy which deliberately modified the traditional policy of free trade was introduced in

1932 under the Ottawa Agreement. The Agreement introduced a system of Imperial Preferences which favoured goods originating from the United Kingdom and its dominion territories. Quotas and duties which discriminated directly against 'foreign' goods were imposed. However, this policy of Imperial Preferences was introduced, not in the interest of Malaysia, but in the interest of British exports, and directed principally against Japanese competition.[6] Protective effects for Malaysian manufacturing industries were only incidental and confined principally to the domestic production of liquor, tobacco leaves, cigarettes, matches, and textiles.[7] Since western control and interests were not as significant in the manufacturing industries as in other industries, and since their interests were confined mainly to engineering works, subsidiary to their main interests in the plantations and tin mines, and in port services, it is not surprising that the British colonial administration had little interest in the protection of Malaysia's manufacturing industries.[8]

In the immediate post war period, customs and excise duty unification was recognised as a prerequisite to the amalgamation of the separate states into a Malayan union. Customs tariffs of the Federated Malay States were extended to the Unfederated Malay States, and the Straits Settlements. The process of amalgamation actually came earlier than anticipated. In 1952 the British colonial administration tabled a plan for accelerated constitutional advance leading to independence.

The move towards independence was marred by a communist led insurrection, termed the 'Emergency', which had as one of its early objectives the disruption of the Malayan economy. In order to finance the cost of the Emergency, which was officially proclaimed in 1948, import and excise duty taxation were raised and extended. This is clearly reflected by the rise in contribution of import duty revenue to total inland revenue.

In the pre-independence period, both customs and excise duties were levied in order to meet public sector expenditure and the costs of Imperial defence. Policies which deliberately modified free trade were discriminatory rules which favoured access of British industry to the Malaysian market. Malaysian industries were expected to thrive on the basis of comparative advantage. Protection for Malaysian industries was always incidental. Nevertheless, these policies of the British colonial administration did produce one lesson. The establishment of branch factories, particularly by Singapore based firms affected by the imposition of customs duties, created a growing awareness of the effectiveness of tariffs as a device which could be used to encourage and foster the development of manufacturing industries in Malaysia.

Second stage: industrial promotion

With self government in 1955 came the first serious consideration on the use of tariffs as a policy instrument for the development of Malaysia's manufacturing industries. At the request of the governments of Malaya, the crown colony of Singapore, and the United Kingdom, the World Bank Mission submitted its report, which suggested a two fold role for the government of Malaya in relation to the development of manufacturing industries in Malaysia.[9] The first role was to provide adequate infrastructure and the second was to foster individual enterprise, according to the principle of *laissez-faire*, but also by the creation of a favourable climate for its development.

Despite its advocacy of *laissez-faire*, among the more specific recommendations of the IBRD report was the use of tariffs as a means of encouraging development.[10] The Mission argued that general concessions along the lines of depreciation allowances and tariff protection were likely to be more effective in encouraging industrial development than would the provision of income tax exemptions to pioneer industries. The Mission made no specific recommendation to increase the existing initial and annual depreciation allowances provided under the original Malayan Income Tax Ordinance of 1947. On the other hand, the Mission recommended that the protective effects and possibilities of the tariff be deliberately recognised and systematically reviewed. To accomplish the task of industrial development, a Division of Industrial Development should be established within the government.[11]

The broad principles of industrial policy as outlined by the World Bank Mission were accepted and incorporated into the election manifesto of the ruling Alliance Party. In the following year, 1956, a working party was formed by the Minister of Commerce and Industry, in order to formulate an industrial development policy, the broad outline of which had already been stated in the government's election platform.

Between 1954 and 1957—the year of independence—import duty revenue from nontraditional commodities almost trebled, and their share of total import duty revenue doubled from 9.6 percent to 18.8 percent; see the row denoted 'other' in Table 4.1. This rapid increase in import duty revenue from nontraditional commodities has been interpreted by Hoffman and Tan[12] as a reflection of the immediate reaction of the government to the World Bank Mission's proposal on the use of tariffs. However, the rise in the relative importance of import duty revenue from nontraditional sources was a consequence of several factors apart from the extension in the coverage of import duty taxation. The Suez Crisis boom in the prices of natural rubber and tin meant an increase in export earnings, an increase in the demand for nontraditional commodities, which were by then subjected to import duty taxation, and a decline in the relative importance of expenditure on the two traditional dutiable commodities, liquor and tobacco. Also, the unchanged rate of duty charged on liquor and tobacco since 1953 contributed towards the decline in the relative importance of import duty revenue from these traditional sources.

However, strong evidence against influence of the World Bank Mission lies in the fact that in 1957 there was still no official policy on the choice of tariffs as a means of promoting industrial development in Malaysia. The Working Party on Industrial Development appointed in 1956 had been toying more with the idea of tax concessions than with tariffs. As it turned out, the Working Party took a distinctly different stand from the World Bank Mission by advocating temporary tax exemptions for pioneer industries with the provision of tariff protection only in selected cases.[13]

The Working Party reasoned that:

1 the establishment of a pioneer industry was in the nature of a gamble for prospective investors who, motivated by profits, were more likely to invest in a territory which guaranteed some relief from income tax since profitability would be significantly affected by the potential size of the tax bill; and
2 the cost of temporary tax exemption was small since it was merely a matter of foregoing tax on profits which would not have risen had the investment not been made.[14]

The decision of the Working Party may have been influenced by the strong opposition to the use of tariffs from at least two sectors of the community. The foreign agency houses, which controlled 70 percent of the nation's foreign trade and 75 percent of the plantations, opposed on the grounds that tariffs would reduce foreign trade and raise the cost of living leading to wage increases, both of which would imply a loss on the returns to their investments. On the other hand, it was strongly felt that the use of protective tariffs would benefit the Chinese and Indians much more than the Malays, since the former controlled the nation's wholesale and retail trade, and were therefore in a position to benefit at the expense of the latter. In both cases, the arguments were weak, reflecting the tendency to equate the general problem of racial imbalances with the particular problems raised by the imposition of protective tariffs.[15] As a final point, although Malaysia had been self governing since 1955 it was not independent and was, therefore, unable to pursue an independent trade and industrial development policy.

Full independence in 1957 marked a major shift in official policy on the development of Malaysia's manufacturing industries. As stated above, in the pre-independence period customs duties served as a principal source of government revenue and all protection for Malaysian manufacturing industries was incidental. If there was a protectionist policy at all, it was not in the interest of Malaysia but in the interest of the United Kingdom and its dominion territories.

The first departure came in 1959 with the establishment of the Tariff Advisory Committee (TAC), the reclassification of the tariff schedule, and the elimination of some categories of Commonwealth preferences. The TAC was not a statutory body, and most of its seven members were from the private commercial and industrial sectors. Its principal function was to investigate and advise, within its limited means and expertise, the Minister of Commerce and Industry on individual applications for tariff protection, tariff exemptions on raw materials, and duty drawbacks. However, final approval lay in the hands of the Minister of Finance, whose ministry adopted a cautious approach with respect to tariff exemptions and duty drawbacks. In its 'Notes for the Guidance of Applicants for Tariff Concessions' of October 1961 the Ministry states that:

> 'the margin of protection granted will in no case be greater than that which will obtain for the local manufacturer the market for goods which can be economically produced in the Federation within a reasonable period. ... The Government will not grant exemption or protection to an extent which would permit the marketing of goods of inferior quality or at excessive prices in comparison with imported goods. It will not grant tariff concessions to industry to an extent which would materially affect public revenue.'[16]

Import duty revenue from nontraditional sources increased by 34 percent between 1959–1960, compared with an average annual rate of 7 percent over the previous three years (see Table 4.1). Hoffman and Tan[17] cite this as an indication of the immediate impact of the TAC. However, 1960 was another boom year for Malaysia. Natural rubber and tin prices were high compared with the dip after the Suez Crisis of 1956. With a highly open economy, movements in Malaysia's export earnings generally have the effect of inducing imports to move in the same direction through the income mechanism.[18] High public sector demand during the period also contributed towards the rapid increase in imports. Increased imports then increased customs duty revenue from nontraditional sources.

Clearer indication of the impact of the TAC lies in the number of items which were accorded protective tariffs. When a second World

Table 4.1 Sources of import duty revenue: Malaysia

Item	1948	1954	1957	1959	1960	1964	1965	1967	1970	1980[a]
Petroleum	17.5	18.3	19.1	22.5	20.0	9.4	8.2	10.1	9.7	5.5
Heavy fuel oils	–	–	–	–	5.2	4.3	3.4	8.0	8.0	1.8
Spirits	7.3	6.8	5.9	5.1	4.8	6.8	7.0	6.2	5.9	3.5
Textiles & apparels	9.3	5.9	6.8	7.0	6.1	6.0	6.5	5.7	7.9	4.6
Tobacco & tobacco products	49.6	47.7	35.5	33.6	29.5	35.0	34.3	30.6	29.7	9.0
Malt liquors	8.4	9.0	8.5	7.4	7.1	4.1	4.2	0.4	0.6	neg.
Sugar	1.0	2.8	5.3	5.3	5.6	7.6	6.7	12.0	7.1	neg.
Others	7.0	9.6	18.8	19.1	21.7	26.7	29.6	26.9	31.0	75.4[b]
Total ($ million)	105.02	199.37	292.68	300.70	353.48	312.06	387.47	352.20	419.20	2060.60
Percentage of total inland revenue	53.7	41.6	44.3	41.6	42.6	31.2	32.9	27.5	23.3	16.1

Notes:
a Malaysia including Sabah and Sarawak.
b Includes import surtax which accounted for 30% of total import duty revenue.
Source: Malaysia, Estimates of Federal Revenue and Expenditure, Kuala Lumpur: Government Printer, various issues.

Bank Mission visited Malaya in 1962 it found that protective rates applied to 25 items (pioneer industry products), including dental paste and powder, plastic articles, footwear with rubber material, cycles, flashlight batteries, wire nails and staples, iron and steel grills, and tyres.[19] By 1963, protective duties had been accorded to more than 200 items.[20]

The second World Bank Mission was invited by the governments of the Federation of Malaya and Singapore to report and make recommendations on the economic aspects of their merger. Among other things, it recommended the establishment of a tariff board, independent of sectoral and political interests, and the 'proper' use of protective tariffs to encourage the expansion of production for the domestic market.[21]

The formation of the Tariff Advisory Board (TAB) on 16 September 1963 constitutes a landmark in the history of protective tariffs in Malaysia (although the TAB became operational and replaced the TAC in the latter part of 1964). However, the TAB, as its name suggests, was only an advisory body with no executive powers. Its principal functions were:

1 to advise the federal government on the establishment of a common market in Malaysia for all goods and products produced or manufactured in significant quantities in Malaysia; and
2 to advise the federal government on the establishment and maintenance of a common tariff of protective duties for the protection of products for which there is to be a common market.

The evaluation of applications for protection by domestic manufacturers was only a secondary function. However, when Singapore left the Federation in 1965, the common market function became less important than the protective tariff evaluation function.

With the departure of Singapore from the Federation, an Action Committee on Tariffs (ACT) was set up to advise the government on the economic implications of separation and on urgent tariff matters. The committee was reconstituted on 18 November 1966 as the Action Committee on Tariff and Industrial Development (ACTID). Though not constituted by any statute of Parliament it possessed extensive powers of jurisdiction above those of the TAB and it was in fact serviced by the TAB. This is reflected by the fact that the committee consisted of fulltime members of the TAB, representatives of the Treasury, the Customs and Excise Department, and the Trade and Industrial Division of the Ministry of Commerce and Industry.

Initially, the committee was charged with an assorted mixture of responsibilities, including the task of considering urgent applications for tariff protection, import duty exemptions on raw materials and component parts, duty drawbacks for raw materials and component parts, and import prohibitions. After its reconstitution, its tasks were extended to include consideration of pioneer status applications, export incentives and, in effect, almost all matters related to industrial development.

The contrast between the ACTID and the TAB lies not only in its composition but, more importantly, in the speed at which decisions could be taken. Since the principal government departments were party to any recommendation of the ACTID, the power to produce efficient and speedy implementation was inherent in the committee. On the other hand, unlike the TAB, it provided very limited scope for public opinion.

The TAB, which received and considered applications for tariff protection from domestic manufacturers, the state or federal governments (e.g. in the case of the motor vehicle assembly industry which the federal government was keen to encourage), and even from within the Board, subjected all applications to public enquiries. These public enquiries were meant as a forum for the expression of all views pertaining to any application for protection. A complete report of the proceedings would be submitted to the Minister of Finance. The latter was required to publish the report and its decisions within six months.[22]

The TAB in its evaluation of any application for protection would take into consideration effects on the following:

1 the need for a balanced distribution of industries throughout the country;
2 the interests of the entrepot trade of Penang and Labuan;
3 the interests of existing industries and of consumers in Malaysia (i.e. including the cost of living, the cost of production in other industries particularly the export industries, and the cost of public sector development works);
4 employment and national income; and
5 the effects on federal and state revenues.

The initial general principles on tariffs were similar to those for the TAC. There was an apparent awareness of the costs of protection. Where the government was concerned, no more protection than was necessary was to be accorded, and for periods no longer than were absolutely necessary. The protected industry was expected to be fully competitive in the due course of time. The quality of the product was to be maintained, and there was to be no adverse effect on public revenue.[23]

In practice, the TAB differed considerably from the TAC. The differences can be gleaned from the outlook which it adopted towards industrialisation in Malaysia. The overall functions of the TAB, according to its chairman, were those of protecting, persuading, cajoling, or even forcing the development of manufacturing industries in the country. The policy was one of deliberately encouraging, assisting, or even coercing industries to be established in order to step up the pace of industrialisation in Malaysia. In line with this outlook, the chairman of the TAB emphasised that the nation could not, in its (present) state of industrialization, be 'fussy' about its choice of industries.[24]

This change in outlook may have been strongly influenced by the 1963 World Bank Mission,[25] which argued for greater economic diversification with emphasis on manufacturing which could generate, in the future, a dynamic and dependable source of private savings and public revenue.[26] However, the actual change coincided with the departure of the relatively more industrialised state of Singapore from the Federation in 1965.

The effect of this change in outlook to active promotion of industrialisation through protection can be seen more clearly in terms of the changes in tariffs which occurred during the period of the TAB's existence (1963–1970). In 1964 and 1965 the TAB held public enquiries on only 30 groups of commodities. However, in 1966 alone it held public enquiries on 23 groups of commodities covering 131 items and subitems under the Malaysian Tariff Schedule. In 1967 it held public enquiries on 29 groups of commodities covering 146 items and subitems. Overall, between 1962 and 1970 (about the time of the dissolution of the TAB) 396 new tariffs were introduced and the rates on 274 of the 811 tariffs existing in 1962 were increased. On the other hand, tariff rates were lowered for only 92 items and subitems during the same period.[27] It should also be noted that the TAB, in 1966, repealed a large number of Commonwealth preferential rates and that by 1967 these preferential rates had been removed completely.

Hoffman and Tan[28] cite the increase in import duty revenue from nontraditional commodities between 1963–1964 as a visible indication of the immediate impact of the TAB. Clearly, the reference years are inappropriate since the TAB, as stated earlier, was not fully operational until June 1964. Nevertheless, a comparison of growth rates in import duty revenue from nontraditional sources shows an increase of almost 58 percent in 1965 compared with an average annual growth rate of 9 percent during the period of the TAC (see Table 4.1). However, the entire increase in 1965 cannot be wholly attributed to the efforts of the TAB. This is particularly the case since 1965 also saw a change in Malaysia's fortunes. Export prices for its primary commodities recorded significant increases, particularly for tin, which recorded an all time high, and import demand increased which generated increased customs duty revenue from nontraditional commodities.

In the years before the TAB and the ACTID, tariffs were generally considered to be at low levels with the exception of a few very high rates imposed for revenue purposes or social reasons, as in the case of tobacco and liquor.[29] However, with the 'aggressive' outlook of the TAB, Malaysia approached the familiar pattern of a medium to highly protected less developed country (LDC) by the end of the 1960s.[30] Although the Investment Incentives Act of 1968 focused on tax incentives and depreciation allowances, almost all companies approved under the Act were accorded tariff protection.

Third stage: protection

Following the racial riots of May 1969, the promotion policy of the TAB received further impetus with the formation of the Capital Investment Committee (CIC) on 6 June 1969. Comprised of Cabinet Ministers, the CIC became the supreme authority responsible for overall policies pertaining to investment and industrialisation in Malaysia. The ACTID would now act on urgent matters along the overall policy objectives of the CIC.

The CIC was charged with:

1 reviewing current policies and programs of industrial development at federal and state levels, where there existed an apparent lack of coordination with individual states continuing to set up their own industrial estates and competing among themselves for investment,[31] with a view to coordinating them into an integral programme of accelerated industrial development, particularly of labour intensive and export oriented industries;
2 identifying development priorities, and guidelines, and incentives for the effective establishment of these industries; and
3 streamlining present government machinery in order to speed up decision-making on industrial development.[32]

The CIC, armed with powers which exceeded those of the TAB and ACTID, felt that '... in the present context of creating more employment opportunities as a national objective, tariff policy should be radically modified to give positive assurance and encouragement not only to existing industries but more particularly to industries not yet established but possessing good prospects for becoming viable ...'[33] In practice, the CIC's positive tariff policy meant the immediate imposition of a protective duty on products which were not yet produced, or for which no productive equipment had yet been installed.

The CIC was clearly not prepared to wait for applications for protection. Rather, it was prepared to impose advance protection (tariffs or quotas) in areas it considered suitable for domestic production. In a number of cases it was prepared to make concrete proposals for investment either to the private sector or to the government.

Within three months of its formation the CIC announced and imposed advance duties on 21 items. To the extent that the objective was to elicit a quick response from the private sector, where fixed investment had been practically stagnant from 1967 to 1969 (with an average annual growth rate of only 2.3 percent compared with 7.9 percent in the previous five years), the result seemed rather successful. Within the first few months several inquiries and serious project proposals were received. However, not all resulted in final production. Furthermore, complaints about increased costs to user industries lead to the removal of some of the duties. In practice, the tendency was for pioneer enterprises to be capital intensive, even though more was done to encourage labour intensive industries. The basic unemployment problem which was a major motivation for the CIC's radical policy remained.

When the social and political climate became more stable, both the TAB and the ACTID, which found the task of handling applications for protection beyond their administrative capacity, were abolished in August 1970. Some of the functions of the TAB were taken over by the Federal Industrial Development Authority (FIDA), now the Malaysian Industrial Development Authority, (MIDA). Under this arrangement, a Tariff Advisory Committee (TAC) and a Special Advisory Committee on Tariffs (SACT) were formed. The SACT advised the federal government on any request considered urgent by the government for measures of protection and promotion of industries referred to it by the government. The TAC took over the functions of the TAB.

The CIC, which was not constituted by an Act of Parliament, ceased to function with the resumption of parliamentary government in February 1971. The SACT currently is, with the exception of the Treasury; the sole tariff making institution with powers to impose or to remove tariff duties. Serviced by the Tariff Unit in MIDA, the SACT retains all the former functions of the ACTID and has taken over the functions of the TAB, which the TAC, now quietly forgotten, was to replace.

Although the CIC was dissolved, its promotional policy remained in the early years of the SACT. The SACT introduced a system of suspended tariffs wherein a protective duty was announced in advance—to be imposed when production started—for selected industries indicating investment opportunities. With the announcement, an import quota would be imposed. When the industry was established, the quota would be reduced progressively and the suspended duty would replace it.

This sytem, like the system of advanced duties, did not prove as successful as had been anticipated. To begin with, there were several inherent shortcomings. Theoretically, if the suspended duty is not overgenerous the government would be freed from sheltering a high cost industry. However, if the suspended duty turns out to be rather high the government may feel constrained to retain it. On the other hand, even if the suspended duty is not overgenerous, once the industry is established the government may be compelled, by social or political pressures, to grant higher protection if the industry turns out to be a high cost producer, a result which could occur because of the presence of protection.

The SACT felt that the system of suspended tariffs did not prove to be an effective means of promoting investment or industrial development.[34] By the middle of the decade, the system was abandoned. The SACT now adopts a less aggressive, perhaps more mellowed, posture by granting protection only to 'deserving industries' on the basis of the working principles originally set out by the TAB.

The government maintains that the extent of tariff protection is still low by international standards.[35] However, it cannot be denied that, under the promotional policy of the TAB, which received further impetus from the CIC and the SACT in its early years, the extent of tariff protection in Malaysia has increased both in terms of coverage and level. As will be seen later, Malaysia seems to have moved from the stage of low level protection to the pattern of medium to high protection found in many LDCs.

It is now unlikely that the aggressive, promotional approach of the TAB, the CIC, and the young SACT will be repeated. In the Third Malaysia Plan (1976–1980), the government stated categorically its intention to reduce the level of effective protection afforded to manufacturing industries and to narrow the differences in effective protection between industries in order to promote a more efficient pattern of industrialisation in the country.[36] This more recent change in approach may be attributed to the findings of two studies conducted within the Economic Planning Unit.[37] The two studies showed that Malaysia was moving from a situation of low tariff protection in the 1960s to high tariff protection in the early 1970s. One of the principal policy recommendations called for a reduction in the 'very high effective rates of protection'.

Nontariff protection

Two other forms of protection used in Malaysia are import restrictions and import duty exemptions.

Unlike in other LDCs, quantitative restrictions as a means of protection have not been particularly important in Malaysia. All quotas and restrictions imposed are classified under three distinct schedules in the Customs (Prohibition of Imports) Order. The first refers to total prohibition on grounds of politics (e.g. all goods from South Africa and Israel). The second covers all goods which are restricted for the purposes of national security or consumer safety (e.g. firearms and substandard consumer products). The third covers goods which are restricted primarily for the protection of Malaysian industries. The number and lists of such goods change continually on the recommendation of the SACT, subject to a joint decision by the Minister of Finance and the Minister of Trade and Industry.

The level of import restriction or quota is in principle based on the industry's intended production capacity relative to the previous year's imports. Since it is not always practicable to conduct a study at the time when the quota is needed, the restriction is often arbitrarily set at 60 to 70 percent of the previous year's imports. The quota is set, however, based on the capacity rather than the production of the industry, and is therefore often generous to the domestic producer. The initial quota period of three months may be renewed and may operate for nine months or longer.

Import restriction is also usually granted where it is feared that large scale speculative imports may take place prior to imposition of tariffs and the commencement of production of an industrial project. Import restrictions may also be granted in exceptional cases of severe disruption of the domestic market caused by overseas dumping, and this will be imposed until the 'threat' passes over.

The extent of import restriction is not large by any standard. In fact, there has been a move away from quantitative restrictions as a form of protection. In 1973, for instance, only 135 items, or 4.2 percent of the total number of import items, were subjected to import restrictions. By 1978 the number of items declined to only 110, or a mere 2.8 percent of the total number of import items (see Table 4.2). This trend has continued into the early 1980s. With the 1982 Budget, the entire list of items subject to quantitative restrictions was reduced to only 12 items.

The decline in the number of items subject to quantitative restriction reflects the basic policy of the government to avoid the use of import restrictions, where possible, as a form of protection. This policy arose from the administrative difficulties encountered in enforcing such quantitative restrictions. The Malaysian experience was that it was easier to collect customs duties at the points of entry than to enforce quantitative restrictions. Furthermore, while the administration of customs duties lay in the hands of only one department—the Royal Malaysian Customs and Excise

Table 4.2 Classification of items subject to quantitative restrictions 1973 and 1978: Malaysia

	1973[a]		1978[b]	
Category	Number	(%)	Number	(%)
Import quotas	119	(3.7)	100	(2.5)
Total prohibition	16	(0.5)	10	(0.3)
Import licensing	86	(2.7)	109	(2.8)
	223		219	
No restrictions	2977	(93.1)	3705	(94.4)
	3200		3924	

Sources: a) Teh K.P. (1977), 'Protection, Fiscal Incentives and Industrialization in West Malaysia since 1957', *Monograph Series On Malaysian Economic Affairs*, Kuala Lumpur: Faculty of Economics and Administration, University of Malaya, Table 4.2, p. 62.
b) Customs (Prohibition of Imports) Act, 1978.

Department—the administration of quotas and quota licences cut across several departments.

Import duty exemptions are firm specific and granted only for raw materials and components. The level of exemption depends on whether the finished products are sold in the domestic market or are exported. Where the final product is meant for export, full exemption will be granted for raw materials and components which are not manufactured locally or, where they are manufactured locally, are not of an acceptable quality and price. Full exemption is normally given if:

1 the final product made from dutiable raw materials and components is not already subject to any import duty; and
2 the price of the final product made from dutiable raw materials and components is not competitive against imports without full exemption.

On the order hand, where the final product is meant for the domestic market, partial exemption for raw materials and components may be granted if:

1 it is ascertained that the industry concerned can compete effectively against imports in terms of cost and can enjoy reasonable profits with only partial exemption; and
2 the raw materials and components for which exemptions are sought undergo very little local processing and, therefore, would achieve only low value added.

No exemption may be granted where it is considered that the industry concerned is already granted adequate tariff protection and the companies concerned are able to enjoy reasonable profits without any exemption.

Like import duty exemptions, duty drawbacks on imported raw

materials and components are also firm specific but apply only where the final product is exported. The conditions for the approval of this facility are that:

1 the goods exported have been manufactured in premises approved by the Director General of Customs;
2 there is proper control and supervision on the movement of the raw materials and components;
3 proper accounts are kept for the purpose of ascertaining the quantity of the raw materials and components used;
4 the raw materials and components have been imported by the manufacturer; and
5 the final products are exported within twelve months of the date on which the import duty was paid.

Concluding observations

Malaysia has apparently travelled a great distance along the bipolar continuum of free trade on the one hand and absolute protection on the other. In the colonial era, when the principle of free trade reigned supreme, protection was viewed as an unavoidable byproduct of a high dependence on customs duty taxation for public revenue. With independence, the establishment of the Tariff Advisory Board and its successors, the Capital Investments Committee and the Special Advisory Committee on Tariffs, resulted in a major shift towards the opposite end of the continuum, with protection being granted in an apparently aggressive manner. In particular, the more aggressive approach of the CIC can be seen as a response to unemployment problems and the racial riots in 1969. However, since the mid 1970s, the SACT appears to be steering back towards a more mellowed, moderate approach to protection.

There are two broad hypotheses on the evolution of protection policy in Malaysia. One is a public interest model, in which the motivation of the policy makers was to achieve a more diversified economy, less dependent on rubber and tin. This required an increase in the size of the industrial sector. This policy was implemented by a system of 'made-to-measure' tariffs.[38]

At times, protection in Malaysia was granted rather cautiously, and at other times somewhat indiscriminately. However, in almost all instances, this was granted with little in depth study of the effects and consequences.

The ideal was that no industry should be granted more protection than is necessary. Under the 'case by case' method, the tariff is based on the excess of the costs of production over the prices of imports, taking into consideration the need for an adequate return on capital.

The tariffs set are often soon rendered inappropriate. Changes in domestic or in foreign costs which spark off a change in one tariff often lead to further adjustments in other tariffs or to duty exemptions and drawbacks in an attempt to avoid antiprotection effects.

Just like tariffs, import duty exemptions require much information which is often not available. It is, therefore, unlikely that a proper evaluation is made in most cases. Exemptions are therefore granted more often on an *ad hoc* basis. Further, since they are granted only to firms which apply for exemption rather than to the entire industry, this results in unfair competition and discrimination between firms. On the other hand, duty drawbacks, which are also firm specific rather than industry wide, confront the government with the problem of checking the allocation of imported inputs.

Quotas are ideally of a temporary nature, being granted to help start off local production and to avoid speculative stockpiling by importers before the new tariff is imposed. However, though this form of protection is typically granted for six to nine months, it has been on occasions granted for twelve months or longer.

With a made-to-measure approach to protection, coupled with inadequate in depth study of the effects and consequences of the various protection devices, Malaysia clearly runs the risk of an inefficient allocation of resources within the industrial sector. However, within the public interest perspective, this is treated as one of the costs of attaining a more diversified economy.

The public interest perspective is only one model of protection policy making. The other, a private interest model, stresses the gains and losses to particular groups of factor owners. An attempt is made below to distinguish these models, but first, empirical work on the structure of protection is reviewed.

Protection structure and trends

Empirical studies of the structure of protection are not lacking in Malaysia. In the last two decades at least six studies have been undertaken.[39] These differ substantially in methodology and data base.[40] However, all rely on data not more recent than 1970. This paper uses more recent data to examine the structure of protection for the years 1973 and 1978 and compares its findings with those in the earlier empirical studies.

The estimates here are based on input-output structures at the five digit International Standard Industrial Classification (ISIC) level contained in the 1973 Census of Manufacturing Industries, a data source used in three earlier studies,[41] and the 1978 Survey of Manufacturing Industries. In order to provide reasonable comparisons with the structures in the 1960s, and given that tariffs form the principal means of protection in Malaysia, this study adopts a similar approach to the estimation of the effective rate of protection (ERP) as that adopted by Power and Ariff and concentrates on tariff protection. Such an approach is not unreasonable since Malaysia remains a very open economy, where imports do compete quite freely with domestic production in large quantities.[42] Table 4.3 shows nominal and effective rates of protection in Malaysian manufacturing industries for the years 1973 and 1978. The following discussion of the structure of protection and the changes over the decades of the 1960s and 1970s are best understood in the context of changes in ideas on protection discussed earlier.

The unweighted average nominal rate of protection (NRP) in 1963 and 1965 were 10.2 percent and 11.4 percent respectively.[43] These averages do not differ significantly. This is clearly not surprising since the TAB—the principal force in protection policies—was formally constituted only in mid 1964, and its policy recommendations could not have taken full effect by 1965.[44]

In subsequent years, with the more aggressive stance of the TAB and the vigorous use of tariffs by its successor, the CIC, as a means of expediting the industrialisation of Malaysia, NRPs rose to greater heights. For the 62 manufacturing industries covered by Ariff (1975) the unweighted average NRP was 15.5 percent in 1970, a significantly higher level than in 1963.[45] After the SACT succeeded the CIC in 1970, the unweighted average NRP rose to 20.8 percent in 1973, but then only marginally to 21.6 percent in 1978. In general the NRPs showed an upward trend in the 1970s.

ERPs, on the other hand, show a somewhat different trend. The unweighted average ERP rose from a low of 21.7 percent in 1963 to 24.9 percent in 1965. By the end of the decade, with the aggressive stance of the TAB and the CIC, the average ERP rose to 43.8 percent in 1970. However, a decline occurred around 1973 when the average ERP fell marginally to 40.4 percent and toward the end of the decade when the average ERP fell to 39.3 percent in 1978. The latter seems to reflect the general policy of reducing ERPs first outlined in the Third Malaysia Plan (1976–1980).

NRPs and ERPs show wide variation. The lowest NRP has consistently been provided to rubber remilling and latex processing. On the other hand, the highest NRP has consistently been granted on tobacco products. The upward trend in the NRP for tobacco products, due to the upward trend in specific duties and the fall in the price of imports, accounts for the widening gap between the two. More specifically, the range has increased from 93 percentage points in 1963 to 210 percentage points in 1978. It should be noted that much of the change in NRP on tobacco products has been for reasons of revenue and public health. The wide variation in NRPs can also be seen in terms of the dispersion in NRPs. The coefficients of variation show no obvious trend. Dispersion in NRPs was least in 1970 with a coefficient of variation of 18.8, and highest in 1973 with a coefficient of variation of 39.6.

ERPs, on the other hand, show an even wider range. Generally, rubber remilling and latex processing have had the lowest ERP. In all the years covered, except for 1970 when the coffee industry had the lowest ERP (−64 percent), rubber processing industries had the lowest ERPs, though this increased considerably from an estimated −50 percent in 1963 to −28 percent in 1978. On the other hand, textiles had the highest ERP in the 1960s (337 percent in 1963), but this declined consistently over the years. By 1970, chemical fertiliser industries were accorded the highest ERP (248 percent), but in the later years the motor vehicle industry became the most highly protected industry; the ERP rose from 192 percent in 1963 to 282 percent in 1973, and then to 317 percent in 1978. This, no doubt, reflects the intentions of the government to establish a motor vehicle industry in Malaysia.

Although the range in ERPs has widened, actual dispersion or variation in ERPs seems to have declined. In contrast to the NRPs which do not show any obvious trend in dispersion, the ERPs show a clear downward trend in dispersion. The coefficients of variation declined consistently from 102.1 in 1963 to 60.0 in 1973, with a slight upturn in 1978 to 63.8. This seems to reflect the tendency to accord more protection to those industries which had negative ERPs, particularly in the 1960s, and perhaps to some extent, the move towards reducing ERPs at the upper end of the scale with the exception of industries which the government intends to promote.

Table 4.4 shows the NRPs the ERPs and by major product group. Except for the 1970 values which have been averaged by supply weights (the contribution to output of each industry group) the values for all other years have been averaged by derived free trade value added weights. This, together with the changes in the classification of the industries—which will become apparent later—means that the averages are strictly not comparable. Nevertheless, a glance at them might provide an indication of the changing pattern of protection across major product groups.

The pattern of nominal protection by product group has changed over the two decades. In 1963 consumer durables industries appeared to be the least protected but in 1970, apart from beverages and tobacco industries, these became the most highly protected in nominal terms. By 1978 the level of nominal protection for the group had increased to 54.8 percent from a mere 1 percent in 1963. This increase reflects, to a large extent, the increase in nominal protection

Table 4.3 Average[a] nominal and effective rates of protection, manufacturing sector, Malaysia, 1973 and 1978

		Nominal	Effective
Importables[b]	1973	4.3	8.8
	1978	12.7	23.0
Exportables[b]	1973	9.1	6.4
	1978	12.5	−17.0
Other	1973	35.4	48.6
	1978	21.7	55.9
Total manufacturing	1973	28.2	37.4
	1978	18.8	34.2

Notes:
a Value added weights.
b Imports or exports at least 20 percent of apparent consumption.
Source: Appendix 1.

for the motor vehicle industry. Machinery industries also show a substantial change in nominal protection over the two decades. In the 1960s, these industries received little or no protection for their products. However, by the turn of the decade, NRPs actually rose for these industries bringing the average NRP for the group up from 5 percent in 1963 to 22.3 percent in 1978.[46]

On the other hand, industries producing intermediate products of lower levels of fabrication (I) continue to be among the least nominally protected compared with industries producing intermediate products of higher levels of fabrication (II). For the former, the average NRP has remained rather constant at around 4 percent.[47] However, for the latter, the average NRP has increased from 14 percent in 1963 to 24.2 percent in 1978. For all other groups, nominal protection has increased by no more than ten percentage points. Beverages and tobacco industries continue to enjoy the highest rates of nominal protection, but protection *per se* has been secondary to reasons of public revenue and public health.

ERPs show a slightly different picture. Consumer durables industries which were accorded the lowest (negative) average ERP in 1963 became the most highly protected group by 1970.[48] On the other hand, the average ERP has increased rather consistently for the group of industries producing nondurable consumer goods, giving the group the second highest ERP in 1978.

Effective protection for intermediate products of lower levels of fabrication has consistently been far less than that for industries producing intermediate products of higher levels of fabrication. In fact, in the 1970s the former appears to be the least protected group. For the latter, the average ERP increased rather consistently during the 1960s and into the early 1970s but appears to have declined by the late 1970s. As for the beverages and tobacco industries—which have been accorded the highest levels of nominal protection—the level of effective protection for the group in 1978 is less than the levels for consumer durables and nondurables industries. Decline for this group is due primarily to the rise in import duty for raw material inputs; particularly for the tobacco industry where import duty has been increased substantially.

Table 4.4 shows that the system of protection in Malaysia discriminates against primary based industries in favour of consumer goods industries. Processed food industries and industries producing intermediate products of lower levels of fabrication, which includes all the primary based processing industries, have consistently received low levels of nominal and effective protection. It must, however, be noted that within the two product groups NRPs and ERPs vary quite widely.

Protection in the Malaysian manufacturing sector does not discriminate only against the primary based manufacturing industries. On a large scale, manufacturing sector protectiom can result in a discrimination against other sectors of the economy, and therefore bias the allocation of resources. Table 4.5 shows the average ERPs for the major sectors of the Malaysian economy for the years 1965 and 1973.[49] For the year 1965, the estimates can be compared directly with the estimates in Table 4.4. However, the 1973 estimates are, strictly, not comparable since they are derived from two different sources. Nevertheless, with due caution, it can be seen that the system of protection seems to discriminate against the primary sectors though, in 1965, consumer durables industries had an average ERP which was less than that for agriculture and rubber planting. In 1973 processed food, construction materials, intermediate products of lower level fabrication (I) and transport equipment industries seem to have enjoyed lower ERPs than the agricultural and livestock sectors.[50] Nevertheless, there was a bias in favour of the manufacturing sector against the rubber planting, fishing, and mining sectors. With the apparent decline in manufacturing sector protection in the latter part of the 1970s it is possible that the extent of the bias against the primary sector may have declined.

Summary

In summary, it appears that while the levels of NRP have increased, particularly during the period of the TAB, CIC, and the early years of the SACT, the levels of ERP appear to be on the decline. This appears to be a reflection of the rise in tariffs on inputs and the general policy of reducing ERPs first outlined in 1976. Further, while there is no obvious trend in the extent of variation in nominal protection across industries, the extent of variation in effective protection

Table 4.4 Nominal and effective rates of protection by product group: Malaysia

	Nominal					Effective				
Product Group	1963[a]	1965[a]	1970	1973[a]	1978[a]	1963[a]	1965[a]	1970	1973[a]	1978[a]
Processed food	9	11	–	17.3	17.2	6	7	–	16.9	24.2
Beverages and tobacco	21	46	n.a.	159.5	146.7	17	73	n.a.	105.3	44.1
Construction materials	6	7	11	16.7	8.9	7	9	73	3.9	22.8
Intermediate products I	4	4	−4	3.2	4.7	8	9	−19	−7.2	0.1
Intermediate products II	14	13	11	21.2	24.2	27	25	52	42.7	41.6
Nondurable consumer goods	14	14	12[b]	28.2	24.2	19	20	17	42.0	85.2
Consumer durables	1	1	26	54.6	54.8	−11	−5	103	193.9	172.6
Machinery	5	5	20	28.9	22.3	6	6	64	82.7	38.6
Transport equipment	–	–	25	0.7	0	–	–	164	−7.3	−5.4

Notes: a Estimates based on free trade coefficient.
b Includes processed food
— Not available, included in another category.

Source: 1963 and 1965 rates from Power (1971), 1970 rates from Ariff (1975), 1973 and 1978 calculated by the author.

Table 4.5 Average effective rates of protection by major resource based sectors: Malaysia

Sector	1965	1973
Agriculture	2	29
Livestock	na	17
Rubber planting	0	1
Forestry	−16	10
Fishing	1	2
Metal mining)	−16	−1
Other mining)		2

Source: Power J.H. (1971), 'The Structure of Protection in West Malaysia', in B. Balassa and Associates, *The Structure of Protection in Developing Countries*, Baltimore: Johns Hopkins University Press, table 9.6, p. 213; and Malaysia (1975), 'Effective Protection and Industrialization Policies, Report of a Research Project', mimeographed draft, Kuala Lumpur, Table 5, pp. 29–30.

appears to be on the decline; in fact, there is a consistent downward trend in the coefficient of variation except for the marginal increase in 1978. The positive relationship between the level of effective protection and the level of fabrication first found by Power (1971) seems to hold true not only in the 1960s but also in the 1970s. The system also appears to remain biased against primary based industries, in favour of consumer durables industries. Although Malaysia's rates of protection are considered moderate on most counts, the system still appears to discriminate against exports and favour import substitution.

Heavy industry policy in the 1980s

A comparison of ERPs by major sectors appears to show a discrimination against the primary sector, though this may have declined with the general decline in manufacturing sector protection in the latter part of the 1970s. However, this may change with the new move towards heavy industries first announced with the formation of the Heavy Industries Corporation of Malaysia (HICOM) in 1980. HICOM was established to initiate, implement, and manage the development of heavy industries requiring large investment outlays and having long gestation periods. The development of heavy industries is seen as a new source of domestic growth which will strengthen backward and forward linkages in the economy. These industries, forming a part of the move towards a second round of import substitution industrialisation strategy, are meant to provide inputs to domestic industries and make use of the country's extensive energy and gas resources. Projects proposed or under execution include two refineries, cement plants, two sponge iron plants, a cold rolling steel mill, a methanol plant, an ammonia-urea plant, a pulp and paper plant, a petro-chemical complex and an automotive plant.

The latest World Bank Country Report on Malaysia has apparently expressed concern over the list of projects estimated to cost between

\$6–\$8 billion but generating employment of only about 14,000 by 1990.[51] The Report apparently emphasised that it would be extremely harmful to the economy in the long term if the projects do not enjoy domestic cost advantages, or if they require permanent protection.

In the midterm review of the Fourth Malaysia Plan, the government noted that improving the structure of the incentive system is an effective way of enhancing the development of the manufacturing sector. The existing system is considered generous but not fully exploited. A review is now underway for the purposes of the Master Plan on Industrialisation. Meanwhile, some of the HICOM projects are already in the execution stage; that is, even before the final outcome of the Master Plan is known. Problems will arise if these heavy industries prove to be inefficient. With so much at stake in terms of capital outlay, the government may be forced, once these industries are operating, to provide high protection to ensure the survival of these public sector projects. This will then result in a new structure of manufacturing sector protection which may ride against the present desire to reduce ERPs. This structure may also mean a new round of discrimination against the primary sector.

Causes of protection

The above discussion lends greater support to the empirical finding that both the levels of NRP and ERP vary widely in Malaysian manufacturing industries. Reasons why some industries receive more assistance than others are now examined. The issue is of importance because wide variations in protection rates affect resource allocation. High protection for some industries *vis a vis* others may lead to inefficient resource allocation within the industrial sector. From a policy viewpoint, an examination of the forces at work may prompt suggestions as to what might be done to reduce the extent of variation, as well as the level of protection. This would be particularly important in the light of the government's explicit intention to reduce the level of effective protection in the 1980s. Also a review of the determinants of the pattern of industry assistance permits a test to distinguish the public and private interest models of policy making discussed earlier.

The model

The level of protection or assistance is determined by the forces of demand and supply of assistance.[52] If industry groups are well informed and economically rational, the ERP would be their principal objective. On the other hand, if they are less well informed, these groups would concentrate their efforts on the level of nominal protection. For this reason this paper examines the determinants of the level of NRP and of the level of ERP.

On the demand side, the forces affecting or influencing the level of assistance can be grouped under two broad categories reflecting employee and employer interests. The former can be proxied by labour intensity (LI), measured in terms of labour's share of value added; value added share of output (VASO); percentage change in the number of employees (PCNOE); and the average wage per employee (AWPE). The latter can be proxied by variables such as the share of domestic supply met by imports (SDSMBM); the number of firms in the industry (NOF); firm size measured by the turnover per firm (TOPF); industrial concentration measured in terms of the share of the largest four firms in the total output of the industry (IC); fixed assests per firm (FAPF); and the number of foreign firms (NOFF) in the industry.

On the supply side, given the emphasis on employment creation and the development of natural resource based industries, the government is more likely to grant higher assistance the larger the number of employees in the industry (NOE) and the greater the use of natural resources. On the other hand, industries which are able to compete without explicit assistance are likely to be granted less assistance. Here, utilisation of natural resources is proxied by a dummy variable (NATRES) which takes unit value for a natural resource based industry and zero otherwise. The ability of the industry to compete without assistance is proxied by the proportion of its total production which is exported (SPX).

The basic model of the causes of protection can, therefore, be specified as follows:

$$\begin{aligned} RP = {} & a + bLI + cVASO + dPCNOE + eAWPE + fSDSMBM + \\ & gNOF + hTOPF + iIC + jFAPF + kNOFF + lNOE + \\ & mNATRES + nSPX + u \end{aligned}$$

where RP = nominal or effective rate of protection;
u = disturbance term; and all other variables are as defined above.

Following Anderson (1980) and earlier studies, for instance, Pincus (1975, 1977) and Caves (1977), the expected relationships of these variables, reflecting the demand for and supply of protection, to the level of protection are as shown in Table 4.6

Empirical results

Table 4.7 shows the stepwise regression results for 1973. The overall explanatory power is rather low but the F-ratios are significant at the 5 percent level. In Table 7 only two variables are consistently

Table 4.6 Expected relationship: dependent variable and independent variable: Malaysia

Independent variable	Expected effect
Demand factors	
Labour intensity (LI)	+
Value added share of output (VASO)	–
Change in employee numbers (PCNOE)	+/–
Average wage per employee (AWPE)	+/–
Domestic supply met by imports (SDSMBM)	+
Number of firms (NOF)	+/–
Turnover per firm (TOPF)	+
Industrial concentration (IC)	+
Fixed assets per firm (FAPF)	+
Number of foreign firms (NOFF)	+
Supply Factors	
Number of employees (NOE)	+
Natural resource based industry (NATRES)	+
Share of production exported (SPX)	–

significant at the 10 percent level or better—IC and NOE—and both have the expected signs. This seems to suggest that, on the demand side, industries which are more highly concentrated tend to demand more protection and are, in fact, given more protection, perhaps because of the greater lobbying power of the few large firms. On the other hand, on the supply side the significance of NOE suggests that the government is more willing to accord higher protection to industries employing more labour. It is conceivable that the insignificance of labour intensity is due to the significance of NOE. NOE is relatively easier to compute than LI, and given the emphasis on employment creation, the government tends to grant protection based on the former rather than the latter.

In the determination of the effective rate of protection, the share of domestic supply met by imports (SDSMBM) has a consistently negative and significant coefficient, contrary to the predicted relationship. This indicates that the more assisted industries are not necessarily those whose domestic markets have been supplied largely by imports. A plausible explanation is that industry assistance has reduced import penetration ratios in highly assisted industries to well below what they would have been in the absence of protection.[53] This may be particularly true in the rush towards industrialisation during the period of the TAB and the CIC. Another factor which may have contributed towards this negative relationship is the government's policy on essential items and consumer choice. Protection under this policy is kept low for essential items in order to maintain stable low prices and to ensure adequate consumer choice in terms of variety and quality. This issue is discussed further in the case studies provided below. Nevertheless, as Anderson (1980) concludes, the

Table 4.7 Determinants of the rate of protection, 1973: Malaysia

	Nominal	Effective
Constant	−13.680	19.713
LI	17.568	–
	(1.390)	
PCNOE	–	0.133[c]
		(1.702)
AWPE	−0.006	–
	(1.575)	
SDSMBM	−228.563	−748.543[b]
	(1.292)	(2.807)
NOF	0.240	–
	(0.995)	
TOPF	–	−4.043
		(1.003)
IC	81.527[a]	38.882[c]
	(4.748)	(1.733)
NOE	1.561[b]	2.475[b]
	(2.229)	(2.531)
SPX	–	−28.050
		(1.075)
$\bar{R}^2$	0.314	0.264
F	3.980[b]	3.330[b]
N	40	40

Notes:
a = significant at 0.01%
b = significant at 0.05%
C = significant at 0.10%
Figures in parenthese are t-ratios

negative sign does not imply that the stronger the import competition, the less assistance an industry is likely to obtain.

Another significant variable in the ERP regression is PCNOE. The coefficient is positive and significant at the 10 percent level. This suggests that faster growing industries may in fact be in a better position to extract more protection than slower growing industries, because of their potential capacity to utilise more labour. The degree of industrial concentration (IC) and the number of employees (NOE) are also significant in the ERP equation.

Table 4.8 shows the regression results for 1978. The overall explanatory power is extremely low in both regressions. Though most of the coefficients included have the expected signs, only a handful are significant at the 10 percent level or better.

The poor explanatory power of the exogenous variables in contrast to the better fit for 1973 may be reflective of several factors. First, it may be reflective of the mellowing of Malaysia's protection policy from the rash of protection during the period of the TAB and the CIC to the relatively more conservative approach of the SACT in the latter half of the decade. Being less hasty, it was probably easier for

Table 4.8 Determinants of the rate of protection, 1978: Malaysia

	Nominal	Effective
Constant	12.426	27.056
VASO	−28.575	–
	(1.061)	
PCNOE	–	0.241
		(1.052)
NOF	–	−0.103
		(1.172)
IC	35.770[a]	23.979
	(3.534)	(1.319)
NOE	0.732[c]	2.044[b]
	(1.799)	(2.565)
NATRES	−6.289	−13.28
	(1.078)	(1.308)
$\bar{R}^2$	0.127	0.071
F	3.825[b]	2.191
N	79	79

Notes:
a = significant at 0.01%
b = significant at 0.05%
c = significant at 0.10%
Figures in parentheses are t-ratios

the SACT to sidestep much of the pressure which may have been exerted on the demand side, though the significance of IC seems to suggest, as in 1973, that larger firms could still exert considerable lobbying power. On the supply side, the coefficient of NOE remains positive and significant, showing the importance which is attached to labour utilisation in the granting of protection or assistance.

Econometric results suggest an industry is more likely to be protected if it has a large number of employees, a growing number of employees, and a higher degree of industrial concentration. These variables are consistent with elements of a private interest lobbying model, or a public interest model which stresses protection for industries employing large amounts of labour. These industries are not necessarily labour intensive. However, the low explanatory power of the equation suggests that a number of variables have been excluded. Some of these may reflect government intentions for a diversified manufacturing sector and a 'made to measure' approach to tariff making.

Also, the method of granting protection, whether in the form of an increase in NRP, duty exemptions, or duty drawbacks remains *ad hoc*. Each case is still examined on its own merits. Any proper assessment of protection would require a great deal more economic data, staff, and expertise than the SACT or the Tariff Unit in MIDA can possibly command at present. With the rapid growth of industrialisation, it becomes increasingly more difficult for the SACT to coordinate the levels of protection accorded to the different industries, particularly in terms of the ERPs. The lack of information combined with a made to measure approach to policy making suggests it is difficult to find any strong patterns in levels of NRPs or ERPs in Malaysia within the manufacturing sector.

Summary of findings and implications

Manufacturing sector protection in Malaysia is most clearly reflected by its tariff policies. Changes in the latter are in turn reflections of policy makers' ideas on protection and the pace of industrialisation. In the initial pre-independence period, the rapid growth of the natural rubber and tin industries on the basis of comparative advantage gave no grounds for any form of protection, let alone for any protection for the manufacturing sector, which was then of minor importance to the Malaysian economy. The principle of free trade reigned supreme and protection was viewed merely as an unavoidable consequence of a high dependence on customs duty taxation for public revenue. Import duties, in fact, became an important source of public sector revenue needed to finance Imperial defence efforts during World War I and the Malayan Emergency.

With the advent of self government in 1955, tariffs featured as a potentially effective policy instrument for the development of Malaysia's manufacturing industries. The first World Bank Mission recommended the use of tariffs as a means of encouraging industrial development, though the Working Party on Industrial Development opted for a system based on tax concessions. When independence finally came in 1957, a Tariff Advisory Committee (TAC) was established to investigate and advise the government on individual applications for tariff protection. The TAC's cautious approach towards protection was subsequently replaced by the more aggressive approach of the Tariff Advisory Board (TAB), which was established following the recommendations of the second World Bank Mission in 1963. However, even the aggressive approach of the TAB pales in contrast to the Capital Investment Committee's (CIC) more aggressive approach, which came into force following the racial riots of May 1969. When the political and social climate of the country improved, the CIC ceased to operate and the Special Advisory Committee on Tariffs (SACT) was established, taking over the functions of the TAB. Initially, it pursued the CIC's aggressive stance but by the mid 1970s it adopted a more mellowed posture granting protection only to 'deserving' industries. Essentially, the approach adopted by each of these institutions reflects the pace of industrialisation which each felt was needed to propel Malaysia along the road towards economic growth and development.

As the study shows, changes in the levels and structure of protection—both nominal and effective—reflect quite clearly the philosophy of each of these institutions. The levels of nominal protection (NRP) increased particularly during the period of the TAB, CIC, and the early years of the SACT, although the levels of effective protection (ERP) appear to be on the decline particularly in the later part of the 1970s. The latter reflects the rise in tariffs on inputs and the general policy of reducing ERPs first outlined in 1976. However, in spite of the latter policy, the structure of manufacturing sector protection shows a large variation in ERP's and also moderately high levels of ERP. These affect resource allocation in the manufacturing sector. Resources are likely to flow to the more highly protected industries resulting in production inefficiencies. This is particularly true in the case of the textile industries (with ERPs ranging from 35 percent to 120 percent in 1978), the tyre and tube industry (170 percent), and the motor vehicle assembly industry (317 percent). This pattern of protection reflects the creation of inefficient industries serving a small market.[54]

On another level, the positive relationship between protection and the level of fabrication implies an incentive for resources to flow to finished products rather than to intermediate products industries. This has prevented the development of strong backward linkages. A comparison of ERPs by major sectors also shows a tendency to discriminate against the primary sector. The extent of discrimination has probably declined over the 1970s but the move towards a new round of import substitution based on heavy industries may see a reversal of this trend to the disadvantage of the agricultural sector.

Strong autonomous action on the part of the government in protection matters coupled with limited means, expertise, and information have given rise to a made to measure, case by case approach to protection. In this setting, the level of protection may not be strongly influenced by the forces of demand for and supply of protection. This is borne out in the study of the causes of protection. On the demand side, larger firms and those which contribute towards manufacturing sector employment growth appear to be able to exert some influence; particularly on the level of nominal protection received. On the supply side, decision makers seem to favour industries providing large employment. However, these forces do not account for a large portion of the variance in protection levels. Other factors omitted in the econometric analysis may be more important. Case studies of the motor vehicle assembly, tyre and tube, textile and tobacco industries show quite clearly that protection is often based on an attempt to satisfy conflicting interests. For example, attempts to develop an industry supplying inputs to another industry which the government also attempts to encourage results in a conflict of interests. Furthermore, the government must balance private interests with public interests in terms of consumer choice and welfare. The development of heavy industries, for instance, the automobile industry, may also introduce further conflict of interests. Production for a small market in small scale, high cost, capital intensive plants conflicts with the objectives of employment and consumer welfare.

Perhaps the main implication of this study concerns possible improvements to the current approach to protection. Although the made to measure, case by case approach is understandable, attempts should be made towards a more coordinated, highly detailed approach. The approach advocated by the United Nations Development Programme, involving the identification of least cost techniques, might be considered. Such a move would clearly require considerably more resources than the Malaysian Industrial Development Authority (MIDA) now has, but the social benefits to the country from a more efficient manufacturing sector would be likely to outweigh the social costs. Indeed, more information is needed than is now available.

A ceiling should preferably be set on the ERP, particularly in the light of the move towards the development of heavy industries. As a first step, the Economic Planning Unit's recommendation of a 50 percent protection ceiling appears appropriate[55] though this will have serious implications for the motor vehicle assembly, tyre and tube, and textile industries, and for the development of heavy industries such as the automobile industry. Nevertheless, the goal should be the development of uniformly low levels of protection which do not discriminate unnecessarily against the primary sector. As a final point, the required level of protection should be considered at the time approval is sought. Once approval is given it might be politically impossible to withhold excessive protection.[56] This is particularly important in the case of the projects being developed by the government's Heavy Industries Corporation of Malaysia (HICOM).

Appendix 1

Table 4.A1 Nominal and effective rates of protection, Malaysia 1973 and 1978

	Industry	Nominal		Effective	
		1973	1978	1973	1978
31					
110	Meat slaughtering	20	20	25	20
121	Ice cream	15	15	15	16
129	Other dairy products	20	20	19	39
131	Pineapple canning	161	116	49	22
139	Fruits & vegetables canning	20	20	35	22
140	Sea foods canning	20	20	112	65
151	Coconut oil	0	0	−4	02
152	Palm oil	−10	−25	−124	−75
153	Palm oil kernel		0		0
159	Other oils and fats	0	0	0	−1
161	Small rice mills	0		0	
162	Large rice mills	0	0	0	0
163	Flour		0		0
164	Sago and tapioca		0		−1
169	Other grain	0	0	0	0
171	Biscuit	22	22	43	45
172	Bakeries	22	22	52	62
180	Sugar		30		42
190	Cocoa, chocolate confectionery	13	13	13	13
211	Ice	0	0	0	0
212	Coffee	0	0	−17	−11
213	Tea		23		23
214	Noodles	11	9	31	14
215	Spices and curry powder	20	20	72	23
219	Other food products	8	3	11	16
220	Animal feed	0	0	−1	−1
310	Spirits		26		27
330	Malt liquors and malt		15		17
340	Soft drinks	38	23	54	22
400	Tobacco	180	210	119	76
32					
111	Textile spinning & weaving	25[a]	15	54[a]	15
112	Textile, dyeing, bleaching & printing		35		66
113	Handicraft spinning & weaving		25		20
114	Batek		50		218
115	Synthetic textile		25		36
119	Other primary textile		25		35
120	Made up textile	50	50	130	120
130	Knitting		15		28
140	Carpets and rugs		50		221
150	Rope and twine		15		19
190	Other textile		25		25
201	Clothing	25	25	34	14
310	Tanneries and leather	25	25	24	22
330	Leather substitute products		20		25
400	Footwear (nonrubber or plastic)	40	40	130	107
33					
111	Sawmill	0	0	−1	−1

112	Plywood, hardboard	20	20	47	42
113	Planing, window and door frames	20	20	62	28
114	Prefabricated wood houses		20		28
115	Wood preservation	20		35	
119	Other wood products		20		27
120	Wood and cane containers	20	20	46	21
190	Wood and cork products	20	20	49	25
200	Furnitures and fixtures	20	20	59	28
34					
110	Pulp, paper, paperboard		15		18
120	Paper and paperboard containers	20	20	84	49
190	Pulp, paper, paperboard articles		20		35
200	Printing and publishing	15	15	27	24
35					
110	Basic industrial chemicals		25		35
120	Fertilisers and pesticides	23	12	95	36
130	Synthetic resins, plastic materials		25		38
210	Paints, varnishes, lacquers	25	25	55	43
220	Drugs	0	0	−5	−4
231	Soap and cleaning preparations	20	20	32	30
239	Perfumes, cosmetics	50	45	70	65
290	Other chemical products	25	25	42	51
300	Petroleum refinery		0		−3
400	Other petroleum & coal products		20		40
510	Tyre and tube	40	66	133	170
591	Rubber processing	0	−17	−34	−28
592	Rubber smokehouses	0	−17	−41	−13
593	Rubber footwear		25		46
599	Other rubber products		25		61
600	Plastic products	40	50	129	114
36					
100	Pottery, china and earthernware	10	15	6	20
200	Glass and glass products	25	25	42	32
910	Structural clay products		15		18
921	Hydraulic cement	23	14	10	15
922	Lime and plaster	0[b]	0	0[b]	0
991	Cement and concrete products	10	10	9	37
992	Stone and stone products	20	20	33	21
999	Nonmetallic mineral products		20		−5
37					
101	Primary iron and steel	15[c]	15	32[c]	21
102	Foundries	0	0	−3	−10
109	Other iron and steel		30		103
209	Nonferrous metal		30		113
38					
111	Cutlery, handtools and hardware	20	20	56	16
112	Tinsmithing and blacksmithing	15		25	
120	Metal furniture and fixture		50		79
130	Structural metal products	15	30	19	59
191	Tin cans and metal boxes	15	15	76	33
192	Wire and wire products	11	6	12	11
193	Brass, copper, pewter & aluminium products	15	15	38	24
199	Other fabricated metal products		15		22
210	Engine and turbines	30[d]	20	63[d]	93
220	Agricultural machinery & implements	0	0	0	8
230	Metal and wood working machinery		35		64
240	Industrial machinery & equipment		20		27

Table 4.A1 Continued

Industry		Nominal		Effective	
		1973	1978	1973	1978
291	Refrigeration, exhaust, ventilating machinery	20	20	76	23
229	Other machinery & equipment		30		64
310	Electrical industrial machinery		25		54
320	Radio, television, communication		25		135
330	Electrical appliances	30	45	94	113
391	Cables and wires		30		66
392	Dry cell and storage batteries		50		178
393	Electric lamp and tubes		10		17
339	Other electrical apparatus		20		38
410	Shipbuilding and repair	0	0	−4	−1
431	Motor vehicle bodies	0	0	−14	−14
432	Motor vehicle assembly	60	60	282	317
439	Motor vehicle parts	0	0	−8	−10
441	Motor cycle assembly		45		54
449	Bicycle, tricycle assembly	12	16	21	31
490	Other transport equipment	25		100	
510	Scientific equipment		25		36
520	Photographic and optical goods		20		20
530	Watches and clocks		20		9
39					
010	Jewellery		5		5
020	Musical instruments		25		27
030	Sporting goods		0		0
091	Brooms, brushes, mops	16	25	61	85
092	Pens, pencils, art supplies		10		17
099	Other manufacturing		15		26

Notes:

a Includes 32115, 32119, 32150, 32190.

b Includes 36999.

c Includes 37109, 37209.

d Includes 38230, 38250, 38299.

Endnotes

* I am indebted to Christopher Findlay for the many comments and suggestions which have greatly contributed towards the exposition in this paper, and to Carol Chan for the arduous task of matching industrial and trade data, and for calculating the rates of protection. Needless to say, I remain responsible for any errors and shortcomings.

1 Unless otherwise stated, the term 'Malaysia' refers to peninsular Malaysia.

2 See Allen G.C. and Donnithorne, A.G. (1954), *Western Enterprise in Indonesia and Malaya*, London: George Allen and Unwin; and Puthucheary, J.J. (1960), *Ownership and Control in the Malayan Economy*, Singapore: Eastern Universities Press, for a discussion of the role of rubber and tin in the development of the Malaysian economy.

3 Teh, K.P. (1977), 'Protection, Fiscal Incentives and Industrialization in West Malaysia Since 1957', *Monograph Series on Malaysian Economic Affairs*, Kuala Lumpur: Faculty of Economics and Administration, University of Malaya, has identified three similar stages.

4 Edwards, C.T. (1965), 'The Structure of Import and Excise Duty Taxation in the States of Malaya and Singapore', *Malayan Economic Review*, X, 2 (October).
5 Malaya, States of, (1930), *Estimates of State Revenue and Expenditure, 1930*, Kuala Lumpur: Government Printing Office.
6 Allen G.C. and Donnithorne, A.G. (1954), pp. 44, 235–239.
7 The textile quotas, introduced in 1934, were designed to safeguard British and Commonwealth textile trade, but the main emphasis appeared to be on the protection of the producers in Lancashire who had suffered badly under intensive Japanese competition; cf. Allen and Donnithorne, (1954), p. 236; and Stahl K.M. (1958), *The Metropolitan Organisation of British Colonial Trade*, London: Faber and Faber, pp. 77, 96–97, 137, 144 and 161.
8 See for instance: Allen and Donnithorne (1954); Puthucheary, J.J. (1960); and IBRD (1955), *The Economic Development of Malaya*, Baltimore: Johns Hopkins University Press.
9 IBRD (1955).
10 ibid, p. 123.
11 ibid, p. 438.
12 Hoffman, L. and Tan, S.E. (1980), *Industrial Growth, Employment, and Foreign Investment in Peninsular Malaysia*, Kuala Lumpur: Oxford University Press, p. 54.
13 Malaya, Government of, (1957), *Report of the Industrial Development Working Party*, Kuala Lumpur: Government Printer.
14 ibid, pp. 37–38.
15 Lim, D. (1973), *Economic Growth and Development in West Malaysia, 1947–1970*, Kuala Lumpur: Oxford University Press.
16 Wheelwright, E.L. (1963), 'Industrialization in Malaya', in T.H. Silcock and E.K. Fisk (eds.), *The Political Economy of Independent Malaya*, Singapore: Eastern University Press.
17 Hoffman, L. and Tan, S.E. (1980), p. 54.
18 Lo, S.Y. (1972), *The Development Performance of West Malaysia: 1955–1967*, Kuala Lumpur: Heinemann Educational.
19 IBRD, (1963), *Report on the Economic Aspects of Malaysia*, Kuala Lumpur: Government Press, p. 30.
20 Teh, K.P. (1977), p. 49.
21 IBRD (1963).
22 Until the dissolution of the TAB, for reasons unknown, none of the reports had been published.
23 Malaysia, Government of, (1966), *First Malaysia Plan, 1966–1970*, Kuala Lumpur: Government Printer, pp. 132–3.
24 Fong, L.S. (1969), 'The TAB and Industrial Development', unpublished graduation exercise submitted to the Faculty of Economics and Administration, University of Malaya, pp. 39–40.
25 IBRD (1963).
26 ibid, p. 11.
27 Teh, K.P. (1977), p. 51.
28 Hoffman, L. and Tan, S.E. (1980), p. 55.
29 Power, J.H. (1971), 'The Structure of Protection in West Malaysia', in B. Balassa and Associates, *The Structure of Protection in Developing Countries*, Baltimore: Johns Hopkins University Press.
30 Lim, D. (1973), p. 258.
31 ibid, p. 258.
32 Federal Industrial Development Authority (FIDA) (1970), *Annual Report, 1970*, Kuala Lumpur, pp. 63–66.

33 Hoffman, L. and Tan, S.E. (1980), p. 56.
34 Ascertained from unpublished source, Malaysian Industrial Development Authority.
35 Malaysia, Government of, (1976), *Third Malaysia Plan, 1970–1980*, Kuala Lumpur: Government Printer, p. 279.
36 ibid, p. 279.
37 Von Rabenau, K. (1975), 'Trade Policies and Industrialization in a Developing Country: The Case of West Malaysia', *Economics Discussion Paper*, No. 55, Regensburg: University of Regensburg, an abridged version can be found in *Malayan Economic Review*, XXI, 1, (April 1976); and Malaysia, Government of, (1975), 'Effective Protection and Industrialization Policies: Report of a Research Project', Kuala Lumpur: mimeographed, Economic Planning Unit.
38 Teh, K.P. (1977), Chapter 4.
39 See, for instance, Power J.H. (1971); Panchamukhi V.R. (1972), *Effective Protection and Intra-Regional Trade: A Case Study of Malaysia*, Bangkok: mimeographed, ECAFE Secretariat; Ariff K.A.M. (1975), 'Protection for Manufactures in Peninsular Malaysia', *Hitotsubashi Journal of Economics*, XV, 2, February; Edwards C.B. (1975), *Protection, Profits and Policy: An Analysis of Industrialization in Malaysia*, 2 Vols., Norwich: University of East Anglia; Malaysia (1975); von Rabenau. K. (1975).
40 For a succinct review of these studies see Hoffman, L. and Tan, S.E. (1980), pp. 36–88.
41 Power, J.H. (1971); Ariff, K.A.M. (1975); Edwards, C.B. (1975).
42 Power, J.H. (1971); p. 211; Ariff, K.A.M. (1975), p. 45.
43 Insufficient information prevents computation of weighted averages for all years.
44 NRPs were actually introduced or increased for about 81 items in the tariff schedule in 1964, and for 194 items in 1965.
45 NRPs were introduced or increased on 279 items in 1969 and on 293 items in 1970.
46 The substantial change for transport equipment is probably a consequence of the reclassification of the motor vehicle assembly industry as a consumer durable.
47 Free trade coefficients produce a NRP of 4 percent while domestic input-output coefficients gave a NRP of −4 percent for the group in 1963 and 1965. This difference is explained by the large weight given to the few export industries which are subject to export taxes.
48 The high ERP for 1970 for transport equipment may be due to reasons stated in endnote 46.
49 The present study based on the 1973 Census of Manufacturing Industries and the 1978 Survey of Manufacturing Industries does not permit a computation of the ERPs for the other sectors of the Malaysian economy.
50 See also Malaysia (1975), Table 5, pp. 29–30 for comparable estimates for manufacturing and the primary sectors. The basic observation here is not changed.
51 The employment figure was cited by the Vice-President of HICOM in the New Straits Times, 1st August 1984, p. 11.
52 Anderson, K. and Garnaut, R. (forthcoming), *Australian Protection and Trade with Developing Countries*, monograph to be published by the World Bank.
53 Anderson, K. (1980), 'The Political Market for Government Assistance to Australian Manufacturing Industries', *Economic Record*, LVI, 153, June.
54 Shepherd G., 'Policies to Promote Industrial Development', in K. Young, W.C.F. Bussink and P. Hasan (eds), *Malaysia: Growth and Equity in a Multi-racial Society*, Baltimore: Johns Hopkins University Press (1980).
55 Malaysia (1975).
56 Shepherd, G. (1980), pp. 194–9.

5 The structure and causes of manufacturing sector protection in Singapore

TAY BOON NGA*
April 1985

SINGAPORE is a tiny island state. The area is about 225 square miles. It has no natural resources except its superb location, yet its growth record has been impressive. Average annual real growth rate from 1960 to 1984 was nearly 9 percent (Table 5.1). Singapore's growth was generally faster than that in other industrialised or industrialising countries.[1]

Many factors have contributed to this rapid growth, two of which have included the success of industrialisation and the trade policy. Although Singapore had advocated a free trade policy since its foundation, it did impose trade restrictions to protect its local manufacturing sector for some periods of time. Also, from time to time, the protectionist effects of other countries' policies have been felt by Singapore. In this exercise, we will first examine Singapore's trade policy, then the struture and causes of protection, the effect of protection and, finally, the Singapore government's policies in response to protectionism in other countries, with special reference to the textile and garment industry.

Early trade development

Merchants have traded in South East Asia for more than two thousand years. Singapore, located at the end of the 'Golden Cheronese' of Peninsular Malaya, became one of the main ports of call for adventurers, navigators, pilgrims, and traders by 1820. But the right to establish a trading post (of the then British East India

* Tay Boon Nga is with the Department of Economics and Statistics, National University of Singapore.

Table 5.1 Real gross domestic product growth rates (at 1968 factor cost), 1961–1984: Singapore

Year	GDP ($m)	Growth rate (%, over previous year)
1960	2,122.3	–
1961	2,303.5	8.5
1962	2,466.2	7.1
1963	2,724.3	10.5
1964	2,607.0	−4.3
1965	2,780.3	6.6
1966	3,074.6	10.6
1967	3,475.2	13.0
1968	3,970.8	14.3
1969	4,501.8	13.4
1970	5,107.0	13.4
1971	5,747.4	12.5
1972	6,514.1	13.3
1973	7,247.2	11.3
1974	7,737.1	6.8
1975	8,043.5	4.0
1976	8,621.2	7.2
1977	9,290.3	7.8
1978	10,088.6	8.6
1979	11,030.9	9.3
1980	12,160.5	10.2
1981	13,369.3	9.9
1982	14,217.9	6.3
1983	15,334.2	7.9
1984	16,584.0^p	8.2^p
Average		8.9

Note: p for preliminary figure only.
Sources: 1960–1981, Singapore, Department of Statistics, *Economic & Social Statistics, Singapore 1960–1982*, pp. 59–60; 1982–1984, Republic of Singapore, *Economic Survey of Singapore 1984*, p. 95.

Company) was politically established when Sir Stamford Raffles signed a treaty with the *de facto* ruler of Singapore, the Resident Temenggong on 30 January 1819. The post was placed under British jurisdiction when a treaty was signed with both the Temenggong and Sultan Hussein on the 6th of the following month.

Raffles insisted that Singapore be a free port. Trade expanded rapidly. During the early period, the India and Peninsular Malaya trade monopoly power enjoyed by the British East India Company and the archipelago trade expansion launched by the Dutch East India Company contributed to the increase of trade handled by Singapore. Singapore's trade was mainly entrepot. By 1824, Singapore overtook Penang and became the main trade centre for the trade between continents (Europe, India, and China) and the Malay Archipelago. Textile traders from India and England as well as merchants for Chinese products used Singapore as their main distributing centre.

There was some setback in the entrepot trade of Singapore when the monopoly of the India and Malaya trade held by the East India Company was abolished in 1867. However, the development of tin mining and the commercial cultivation of rubber in peninsular Malaya in the early 1920s made Singapore the main tin and rubber export outlet of the world.

Several developments have sustained Singapore as the entrepot trade centre for this region. These include[2] the establishment of Hong Kong in 1842, the opening of the Suez Canal in 1869, the invention of steamships and international communication in the 1870s, and the so called 'colonial legacy'. This intermediary trade pattern remained unchanged even after the end of World War II. However, since 1960, the entrepot activity has not grown as rapidly as the rest of the economy[3]. A contributing factor was that the Indonesians and the Malaysians preferred direct trade with other countries, purportedly to cut down on middlemen.

Changes in development strategies

Revenue raising

The free trade policy remained practically unchanged until early 1960.[4] There were no export duties, and import duties were very low. According to Edwards,[5] in the early 1900s the main purpose for imposing import (and excise) duties was to control the import and production of intoxicating liquor and opium. It was also stated that 'The objective of the imposition of the duties was the raising of additional revenue for the purpose of making an annual contribution to the Imperial Government towards the expenditure incurred in the war then being waged.'[6]

Duties on liquor and petroleum were first imposed in 1909. In 1916, tobacco imports were subject to import duties. This tariff system did not extend beyond these three products (i.e., excluding opium) until early 1960.

Unemployment and industrialisation

With this free port policy, entrepot trade contributed greatly to the development of Singapore. For example, in 1957, the percentages of labour force employed in the trade and services were 24 and 27 percent respectively. To contrast, only 16 percent of the labour force were engaged in small scale manufacturing.

In 1960, the unemployment rate rose to 13.5 percent, from 4.9 percent in 1957. Demographic factors contributed to the higher unemployment rates.[7] Also, by 1960, it was clear that entrepot trade may have limited possibilities of expansion because of more direct trade between some South East Asian countries and the more industrialised western nations. In these conditions, Singapore then began to consider an industrialisation development strategy, rather than one based on entrepot trade. Many of the nations in the region also embarked on industrialisation programmes at this time.

An Industrial Survey Mission was assigned by the then United Nations Economic Commission for Asia and the Far East to undertake economic investigations into industrialisation strategies, to draw up a list of economically feasible industries, to prepare an outline plan for the development of industrial estates, and to advise on general economic policies. The Mission, headed by Albert Winsemius, visited Singapore from October to December 1960. A Preliminary Report was submitted to the government at the end of December 1960.

According to the Mission's estimate, a maximum industrial progam required funds of about S$820 million during the period 1961–1970.[8] But domestic private investment of this scale was not feasible. To help to reach the target, measures were taken to attract foreign private capital inflow. The Economic Development Board (EDB) was established in August 1961. Policy measures included the creation of industrial estates, tax relief for pioneer industries, and selective protection by tariffs.

Tariff protection

Tariff protection was used as a means of stimulating industrialisation in Singapore. The earliest known case for protection was in 1959, when a committee was formed to consider the imposition of duties on imports of hard soap and detergents. The imposition on these two commodities was a move to protect Singapore made goods but was also to retaliate against Malaysia which, during that period, had introduced high tariffs against Singapore goods.[9]

The revenue generated from duties on these commodities was negligible.[10] What is important to note here, however, is that the imposition of duties in 1959 signified a change of trade policy.

In August 1961, the Protection Advisory Committee (Paints) recommended that there should be import duties on imported paints of certain categories so as to protect the local paint industries. Subsequently, a protective duty was imposed on certain classes of prepared paints, enamels, varnishes, and distempers in April 1962.

The UN Industrial Survey Mission argued that although import protection could only be applied selectively, because on a larger scale it might damage entrepot trade, Singapore could not do without infant industry protection and recommended (1) to continue giving import protection to selected industries; and (2) to immediately introduce anti-dumping legislation within the provision of Article VI of GATT. Accepting one of the recommendations made by the Mission, the Tariff Advisory Commission was established in November 1962 to advise the government on the imposition of tariffs on imported commodities. The Commission named 30 local products it was considering for tariff protection on 1 July, 1963.[11] The Government found that it was necessary to give protection to certain industries which were subject to unfair competition and dumping from abroad.[12] Thus tariffs were imposed, mainly, for the protection to infant industries. However, to ensure the continued development of the entrepot trade, 'tariff duties can be imposed on imported goods only if they have slight or no unfavourable effects on the entrepot trade. Moreover, a genuine need for such tariff duties to protect local industries and encourage their expansion would have to be demonstrated in each case'.[13]

Another reason given by the then Minister for Finance, Dr Goh Keng Swee, for imposing import duty on goods was 'that once we start protecting industries by import duties, it is an advantage to have the largest possible protected market. This means that your industries can then be larger and more efficient. The additional cost to you as consumer will be less'.[14]

However, soap and detergents, and paints were the only two product groups actually protected by import duties until 1965. The reason for the absence of protection for other products was related to the establishment of Malaysia. Separate duties on imported goods through Singapore would not be appropriate as they would be entering, if realised, a custom union or a common market in the Federation of Malaysia. Malaysia was formed on 9 July 1963 and 25 months later, Singapore separated from the Federation. During this period, neither a custom union nor a common market was realised.

When Singapore ceased to be a member of Malaysia, on 9 August 1965, and to accelerate industrial programmes, it was felt that 'the incentives offered in the form of financial and technical assistance, temporary reliefs from corporate income tax, the alienation and development of suitable industrial sites and so on, would not be sufficient to induce investors to go in for large scale industrial production in the absence of tariff protection for the domestic market'.[15] On the day of independence, protective tariffs were extended to electric filament bulbs and rubber tyres and tubes. There was a tremendous increase of items under tariff protection later the same year. The number of these items grew from 166 in October 1965 to 183 by December.

Table 5.2 Number of items subject to customs duties as at end of year, 1910, 1916, 1932, 1951, and 1959–1081: Singapore

Year	Total		Protection		Revenue	
1910		(8)[a]		(0)		(8)
1916		(14)		(0)		(14)
1932		(17)		(0)		(17)
1951		(20)		(0)		(20)
1959		(24)		(0)		(24)
1960		(29)		(5)		(24)
1961		(29)		(5)		(24)
1962		(48)		(8)		(40)
1963		(48)		(8)		(40)
1964		(79)		(8)		(71)
1965						
Aug 9			11[b]			
Oct 11			166[b]			
Dec 31		(225)	183	(183)		(72)
1966		(265)	199	(199)		(66)
1967		(281)	229	(230)		(51)
1968		(378)	295	(293)		(85)
1969	398	(401)	312[c]	(315)		(86)
1970	388	(390)	301[c]	(303)		(87)
1971	352	(355)	258[c]	(266)	88[c]	(89)
1972	346	(345)	259	(255)	87	(90)
1973	197	(197)	106	(106)	91	(91)
1974	188		97		91	
1975	211[d]					
1976	208[e]					
1977	203[e]					
1978						
1979						
1980			0[f]			
1981						

Notes:

a Numbers in parenthese are figures as stated in T.S. Lim, S. Nair, and S.M. Razak, *The Impact of Customs*, (Singapore: Singapore National Printers (Pte) Ltd., 1974, Appendix 5, p. 117).

b Based on the assumption that the number of items under protection at the end of 1964 was 8 and on the information in the *Annual Report of the Customs and Excise Department*, 1965, p. 1 and p. 5.

c Based on *Customs and Excise Department Annual Report*, 1972, Appendix "L".

d As a result of the conversion of United Nations' Standard International Trade Classification (SITC) (revised) to the Brussels Tariff Nomenclature (BTN) on January 1, 1975, 210 dutiable BTN items were created out of 188 SITC items. Thus, during the year, only one new tariff item was introduced.

e Based on figures in *Singapore Annual Reports*.

f According to the *Monetary Authority of Singapore Annual Report*, 1979/1980, p. 61.

Sources: T.S. Lim, S. Nair, and S.M. Razak, *The Impact of Customs*, p. 117; *Annual Report of the Customs and Excise Department*, various years; *Singapore Year Book or Singapore Annual Report*, various years.

The number of items under tariff protection is given in Table 5.2. The number in most years is not certain. For the years 1965, 1967, 1973, and 1974 the number of items under protection was unambiguous as the words 'protective tariffs (or duties)' were used; in other years, the number was uncertain or unknown either because the words 'dutiable items' sometimes referred to protective items, and sometimes referred to both protective and revenue items, or because only the total number of items 'on the tariff' or 'subject to customs duties' was given. Part of the reason was that it was difficult to separate the items. For example, '. . . the duties on refined sugar, cosmetics, and clothing . . . are partly protective and partly revenue in intent'.[16] However, since the number of revenue items varied little over the years, changes in the total number of items subject to customs duties did reflect the changes in the number of protective items, and thus the changes of protection policy.[17]

The number of commodities under tariff protection started to reduce from 1970. According to the Monetary Authority of Singapore there were no items subject to a protective tariff in 1980. This reflected changes of development strategies in the late 1960s and early 1970s and other factors. For instance, 'in an effort to counter inflation and to give local consumers a wider range of imported products at competitive prices, protective duties on 153 items were abolished',[18] and 'taxes on cosmetics, perfumeries and toiletries were abolished to encourage the tourist trade'.[19]

Between 1961, the year when a comprehensive development plan was conceived, and 1967, two years after Singapore separated from Malaysia, the industrialisation programme and supportive measures were basically aimed at import substitution in the home market. But by the late 1960s, it was believed[20] that with a population of about two million and lack of natural resources, Singapore would not be able to develop or survive unless a more outward looking attitude was adopted.

Another event that led to a change of industrial policy in Singapore in the late 1960s was the decision of the British government to advance the date of military withdrawal from Singapore from the mid 1970s to 1971. The accelerated withdrawal would have far reaching adverse economic effects on the Singapore economy. For example, in 1965, it was estimated that British military spending in Singapore contributed 17.6 percent of Singapore GDP.

Several policies were implemented in response to these changes. The first was the introduction of the Employment Act and Industrial Relations (Amendment) Act in 1968 to secure industrial peace. The second was the acceleration of establishment of technical and vocational schools to increase the future supply of skilled labour. The third was the increase in the rate of contribution to the Central Provident Fund from 10 to 13 percent to increase the volume of savings. The fourth was the introduction of the Economic Expansion Incentives (Relief from Income Tax) Act in 1967 to encourage export oriented industries. With the implementation of export incentives embodied in the Incentives Act, Singapore entered into an era of export orientation. There was an influx of multinational corporations to Singapore, partly in response to these measures. These corporations brought with them new technology, modern management, and foreign markets. The reduction of protective measures and import restrictions in the late 1960s was to complement the export oriented strategy. It was hoped that by doing so foreign markets could be expanded through reciprocal access to Singapore market.

The change from import-substitution to export oriented industrial policy can be summarised by the statement that

> '... the small domestic market of a country with a population of two million limited the number of products that could be produced to replace imports—a favourite method of stimulating industrial growth in developing countries.[21] In the latter part of the sixties, we began to shift away from import substitution industries. We recognised that a policy of import substitution behind tariff walls will at best give us short term gains but will stifle our growth in the longer term. Protected industries, sheltered from competitive pressures, will become inefficient and stunted in their growth potential. Moreover, as we depend on the world markets for large scale expansion of manufacturing activities, our products must be internationally competitive. This cannot be achieved with tariff protection.[22] We ... (emphasise) ... the development of export oriented manufacturing industries for countries wishing to attain a rapid economic growth. Exceptions would, of course, be countries with rich endowments of exportable natural resources such as oil and minerals. Larger countries also can depend more on domestic markets as compared to smaller countries. For us, however, there is no choice'.[23]

Duty exemption

Tariff and non tariff barriers were erected to protect certain domestic producers. However, commodities using imports which were subject to these restrictions as inputs, would, other things constant, face higher costs of production and weaker competitiveness in the world market. In view of this, domestic factories were given duty exemptions on raw materials used in manufacturing. For instance, when certain types of paints were granted protection in 1962, duty free paints were accorded to eight local shipyards and three other local industries for special types of paints which could not be produced domestically but were needed in production. In 1967, duty exemptions were extended to 494 manufacturers.[24]

Quotas

As in the case of tariff protection, quota protection was applied with great caution and was temporary. In the *Trade Division Annual*

Table 5.3 Number of items subject to quota restriction, 1962–1981: Singapore

End of Year	Number
Before 1963	0
1963	3
1964	7
1965	9
1966	13
1967	15
1968	15
1969	10
1970	9
1971	5
1972	3
1973 to 1979	0
1980 to date	1

Sources: Republic of Singapore, *Annual Report of the Trade Division*, 1965–1970, Republic of Singapore, *Trade Division Annual Report*, 1971, Republic of Singapore, Department of Trade, *Annual Report*, 1972–1974, Republic of Singapore, Ministry of Finance, *Department of Trade, Annual Report*, 1975, 76/77 to 77/78, and Republic of Singapore, Ministry of Trade and Industry, *Department of Trade, Annual Report*, 78/79 to 79/80, *The Straits Times*, May 28, 1980, p. 7.

Report, 1968, it was stated that '. . . where the particular industry was found not making any serious effort to succeed in its ventures, import quota protection was withdrawn. Quotas were also removed when industries were able to function without further protection from Government'.[25] In this connection, it is worth noting that protection measures were reduced—the import quota was increased from 10 percent in 1962 to 50 percent in 1963—for the wire nail manufacturing industry, as the Singapore government found that the two factories manufacturing wire nails increased the prices of their products on several occasions without the knowledge or approval of the government.[26] Also, all quota controls are subject to half yearly reviews.[27]

The earliest case of quantitative restrictions on imports was in 1963 when import ceilings were imposed on batteries, monosodium glutamate, and flour. The rationale for placing quotas on these commodities, as it appeared in the *Trade Division Annual Report* 1963 and 1964, was to promote the growth of industrialisation in Singapore,[28] an infant industry argument. The numbers of items subjected to quota restriction are summarized in Table 5.3.

It is worth noting that the government granted quota protection to air conditioner manufacturers in April 1980 after it had removed duties on the product. The duties were imposed in 1968, and the government felt that the industry should be able to stand up after 12 years of protection. After the government removal of tariff protection the manufacturers complained that they could not stand the competition. The government agreed and granted quota protection.

This protection was to be abolished in 1983, it being hoped that by then this industry could upgrade its production.

Antidumping

As mentioned, the UN Industrial Survey Mission recommended the introduction of antidumping legislation in its report. As a result, the Custom (Dumping and Subsidies) Ordinance came into force in April 1962 with the intention of protecting local industries from dumping. However, antidumping duties were only imposed for short periods of time on plastic laminated sheets in 1969 and on frozen chickens in 1970. Since then they have been abolished. Another official reason for antidumping legislation was to strengthen Singapore's bargaining power in international trade negotiation.[29]

Other import restrictions

There were some other minor import restrictions, namely, the restriction on US dollar goods, on Japanese goods, and on goods from Communist countries and countries practicing racial segregation policies.

On dollar goods

Prior to 1959, special licences were required for imports of goods from US dollar sources, i.e., goods originating in or consigned from the US, Canada, and other countries in the American Account Area. When Singapore was a member of the Sterling Area, restrictions on imports from nonsterling sources were 'for the purpose of conserving foreign exchange'.[30] These restrictions on imports from American account countries were lifted in 1959.[31]

On Japanese goods

Before 1948, trade with Japan was on a government to government basis. After the signing of the Sterling Area-Japan Trade Arrangement, private traders were allowed to conduct business in Japanese trade on a limited scale. The export of rubber to Japan was still under the control of the British government.

Also under this arrangement, all payments were to be made in the pound sterling. In the second half of 1953, an improvement in the balance of payments between the Sterling Area and Japan allowed a great liberalisation in trade with Japan. Although individual licences for certain items were still needed they were granted very liberally. All other items were freely imported under the Open General Import

Table 5.4 Number of commodity groups from Japan under special import licence, 1970–1981: Singapore

Year	Number of commodity groups[a]
1970	26
1971	25
1972	25
1973	24
1974	5
1975	4
1976	1
1977 to 1981	0

Note:
a Plus all items under special licensing for import from any other country.
Sources: Republic of Singapore, *Annual Report of the Trade Division*, 1970; *Trade Divisional Annual Report*, 1971; Department of Trade, *Annual Report*, 1972–74; and Ministry of Finance, Department of Trade, *Annual Report*, 1975, 1976/77 to 1980/81.

Licence. The last item on the Japanese licensing list, the standard type of textile fabrics, was removed in June 1977. Table 5.4 shows the changes of the number of items on the Japanese licensing list.

On other countries

From time to time, restrictions on imports were for political reasons. For example, a ban on the imports of goods of South African origin was imposed in May 1965, and on Rhodesian goods in December of the same year.

Special licences for imports originating from some Communist countries were also required. These countries were: Albania, East Germany, Socialist Republic of Vietnam, People's Republic of Mongolia, Laotian People's Democratic Republic as at November 1984, but not the People's Republic of China.

Special licensing

There are certain goods that have been placed under special licensing restrictions for security, health, and foreign exchange reasons. The numbers of items in this category were also very limited. The maximum number was 23 in 1972 and in 1973. In July 1974 there was only one item under special licensing; about 18 items were delicensed but were subject to endorsement by other authorities. To date, there have been marginal changes in these numbers.

Export control

To ensure supply for domestic uses at prices regarded as reasonable, certain exports of goods from Singapore were under control. However, the items were very limited. The largest number of items of goods that needed export licences was 16 in 1974. In 1979, the number was reduced to 7.

Preferential tariffs

There are many forms of economic cooperation, one of which is preferential arrangement of tariffs and other trade restrictions among nations. Examples of preferential tariff arrangements are the Commonwealth preferential system and the ASEAN preferential trading arrangements.

Under the Commonwealth preferential system, Commonwealth countries grant preferential tariff treatment on imports from member countries. The preferential treatment takes the form of imposing lower rates of tariff on imports from member countries than on those from nonmember countries.

The argument for preferential tariffs was that while it is unrealistic to have completely free trade among all countries, the volume of trade may increase and thus benefit Commonwealth nations if reduced rates on imports could be given among these nations.

However, this preferential arrangement did not seem to be advantageous to Singapore. Singapore's long standing free trade and 'trade with all nations; doctrine was in contradiction to this system. It became acute when Singapore began its export orientation industrial programme aiming at exporting to all nations in 1967 that led to the elimination of preference for some commodities in the second half of the 1960s. Very few items at preferential rate have remained since 1973.[32]

ASEAN was formed in August 1967. In terms of economic cooperation, there was nothing significant in the first several years of existence. In 1972, a United Nations team submitted a report and recommendations on economic cooperation for ASEAN. The recommendations, which, among other things, included recommendations for trade cooperation, were not seriously considered by the ASEAN nations.

One major step in trade cooperation was made at the first Summit in Bali in 1976. The ASEAN Heads of State agreed to establish the Agreement on ASEAN Preferential Trading Arrangements to gradually liberalise trade among ASEAN countries.

A Committee on Trade and Tourism (COTT) was formed mainly to decide on commodities that could be accorded preferential treatment under the Agreement. After much negotiation, in June 1977, the Fourth COTT meeting finally agreed upon 71 items on the first list of products for preferential treatment. In June 1978, an additional

755 products were added to the ASEAN Preferential Trading Arrangements (PTA) list. By the end of 1982, a total of 8,560 items had been given preferential treatment. By the end of 1984 the number of eligible items increased to 15,715. Singapore's PTA trade value reached S$40 million in 1982, and increased to S$107 million in 1984.

Free trade zones

The free trade zones were established under the Free Trade Zones Act, 1966. Dutiable or quota restricted items pending reexport or transshipment could be entered free from customs control. To date, there are four free trade zones. The zones are an integral part of Singapore's entrepot trade strategy.

Government procurement policies

When members of the United Nations Industrial Survey Mission examined Singapore's building materials industries in 1960, they found that nearly all of the items of building materials imported could be manufactured domestically. The two most important items were cement and steel. The team also felt that government should give, in some way or other, in its purchasing programmes a priority to local building materials industries. Towards this end, it recommended that the Government should declare its intention of requiring building contractors to purchase locally manufactured products as one of its conditions for awarding its contracts.[33]

In addition, the Mission further suggested that if there was no local manufacturer for some of the building products,

1 preference should be given to those suppliers who were planning to produce these products in Singapore in the near future, or
2 preference should be given to those manufacturers who were planning to produce these products in the near future and who might supply part of the immediate need through imports.[34]

Preferential government purchase of domestically produced products was (and is) practised, provided that the prices were less than 10 percent higher and the quality of these products was comparable to those of imported products. However, no data are available to assess the impact of preferential purchasing by government.[35] But as there was virtually no trade restriction with any country, the protective effects of procurement policies may not be substantial, as prices of local products (ignoring the quality aspect) have to be compared with the lowest possible prices of foreign products before preferential purchases could be granted. Today the most substantial practice in procurement policies can be found in the building industry. In 1982, a 5 percent preferential margin was given to Mass Rapid Transit contracts. The ceiling of the margin of $2.5 million and $5 million was awarded to joint ventures with equal Singaporean participation. This preferential treatment was later extended to all government construction projects.

Government policy on export promotion

As a small country, Singapore has had to adopt an outward looking policy. The government started the export orientation programme from 1967. Initially, it took the form of fiscal incentives of granting tax concessions for export earnings and expenditures. Established in September 1976, the Export Credit Insurance Corporation provided fiscal assistance to manufacturers to explore overseas markets. Commercial Secretaries Services, which helps to identify new export markets, was established and expanded. Overseas trade offices were set up to answer trade enquiries and to coordinate trade missions and trade fairs. Through the assistance of the Department of Trade, companies are encouraged to participate in international trade fairs, trade missions, and trade promotional trips to venture into new world markets. The Department of Trade also conducts export promotion training courses, marketing seminars, and workshops.

Perhaps the most significant change in the institutional organisation to expedite export promotion is the reorganisation of the Department of Trade into a statutory board, the Trade Development Board (TDB) in January 1983. It is believed that the TDB could function more efficiently than the Department of Trade for the following reasons:

1 With greater flexibility in staff recruitment, administration, and program implementation, the TDB would be more efficient in the development and implementation of a trade development programme.
2 The TDB will be in a better position to assist local manufacturers in their search for export markets.
3 The private sector can be more closely involved in formulating the direction of trade promotion and taking part in trade promotion and activities.[6]

A 10 year Economic Development Plan for the 1980s was announced in the 1981 Budget Statement. Prepared by the Ministry of Trade and Industry, 'the main objective of this plan is to develop Singapore into a modern industrial economy based on science, technology, skills and knowledge'. Through the modernisation of the industrial sector, workers will be better paid, and the whole economy will be less vulnerable to protectionism. In fact, the second aspect of industrial modernisation, i.e., a less recession and less protection prone economy, is the more important one. Since the 1981 Budget Statement, this fundamental change of strategy to counter protectionism has been stated repeatedly in the yearly Budget Statement. It may be worth quoting a passage in he 1983 Budget Statement relating to this belief:

> 'As I said last year, and it bears repeating, our best defence against trade protectionism is to increase our productivity so that our goods will still be competitive in price and quality despite protectionism barriers. We will need to expand our domestic exports by pursuing more aggressive strategies in existing markets and by penetrating new markets. In the longer term we can become less vulnerable to protectionism only by restructuring our economy and producing products and services of increased value added content'.[37]

Another major policy proposed by the government to reduce the negative effects of protectionism is to diversify the economy. 'We must continue to diversify our economy to achieve the economic security objective as we are to make our economy more competitive both within and in the export market. We must diversify: . . . (b) our market: to get around the problem of protectionism and to expand our export to developing countries'.[38]

The government has identified and given priority to eleven industries. These industries will produce automotive equipment, machine tools and machinery, medical and surgical instruments, computers and computer peripheral equipment, and electronic instrumentation. Two main reasons were given for this change of policy. First, these industries were not labour intensive and would reduce the pressure on the already tight labour market. Second, these industries were less prone to protectionism by industrialised countries.

The political economy of protection

In 1976, Caves proposed three hypotheses on protection. They are: the national interest, the adding machine, and the vested interest group hypothesis. Compared to some countries, the political economy of protection in Singapore is comparatively less complicated.

For example, in the Australian case, the trade policy of the government, either by the Liberal or the Labor Party, is constrained by the existence of the other almost equal rival. The different interests of industrial groups and the electorates associated with industries in different regions influence directly or indirectly the decision of the government in power.

In contrast, the present government of Singapore (formed by the People's Action Party) has been in power since the first state election in 1959. Between October 1966, when the Barisan Socialist party, the only other strong party in that period, resigned from Parliament, and December 1981, when the Secretary General of the Labor Party won a byelection, there was no Member of Parliament from the opposition party in the Parliament.[39] As a matter of fact, there has been no immediate threat of power from any political party since 1966.

Singapore is a tiny city state with an insigificant agricultural sector and small industrial establishments in the manufacturing sector. Constituencies were not characterised by any particular industrial activity and the electorates hardly pressed their representatives in the Parliament for protection.[40] In fact, major economic policy decisions have been determined by top policy makers, who are convinced that these policies or policy changes are beneficial for the country as a whole.[41] When policies or policy changes are made in the Parliament or elsewhere, Members of Parliament are usually supportive of the government policies.

In the case of granting and withdrawal of protection, the main issues were whether protection would have possible unfavourable effects on entrepot trade and the general level of prices, and timing of changes of policy, and efficiency.[42]

The criteria for choosing the types of industry for protection and the bargaining process of vested interest groups were not made clear to the general public. For example, when the Tariff Advisory Commission Bill was tabled for second reading Parliament on 17 April 1962, a Member of Parliament from an opposition party commented that:

> 'Clause 15(2) of the Bill confers on the Commission unwarrantable powers of secrecy with regard to the proceedings of the Commission. Furthermore, there is no provision to publish the reports of the Commission upon which the Government's decision of protection for industry is based ... We do not like to see a repetition of what happened recently when the Government took a decision to give protection to the paint manufacturing industry in Singapore. The Government did not publish the report of the Advisory Committee although action was taken'.[43]

It is also not possible to conduct a realistic exercise to determine the factors that entered into the tariff protection decision equation. The main reason is that published data on industries are highly aggregated. The Department of Statistics has been publishing principal statistics of industrial production since 1959. These statistics provide information on the number of establishments, number of workers, input, output, value added, employees' remuneration, and capital expenditure for major industry groups, but the groupings are highly aggregated. In the 1965 principal statistics of industrial production, there were 18 major industry groups. Between 1960 and 1965, there were about 183 items[44] on the tariff protection list. We would not therefore be able to determine the differences in industrial characteristics of the protected and unprotected industries (and thus the rationale of protection).

For the same reason, it would be fruitless to try to determine the factors that affect the withdrawal of protection for certain industries and the rationale of granting tariff (quota) protection rather than quota (tariff) protection. Besides the problem that although products under tariff and quota protecion may be quite distinct but nevertheless fall into the same industrial group, the additional difficulty is that products under both types of protection are very similar or identical in some instances. In 1966, for example, out of the 13 items under quota restriction, 7 items are indistinguishable from some of products on the tariff protection list.[45]

Given the above constraints, perhaps the safest thing we can say is that the types of industry which were most likely to be protected in the import substitution phase of industrialisation (mid 1960s) were those which were considered to have the greatest potential of creating employment. The plausible reason is that unemployment was one of the major problems in Singapore in the mid 1960s. Unemployment rates were 8.9 and 8.1 percent in 1966 and 1967 respectively. Coupled with a large increase of new entrants into the labour market during that period, the most immediate task of the government was to create jobs for the unemployed.[46]

Other factors of protection could be summarised in the statement quoted below:

> '. . . The Government is prepared to grant protection to industries if several factors are satisfied. Firstly, the industry must be in a position to supply a major part of the local market. Secondly, the industry should result in a net gain to the economy by way of foreign exchange savings, local income generated and local employment created. In addition, the industry should as far as possible be one with some export potential.'[47]

Table 5.5 Nominal rates of protection (NRP), selected industry groups, Singapore, 1979

SI-0 1973 Code[a]		NRP (%) 1979
01	Livestock	26
05	Rubber processing	6
08	Milled cereal and cereal preparations	5
09	Sugar and sugar confectionery	15[b]
13	Beverages	42[b]
14	Tobacco products	166[b]
18	Clothing	15
24	Furniture and fixtures	20
32	Other chemicals	76
33	Petroleum and petroleum products	0[b]
46	Electrical goods and machinery	25
50	Vehicles	22
55	Other manufactures	20
56	Gas	2

Notes:
a SI-0 1973 stands for Singapore Input-Ouput Table 1973.
b Adjusted by excise taxes. For example, although import duties on petroleum are extremely high, they have no protective effects, as excise duties are applied at the same rates as import duties. For other commodities, excise duties are less than import duties. The government only began to harmonise excise duties with import duties from March 1980. The two commodities affected were cigarettes and tobacco. Yet the process was not completed. (see Bela Balassa, and Associates, *The Structure of Protection in Developing Countries*, (Baltimore: Johns Hopkins Press, 1971), p. 13).

Rates of protection

Nominal rates

Nominal rates of protection are rates imposed on imports entering the Singapore market. They are stated in the Custom Duties Order in terms of percentages of import prices, or some fixed dollar values per unit of import, or both when the rate which gives the greater revenue would be applied. Specific rates could be translated into *ad valorem* rates if we know the import prices. As a first approximation of degree of protection for certain commodity groups for 1979, the nominal rates of protection are given in Table 5.5.

ERPs based on the Corden and Balassa methods are reported for comparison. The effective rates are calculated for 1979, the most recent year when a complete listing of commodities subject to tariff is available. At the time of writing, the lastest input-output table available was drawn up in 1973. Input-output coefficients are taken to be fixed between 1973 and 1979.

It was noted above that to promote industrial expansion, local manufacturers were given generous duty exemption on raw materials where local substitutes were not available. However, as we have no information on duty exemption for these inputs, the ERPs reported

Table 5.6 Average[a] nominal and effective rates of protection, Singapore, 1979

Industry group	Nominal	Effective	
		Balassa	Corden
Importables[b]	0.5	5.1	4.5
Exportables[b]	0.2	3.9	3.4
Other	0.9	0.2	0.04
Total	0.5	2.7	2.3

Notes:
a Value added weight.
b Imports or exports more than 20 percent of apparent consumption.
Source: Appendix 1.

in this report are biased as they are calculated on the assumption that no duty exemption was extended to inputs.

ERPs are estimated for industries defined at the Singapore Input-Ouput Table 1973 2 digit level. The economy was divided into 74 industry groups. The ERPs for the 74 industry groups in 1979 are shown in Appendix I. From Appendix I, we observe that ERPs for most industries are relatively low. Average rates are reported in Table 5.6. The average ERP in 1979 was less than 3 percent.

Conclusion

There have been some changes in protection policy in Singapore since the 1960s when the industrialisation programme was started. However, the policy played a rather passive role in the package of economic policies. In the early stage of implementation, it was used to help solve umemployment problems. It was then realised that a protective trade policy would hamper the growth of Singapore's economy, an economy in which an export promotion strategy was more viable than the import substitution policy. To gain access to the world markets, trade policy was liberalised, it being hoped that reciprocity from exporting countries would be received.

The export promotion policy was quite successful. Singapore, for the past decade or so, has achieved full employment. Foreign workers were recruited to relieve the labour shortage. However, these guest workers could bring in social problems to this country. The government also felt that in order for Singapore to continue to remain prosperous, the economy should produce more high value added products. The recent changes in trade policy could be partly explained by this switch of policy. Labour intensive industries would receive less protection from the government in the future. New high skill products and services would be identified to break through the protective barriers of the more developed countries.

Protectionism has not been a political issue in Singapore. This is related to the domination of politics by a single party and to the lack of identifiable industry characteristics in any single constituency. In other countries, the different interests of industrial groups, and electorates associated with industries, have influenced government decisions. This has not been so evident in Singapore.

To conclude, the whole economic philosophy of development and protection of the Singapore government can be summarised by the following quotation:

> 'Another lesson (for the Eighties) is that we must continue with our economic philosophy of open and free competition. It is a harsh but effective approach of rewarding the capable and punishing the incapable. We have no choice, given our small domestic market, lack of natural resources and limited leverage in the world market. We cannot follow a self-sufficiency or self-reliance economic ideology. Our philosophy must be to import from the cheapest reliable source of raw materials, machinery, parts and components; we cannot close up our domestic market to shelter local industries'.[48]

Appendix I

Table 5.A1 Effective rates of protection (ERP), 1979: Singapore

SI-O 1973 Code	Industry	ERP(%) Corden	Balassa
01	Livestock	46.25	55.05
02	Other agriculture	−1.98	−2.33
03	Fishing	−3.96	−4.17
04	Quarrying	−0.08	−0.10
05	Rubber processing	6.03	0.01
06	Meat and meat preparations	−3.33	−4.70
07	Dairy products	−4.23	−4.83
08	Milled cereal and cereal preparations	4.31	2.06
09	Sugar and sugar confectionery	16.47	18.49
10	Oils and fats	−0.29	−0.31
11	Other food	−2.24	−2.55
12	Animal feed	−0.06	−0.07
13	Beverages	56.67	74.71
14	Tobacco products	178.08	186.93
15	Textile yarn and thread	−0.16	−0.18
16	Textile fabrics	−0.53	−0.59
17	Textile articles	−0.53	−0.59
18	Clothing	−15.84	17.10
19	Footwear	−2.58	−2.92
20	Leather and leather products	−4.74	−5.67
21	Sawmilling	−0.37	−0.41
22	Wood products except furniture	−0.61	−0.69
23	Ratan processing	−0.05	−0.06

24	Furniture and fixtures	26.31	32.11
25	Paper and paper boards	−0.73	−0.93
26	Products of paper and paper boards	−2.09	−0.93
27	Printing and publishing	−0.11	−0.12
28	Paints and dyeing materials	−0.50	0.57
29	Medicinal and pharmaceutical products	−1.25	−1.50
30	Toilet preparations	−5.41	−6.50
31	Plastic and plastic products	−0.45	−0.49
32	Other chemicals	51.12	66.67
33	Petroleum and petroleum products	0	0
34	Rubber products except footwear	−1.38	−1.54
35	Pottery, china and earthenware	−2.59	−3.91
36	Glass and glassware	−4.78	−6.20
37	Bricks, tiles & other structural clay products	−2.00	−2.45
38	Cement	0	0
39	Structural cement & concrete products	−2.24	−3.03
40	Nonmetallic minerals	−1.09	−1.30
41	Metal ores and scrap	−0.32	−0.38
42	Basic iron and steel	−0.12	−0.14
43	Nonferrous basic metals	−5.40	−6.45
44	Fabricated metal products	−0.52	−0.61
45	Nonelectrical machinery	−0.40	−0.46
46	Electrical goods and machinery	27.76	32.97
47	Insulated wires and cables	−0.08	−0.09
48	Electronics and telecommunication	−0.10	−0.11
49	Musical instruments	−1.16	−1.40
50	Vehicles	22.66	26.65
51	Ships and aircraft	0.37	−0.43
52	Watches and clocks	0.39	0.45
53	Scientific and precision equipment	0.12	−0.13
54	Jewellery	−0.03	−0.03
55	Other manufactures	13.74	14.22
56	Gas	0.61	0.70
57	Electricity	−12.07	−13.15
58	Water	−2.31	−2.75
59	Construction	−0.49	−0.52
60	Wholesale and retail trade	−0.06	−0.07
61	Restaurants and other eating places	−10.89	−12.35
62	Hotels and lodging houses	0	0
63	Transport and storage	−1.28	−1.73
64	Communication	0.05	−0.06
65	Financial services	−0.04	−0.05
66	Insurance	0	0
67	Real estate	0	0
68	Other business services	−0.06	−0.08
69	Ownership of dwellings	0	0
70	Social and community services	0	0
71	Recreational and cultural services	−0.05	−0.06
72	Personal and household services	−0.87	−1.16
73	Producers of government services	−0.78	−0.85
74	Domestic services and private nonprofit organisations	−0.13	−0.16

Endnotes

1 See Table Intro. 1; World Bank (1985), *World Development Report, 1985*, New York: Oxford University Press for the World Bank.

2 These factors are discussed in State of Singapore, (1961), *Annual Report 1961*, Singapore: Government Printing Office, p. 23.

3 Entrepot exports constituted about 73 percent of total exports in 1960. Their share reduced to 15 percent in 1981. The decline in the entrepot trade, in relative terms, was also partially reflected by the decline of the trade sector in terms of the GDP composition. As a component of GDP, trade declined from 35.9 percent in 1960 to 21.4 percent in 1983. However, in terms of value added, the absolute contribution of the trade sector rose from S$713 million in 1960 to S$6,908 million in 1983 (Lim Chong-Yah, 'Economic Development Experience in Singapore Since Self-Government in 1959', paper presented at the Asian Development Consultation of the Christian Conference of Asia in Singapore, 26 November, 1984, p. 10).

4 Control on imports was only exercised on those which would affect security or health, or the exchange control system, and on rice, in which import is conditional on purchase from the government stockpile.

5 Edwards, C.T. (1965), 'The Structure of Import and Excise Duty Taxation in the States of Malaya and Singapore', *Malayan Economic Review*, X, 2, October, p. 84.

6 Singapore, Colony of, (1949), *Customs and Excise, Annual Report, 1949*, p. 1.

7 The average annual increase in population between 1947 and 1957 was 4.3 percent and about 43 percent of the population was 15 years old or younger in 1957. This period is also discussed by Helen Hughes in 'The Singapore Economy: The Next 25 Years', Lim Tay Boh Memorial Lecture, Singapore, December 1984 (revised January, 1985).

8 The United Nations Industrial Survey Mission (1963), *A Proposed Industrialization Programme For The State of Singapore*, United Nations, Commissioner of Technical Assistance, Department of Economic and Social Affairs, 1 September 1963, p. 14.

9 Report of the Committee to consider the Imposition of Duties on Imports of Hard Soap and Detergents, pp. 2–5; and Hughes H. and You Poh Seng (eds), (1969), *Foreign Investment and Industrialization in Singapore*, Canberra: Australia National University Press, p. 32.

10 During the 1930–1960 period, import duties were imposed on only three commodities, ie, liquor, tobacco, and petroleum mainly for revenue purposes.

11 *The Straits Times*, 1 July 1963.

12 Quoted from State of Singapore, *Annual Report, 1962*, Singapore: Government Printing Office, 1964, p. 127. Note however that the Committee on Tariff Matters could not or would not give exact reasons for the percentage of duty imposed on imports; see Khoo Saik Liew, 'A Study of the Singapore Footwear Industry and Its Protection', unpublished academic exercise, University of Singapore, 1970/71, p. 24.

13 Singapore (1961), *Annual Report, 1961*, p. 93.

14 *The Straits Times*, 1 July 1963.

15 Singapore, *Customs and Excise Department, Annual Report, 1965*, p. 5.

16 Singapore, *Customs and Excise Department, Annual Report, 1969*, p. 8.

17 Although the number of protective items was sometimes stated in the *Singapore Year Book*, the number may not be the same as that in the *Customs and Excise Department, Annual Report*.

18 Singapore, *Customs and Excise Department, Annual Report, 1973*, p. 2.

19 Singapore, *Customs and Excise Department, Annual Report, 1977*, p. 2.

20 See for example, quotations by Dr Goh Keng Swee (1977), in his *The Practice of Economic Growth*, Singapore: Federal Publications, pp. 7–8; and Augustine H.H. Tan and Ow Chin Hock, 'Singapore', in Bela Balassa and Associates, *Development Strategies in Semi-Industrial Economies*, World Bank, 1982, p. 282.

21 Goh Keng Swee (1977), p. 7.
22 Singapore, Republic of, *Parliamentary Debates, Official Report*, Vol. 39, pp. 622–23.
23 Singapore, Republic of, *Parliament Debates, Official Report*, Vol. 30, p. 561.
24 Singapore, *Customs and Excise Department, Annual Report, 1967*, p. 17.
25 Singapore, *Trade Division, Annual Report, 1968*, pp. 39–40.
26 *The Straits Times*, 28 January 1967.
27 Singapore, *Department of Trade, Annual Report, 1972*, p. 32.
28 Import licences for these commodities were issued on the basis of an applicant's imports for the preceding year: *Trade Division, Annual Report, 1963*, p. 21.
29 Singapore, State of, *Legislative Assembly Debates, Official Report*, Vol. 16, p. 16.
30 Singapore, Colony of, *Customs and Excise, Annual Report, 1947*, p. 56.
31 Singapore, *Annual Report 1960*, p. 189.
32 Singapore, *Customs and Excise Department, Annual Report, 1973*, p. 2.
33 The United Nations Industrial Survey Mission, *A Proposed Industrialization Programme for the State of Singapore*, p. 178.
34 ibid, p. 179.
35 Balassa, B. and Associates, *Development Strategies*, p. 307.
36 Tony Tan Keng Yam (1982), 'How We Can Put the Economic Recession to Good Use,' in S. Jayakumar (ed.), *Our Heritage and Beyond*, Singapore: Singapore National Trade Union Congress, 1982, p. 70. The main functions of the Board are, 1 to promote Singapore's trade and services, in particular to promote the exports of Singapore manufactured products and services, by developing and implementing specific export promotion programmes, 2 to establish trade offices overseas to assist Singapore manufacturers and exporters to promote and penetrate overseas markets, 3 to develop an effective system to provide current trade and commercial information of overseas markets to Singapore manufacturers and exporters.
37 Singapore, *1983 Budget Statement*, pp. 13–14.
38 Appendix 1 to *1981 Budget Statement*, mimeographed, p. 6.
39 Even at the height of struggle for political power in the early 1960s protectionism was not a political issue. Rather, the issues were of mergers and Malayanisation.
40 However, one exceptional case to bring out the question of protection for a particular group of producers was recorded in Parliament on 25 March, 1974. A Member of Parliament from a rural constituency expressed concern that the lifting of import duty on frozen chicken and the lifting of the import ban of pigs would cause hardship to farmers in Singapore.
41 A most recent case of policy change made by the government is the phasing out of pig farms in Singapore. The government decided that pig farms have to be phased out because pig farming is highly pollutive and uses up too much land and water; *The Straits Times*, 24 April, 1985.
42 See for example: Singapore, *Legislative Assembly Debates, Official Report*, Vol. 17, pp. 736–38 and p. 741; Singapore, *Parliamentary Debates, Official Report*, Vol. 26, p. 799; Vol. 30, p. 695; Vol. 32, p. 1003; Vol. 33, p. 827; Vol. 38, p. 610; Vol. 39, p. 954, p. 1017, p. 1280 and p. 1282.
43 *Legislative Assembly Debates*, Vol. 17, p. 742.
44 A complete list of items is given in Lim et al., (1974), *Impact of Customs*, Appendix 4, pp. 110–13.
45 These items are: iron and steel wire; hardware, bolts, hinges, hooks and flush pulls; monosodium glutamate; batteries and electrical lamp bulbs; printing ink and ink for duplication; washing and bleaching preparation; and pig meats and colouring materials.
46 The urgent need for industrialisation and employment creation was repeatedly

presented in Parliament debates and in Ministerial Statements. See, for example, *Legislative Assembly Debates*, Vol. 17, p. 741; *Parliamentary Debates*, Vol. 25, pp. 836–37; and Goh Keng Swee, 'Some Problems of Industrialisation', in *Towards Socialism*, Vol. 7, pp. 1–2, Singapore: Government Printing Office.

47 Singapore, Republic of, *Preliamentary Debates, Official Report*, Vol. 26, pp. 181–82.

48 *1981 Budget Statement*, Appendix 1, Highlights of 'Singapore's Economic Development Plan for the eighties', mimeographed, pp. 4–5. See also an equally strong and unambiguous economic policy statement by Dr Tony Tan Keng Yam, Minister for Trade and Industry, entitled, 'Protectionism Breeds Inefficiency in the Long Run' in Singapore Ministry of Culture, *Speeches*, Vol. 6, no. 4, October 1982, pp. 42–43.

6 The political economy of manufacturing protection in Australia

KYM ANDERSON AND ROSS GARNAUT*

July 1985

THROUGHOUT the past half century manufacturing has been more heavily protected against import competition in Australia than in perhaps any other industrial country except New Zealand. Moreover, during the 1970s, Australia increased substantially its barriers against the most important manufactured exports of developing countries, namely, textiles, clothing and footwear. It has lowered its import barriers for most other manufactured goods in recent years, but the overall level of protection to manufacturing by the mid 1980s was still very high in Australia compared with other industrial countries.

This chapter seeks to explain Australia's protectionist stance and its recent changes. It begins with a brief summary of the evolution of Australian protectionism which includes evidence on the size of manufacturing tariffs and the extent of non-tariff barriers in Australia as compared with other industrial countries. It then discusses the public-interest arguments that have influenced Australia's protection debate at various times, and goes on to combine these arguments with explanations in terms of the pressures exerted by private interest groups on the policy-making process. The paper concludes with a discussion of recent and prospective future changes in the influences operating on protection policy decisions, and their implications for trade with developing countries.

* Kym Anderson is with the Department of Economics, University of Adelaide, Adelaide and Ross Garnaut is with the Department of Economics, RSPacS, Australian National University, Canberra.

The evolution of protection[1]

Protection emerged as a political issue in Australia well before Federation in 1901. The discoveries of gold in the 1850s in the two most populous colonies (later States), Victoria and New South Wales, caused the population of Australia to treble in less than a decade.[2] Then as the alluvial gold deposits were depleted during the 1860s there was concern about the likelihood of unemployment among the greatly enlarged workforce. Tariff protection for manufacturing was advocated as a way to increase the demand for labour, but its popularity was by no means universal. There were fierce parliamentary confrontations and narrow voting margins over protection policy in both colonies in the 1860s. As it happened, the protectionists won the day in Victoria while the free traders won in New South Wales. When the first federal parliament met in 1901 these opposing views were clearly expressed. Within a few years, though, the free traders had merged with the protectionists to form an anti-Labor-Party coalition, and both the coalition and Labor became sympathetic to calls for protection.

A variety of considerations were important in explaining the emergence of a protectionist consensus in the first decade of the Federation. Forty years of high protection in Victoria, and considerable protection in South Australia and Tasmania, had created many vested interests which lobbied hard for the retention of protection under Federation. As well, import duties were seen as more feasible politically and administratively than were income taxes as a source of government revenue. And thirdly, the protectionist leadership from Victoria was able to woo the support of the Labor Party. It did this by developing the concept of 'New Protection', whereby the continued provision of protection was made conditional on employers providing superior conditions of employment, including high wages. When members of the Free Trade Party were defeated decisively in the 1906 election, the Party's survivors announced that they interpreted the election outcome as an expression of the electorate's preference for protection.

Thus the first substantial protective tariff, the Lyne tariff, was imposed on manufactured imports in 1907/08. It doubled many of the existing revenue duties, and the average increase was almost a third.[3] But its real significance was not in the height of the duties but in the victory of protectionist forces.

Manufacturing was boosted somewhat towards the end of World War I under the stimulus provided by the disruption of normal import supplies. The disappearance of that stimulus after the war and the decline in ocean transport costs led the government to raise the level of tariff protection in 1920/21 to about one third above the height of the Lyne Tariff. Its purpose was to support infant industries, such as the iron and steel and the motorbody building industries, in the wake of the sharp decline in natural protection. This Greene Tariff met with little criticism, for by this time opposition to protection had virtually disappeared. Defence became an additional justification for protection, along with employment considerations and national pride in watching the industrial sector develop.

Tariffs were raised further in the mid-1920s as production costs overseas fell and as the postwar boom ended. They were increased again for a wide range of commodities by the Scullin tariff adjustments of 1929/30. These increases, which were part of the emergency measures of the depression, applied to British as much as to other goods. The Scullin Tariff was accompanied and followed by some quantitative restrictions on imports as the depression deepened, and by further tariff increases in 1930/31.[4] The competitive position of manufacturers was further strengthened by the 25 percent devaluation of the Australian pound against sterling in 1931.

By the mid-1920s Australia had established itself as one of the highest tariff countries in the world (Table 6.1). It is true that United States' tariffs were higher in 1925, that other countries also raised their tariffs during the depression of 1929–33, and that Australia lowered its British Preferential tariff rates following the Ottawa Agreement of 1932, but the overall level of protection to Australian manufacturers remained high through the 1930s, whereas the United States and a number of its trading partners began to reduce their tariffs after the U.S. reciprocal Trade Agreement Act was passed in 1934.

Manufacturing activity received another increase in natural protection with the disruption of supplies during and immediately following World War II. Again the government raised artificial barriers to imports as this stimulus disappeared following postwar reconstruction. This time, import licensing rather than the tariff was used extensively, at least until 1960. This was seen as a policy instrument also capable of helping to achieve balance of payments equilibrium during this period of fixed exchange rates.[5] Licensing provided even more protection than the tariff so that, when it was abandoned in 1960, numerous tariffs were raised to maintain protection levels.

Australia did not participate in the multilateral tariff reductions negotiated in the Dillon and Kennedy rounds of GATT talks in the 1950s and 1960s,[6] just as it had not participated with the United States in reducing tariffs bilaterally in the 1930s. As a result of that abstention, the manufacturing tariff differences between Australia and other industrial countries widened during the two decades to the

Table 6.1 Average manufacturing tariff rates, selected industrial countries, 1902, 1913, 1925, 1970 and 1975 (per cent)

	1902	1913	1925	1970[b]	1975[c]
Australia	6	16	27	23	13
Belgium	13	9	15		
Canada	17	26	23	14	8
Denmark	18	14	10		
EEC[a]				8	8
France	34	20	21		
Germany	25	13	20		
Italy	27	18	22		
Japan	10	20	13	12	11
Netherlands	9			23	
Norway	12			11	
Sweden	23	20	16	7	5
Switzerland	7	9	14	3	
United States	73	44	37	9	6

Notes:

a Belgium, France, West Germany, Italy, Luxembourg and the Netherlands in 1970; also Denmark, Ireland and United Kingdom in 1975.

b The 1970 estimates are not comparable with earlier estimates because the earlier estimates used f.o.b. export prices of other countries as indicators of border prices, instead of c.i.f. import prices for the country concerned. Thus the earlier rates are somewhat overestimated. This has little effect on the inter-country comparison for each year, however.

c The 1975 estimates, which were compiled by the Industries Assistance Commission from information supplied by the GATT, may not be perfectly comparable with the 1970 and earlier estimates.

Source: League of Nations, *Tariff Level Indices*, Geneva, 1927, as quoted in Little, Scitovsky and Scott (1970, p. 162–3); General Agreement on Tariffs and Trade, *Basic Documentation for the Tariff Study*, Geneva, 1972; and Industries Assistance Commission (1978, Table 1.3.3).

mid-1970s, by which time Australia, with New Zealand, had by far the highest manufacturing tariffs among industrial countries (Table 6.1).

However, in 1973, Australia reduced all tariffs unilaterally by 25 percent. This was Australia's first major general liberalisation of imports, an historic point in protection policy. Having taken that decision, the government cut tariffs further on some items in the mid-1970s, as part of an overall review of tariffs, and on other items in early 1977 following a currency devaluation. The latter cuts brought the average tariff reduction since 1973 to 40 per cent, sufficient for Australia to qualify as a participant in the Tokyo Round of GATT negotiations.

Thus the mid-1970s saw a major turnaround in assistance to Australian manufacturers. This occurred despite very considerable structural adjustment pressures on the sector at the time, due to a boom in mineral exports, increased import competition from newly-industrialising Asian countries, and sharp increases in real wages.[7] The only industries exempted from assistance cuts were textiles, clothing and footwear, and motor vehicles and parts. For other manufacturing industries, the effective rate of protection was reduced

Table 6.2 Effective rates of protection to manufacturing industries, Australia, 1968–69 to 1981–82[a] (per cent)

	1968–69	1973–74	1981–82
Textiles	43	35	54
Clothing & footwear	97	64	204
Motor vehicles & parts	50	38	124
Other manufacturing	32	23	14
Total manufacturing	36	27	26

Note:
a The Industries Assistance Commission estimates are described as effective rates of assistance. However, in the case of manufacturing, virtually all of the assistance is provided via tariffs and quantitative restrictions on imports.
Sources: Industries Assistance Commission, (1976, 1980a, 1983).

from 32 to 23 percent in 1973, and fell steadily thereafter to 14 percent by 1981/82 (Table 6.2). On the other hand, quotas were applied to imports of textiles, clothing, footwear and motor vehicles from 1974, which became increasingly protective over time as Australia's competitiveness in these industries declined. The average rate of effective protection to these quota-protected industries trebled from 45 percent in 1973/74 to 130 percent by 1981/82. Thus the average effective rate of protection for manufacturing as a whole, which remained around 27 percent during the decade from 1973, masks both the tendency for protection to be reduced considerably in most industries and a phenomenal increase in quota protection for two industry groups.

In short, the period from Federation to 1973 was one of periodic increases in protection from manufacturing import competition, while the decade since then has been one of protection cuts for most manufacturing industries. This Australian experience contrasts markedly with that of other industrial countries. The evidence in Table 6.1 suggests that manufacturing tariff reductions have been the norm in most other industrial countries throughout much of the period since 1900. Moreover, according to the evidence summarised in Table 6.3, Australia also subjects a relatively large proportion of its manufactured imports to non-tariff barriers. This evidence is consistent with that in Table 6.4, showing the decline in the openness of the Australian economy relative to other industrial economies, as measured by exports plus imports as a percentage of GDP.

No estimates are available on the extent to which protection has influenced the size of Australia's manufacturing sector.[8] However, we know that at the turn of the century manufacturing contributed less than 15 per cent of GDP and employment in Australia compared with 20 to 25 per cent in the United States and Canada and 25 to 35

Table 6.3 Share of industrial countries' imports of manufactures subject to non-tariff barriers, 1983 (per cent)

	Imports from:		
	World	Industrial	Developing countries
Australia	25	24	26
Austria	2	2	3
EC-10	18	16	23
Finland	9	7	26
Japan	5	5	5
Norway	3	2	19
Switzerland	16	16	18
United States	12	11	18
All industrial countries	15	13	21

Source: Norgues, Olechowski and Winters (1985).

Table 6.4 Share of national product traded, selected industrial countries, early 1900s to 1970s[a] (per cent)

	Early 1900s	1920s	1950s	1960s	1970s
Australia	42	37	35	32	31
Austria			45	54	68
Belgium			62	75	93
Canada	32	42	39	41	47
Denmark	62	57	66	61	61
Finland			46	48	58
France	35		27	28	39
Germany	38	31	31	37	45
Italy	28	26	27	34	49
Japan	30	36	20	20	24
Netherlands			96	92	100
Norway	69	64	87	84	89
Sweden	40	42	46	45	56
Switzerland			56	61	65
United Kingdom	44	38	44	42	55
United States	11	11	9	10	15

Note:

a For the period since 1950, the World Bank's data refer to the ratio of exports plus imports of commodities and nonfactor services to GDP. The earlier data from Kuznets refer to the ratio of exports plus imports of commodities to GDP (or GNP in the case of Australia, Canada, Italy, the U.K. and the U.S.) It has not been possible to get data for years which correspond exactly across countries. The percentages in the first two columns, from Kuznets, refer to the following years for each country: Australia (1911–13, 1924–29), Canada (1911–13, 1926–29), Denmark (1910–14, 1921–29), France (1908–10), Germany (1910–13, 1925–29), Italy (1911–13, 1925–29), Japan (1908–13, 1918–27), Norway (1905–14, 1920–29), Sweden (1911–13, 1921–30), United Kingdom (1909–13, 1924–28), and United States (1904–13, 1919–28). The final three columns, from the World Bank, refer to 1950–60, 1960–70 and 1970–77.

Sources: Kuznets (1966, Table 6.4) and World Bank (1980).

per cent in the major countries of western Europe.[9] By the 1950s this difference between Australia and other industrial countries had virtually disappeared, with manufacturing contributing around 30 per cent of GDP and employment.[10] Yet, as in 1900, manufacturing still contributed no more than 10 per cent of exports in Australia, compared with 70 or 80 per cent in other industrial countries. Thus even though Australia's manufacturing sector expanded substantially over the twentieth century, there appears to have been no increase in its competitiveness internationally. This suggests that much of the expansion was induced by protection.[11]

In seeking to explain Australia's protection policy, it is helpful to consider first the reasons for Australia adopting a strongly protectionist stance until 1973, and then to look at why it altered that stance.

Reasons for high protection levels to the 1970s

One of the strongest conclusions in neo-classical trade theory is that protection against imports reduces the average level of income in countries like Australia whose economies are too small to influence international prices.[12] Nonetheless, a number of public-interest arguments have been used to support the case for protection. These include the infant-industry argument as well as the arguments that protection could boost real wages and thereby redistribute income and attract more immigrants, could alleviate unemployment problems, could overcome balance of payments problems, and could ensure a more 'balanced' economy that was less specialised in primary production.

Infant-industry protection, wages and immigration

Supporters of protection have claimed that there is a dynamic economic case for protection aimed at stimulating new manufacturing activity because the general atmosphere engendered by manufacturing is more conducive to economic growth than that surrounding primary activities. Protection was necessary, it was argued, to ensure that entrepreneurs would risk their capital during the early years before an industry could stand on its own feet. This argument was used in Australia during periods of high growth optimism, such as from the 1860s to the 1880s, and in the two decades following World War II, but in recent times it has been taken much less seriously.[13]

In Australia, a belief that protection raised real wages, or allowed a larger workforce to be employed at a given real wage, played an extremely important part in moulding public opinion in favour of protection. This proposition, first espoused formally by Brigden was incorporated into a major official report on the Australian tariff and was further clarified by Reddaway.[14] It was argued that protection of the import-competing manufacturing sector would raise the demand for labour. This would have two effects that were judged to be socially desirable: it would support greater employment at given real wages, and so attract immigrants, and it would implicitly tax some of the rents of the relatively rich landowners. That is, protection was seen as an instrument capable of both redistributing national income more equitably and increasing Australia's population of European settlers.[15]

Brigden's proposition survived despite considerable controversy and refinement.[16] Its supporters knew they were arguing a second-best case, and they were suspicious that average incomes would be lower because of protection. However, the first-best redistribution policy, namely taxing land rents for the purpose of raising living standards of wage earners, was seen by them as politically impractical because it was more overt than the tariff, as well as being difficult to administer. Critics, on the other hand, argued that if direct income redistribution was necessary it could be achieved most efficiently through rent or income taxation.[17] But their arguments were insufficient to win the public or even the economists' debate at the time.

Had the Australian discussions continued in the Brigden-Reddaway tradition, the protectionists' case would have soon weakened with the increased feasibility of using superior instruments of taxation for redistributing income to wage earners. Income taxation, for example, has been a much more important source of government revenue since the 1940s.

As it happened, though, the relationship between protection and income redistribution increasingly came to be understood in terms of a model which invited conclusions excessively favourable to protection. This was the model presented in the celebrated article by Stolper and Samuelson (1941).[18] The model had two tradeable sectors, agriculture and manufacturing, and two homogeneous and domestically mobile factors, which were called land and labour. The authors concluded, in reference to the Australian situation, that protection of the relatively labour intensive tradeable sector, manufacturing, would unambiguously raise wages and reduce the real return to land. By ignoring the fact that land is specific to the primary sector, this model inappropriately distracted attention from the more efficient means of redistributing income mentioned above.

In recent years the belief that protection is desirable because of its effect on relative factor rewards has waned considerably. The anxiety to stimulate immigration of workers has diminished, and the real incomes of wage an salary earners, at least since the mid-1960s, have been above the average of farm household incomes. Also, it has become increasingly apparent that the real income of a household of workers may not be enhanced by protection if their particular skills are not used intensively in the protected industries, or if a relatively large share of their expenditure is on goods which are highly protected.[19] The fact remains, though, that this belief was an important contributor to the raising of Australia's manufacturing tariffs in the first half of this century.

Employment protection

A variant of the Brigden proposition—that protection for manufacturing allows higher levels of employment to be maintained at a given real wage—has been influential during lengthy periods of high unemployment. The argument is persuasive to the layman, especially in the Australian environment where, for institutional reasons, real wages tend to be more than usually rigid in the face of market pressures for their reduction.

The reality is that general unemployment can be addressed effectively only by measures which either raise aggregate demand (which is feasible only when the economy is operating within its balance of payments constraint), or raise productivity relative to real wages (through raising average productivity, reducing real wages or a combination of the two).

In practice, protection is likely to have very little effect on employment, except in the short run. This is because protecting one set of industries is equivalent to taxing all other industries via the effect of protection on the general cost level and the exchange rate, so that any employment gains in the first set of industries will be more or less offset by employment losses in the second set. The reduced job opportunities in agriculture, however, might take some time to be reflected in the employment statistics; it might, in the interim, take the form of underemployment on farms, which is not measured in the statistics. What is observed therefore, namely less threats of retrenchments in the protected industry itself, is only part of the story. But it can be the part which impresses the voter in times of higher unemployment such as around 1930.

More often, protection is sought for the maintenance not of overall employment but of employment in certain industries or regions. Society is often concerned to cushion workers and owners of enterprises facing large declines in their fortunes, especially when those declines coincide with high general levels of unemployment. Protection is never the first-best policy instrument for this purpose, even if the cause of the problem is increased import competition. Direct subsidies would be a less wasteful means of assisting the industry until the industry's problem disappears.[20] If the cause of the problem is a permanent (rather than temporary) decline in competitiveness, once-only assistance for structural adjustment would be more sensible than handouts every year to prop up the industry. However, protection tends to be used because, unlike subsidies, it does not involve budgetary outlays. As already mentioned, this argument for protection to maintain an industry's existing level of output was used in Australia following the first and second world wars.

Balance of payments

Prior to the movement towards more flexible exchange rates in the 1970s, it was often argued that protection was necessary at times for the maintenance of equilibrium in the balance of payments. This was especially so in the 1950s in Australia when the Sterling Area currencies were pegged to the pound which in turn was pegged to but not freely convertible into the US dollar.

This argument was never strong, since a country's competitive weakness could be corrected more effectively by demand policies which led to lower rates of inflation than those ruling in other countries. It lost all credibility among economists when exchange rates became adjustable in the early 1970s. If a deficit in external payments results from excessively high domestic costs and not solely from excessive domestic demand, a real devaluation is more appropriate than increased protection. This is because the former boosts export industries as well as discouraging imports whereas the latter discourages both exports and imports. Conversely, the first-best response to a surplus in the balance of payments of an economy free of distortions is real currency appreciation, through inflationary demand policies or through nominal currency appreciation. However, if the economy were distorted by import protection (in the absence of offsetting export subsidies), then the most efficient strategy would be to reduce that protection first, and to appreciate only if the payments surplus persisted after all import barriers had been removed.

In practice, when the balance of payments is tending towards surplus, the correction of this tendency is often associated with pressure for some tradeable goods industries to contract. Perhaps the best known example is where there has been strong growth in mineral exports—whether as the result of the discovery of minerals or of an increase in their relative prices on world markets.[21] The consequent rise in the real exchange rate reduces the competitiveness of other export and import-competing sectors, and so may generate public sympathy for their assistance. A protection increase in these circumstances would be an inefficient means of providing assistance, however. Apart from the loss to consumers of importables, protection only assists industries which compete with imports and not other export producers such as farmers. The burden of adjustment for the established industries producing tradeable goods could be eased as well by slowing the rate of real appreciation of the currency. This provides what Corden[22] has called 'exchange rate protection' to industries producing tradeable goods relative to others. While it still has welfare costs, such action may provide short term assistance more efficiently than conventional protection because it is easily reversible and does not distort resource allocation within sectors.

A more 'balanced' and 'self-sufficient' economy

Many societies have protected import-competing industries in order to have a more 'balanced' economy, despite the cost in terms of income foregone. This desire has resulted from anxieties about the availability of imported supplies of essentials in times of war, or from an idea that it is somehow wrong or unnatural for the intersectoral composition of economic activity to diverge markedly from that of other countries, or at least from that of the advanced industrial countries.[23] It has also been argued that protection would allow an economy to be less specialised and hence less vulnerable to changes in the fortunes of particular commodities on world markets. This is the argument used in Europe and Japan to justify protection of their agricultural industries.

The 'balanced economy' argument for protection of manufacturing has always been influential in Australia. Without protection, it is probable though by no means certain that the Australian manufacturing sector would have been smaller and would have accounted for a smaller proportion of employment and domestic product than the manufacturing sectors of the major industrial countries.[24]

Public sympathy for protection on these grounds arises because many people's perception of the Australian economy without manufacturing protection is of many farms and mines and few cities. By far the largest sector, services, is usually ignored or viewed as unproductive ('bringing in each other's washing') and therefore unimportant. Also overlooked is the fact that protection reduces the share of GDP that is traded, so that exports are from a narrower range of industries—are *more* specialised—than would be the case without protection.

Private-interest groups and inter-industry differences in protection

This discussion reveals flaws in each of the arguments that traditionally have been used to justify protection in Australia from society's point of view. However, it also shows how these ideas nonetheless at

Table 6.5 Dispersion of assistance to Australia's manufacturing sub-sectors, 1968–69 to 1980–81

	Average effective rate of assistance (per cent)				Standard deviation of effective rates of assistance[a] (percentage points)			
	1968–69	1973–74	1977–78	1980–81	1968–69	1973–74	1977–78	1980–81
Food, beverages & tobacco	16	18	10	9	26	29	23	9
Textiles	43	35	52	57	25	25	45	69
Clothing & footwear	97	64	142	144	36	26	44	52
Wood, wood products & furniture	26	16	18	14	15	10	13	10
Paper & paper products, printing	52	38	26	25	33	24	14	15
Chemical, petroleum & coal products	31	25	19	15	24	14	13	1
Non-metallic mineral products	15	11	5	6	20	13	8	6
Basic metal products	31	22	12	10	20	18	8	11
Fabricated metal products	61	44	32	34	23	18	12	17
Transport equipment	50	39	57	94	14	13	49	100+
Other machinery & equipment	43	29	20	21	17	10	11	10
Miscellaneous manufacturing	34	24	26	29	14	10	9	15
Total manufacturing	36	27	23	24	30	23	34	41

Note:
a Refers to the effective rates of assistance to 4-digit ASIC industries within each of the sub-sectors shown, based on 1974–75 production weights (except for 1968–69 and 1973–74, which are based on 1971–72 production weights and 1980–81, which are based on 1977–78 production weights). See Appendix Table 6.A1 for details of the 4-digit effective rates of assistance for other years.

Sources: Industries Assistance Commission (1979, 1980a, 1982).

times have supported and encouraged a climate of public opinion favourably disposed toward protecting the import-competing manufacturing sector in a country so richly endowed in natural resources as Australia.

Most of these public-interest arguments for protection are directed towards protection for the import-competing manufacturing sector as a whole. The only important argument concerned with protection for particular industries *within* the manufacturing sector is that protection should be provided to industries facing a large and sudden decline in their fortunes, for example due to increased import competition in the course of post-war reconstruction. Even in this case, however, it can be demonstrated that protection is not the first-best policy instrument for assisting those hurt.

Yet there is a very wide dispersion in the rates of protection and other forms of government assistance to various manufacturing industries in Australia. This is clear from Table 6.5, which reports estimates of a summary indicator of the degree of industry assistance, namely, the effective rate of assistance or the percentage by which government policies have directly raised value added in the industry above what it would be without those policies. The rates of assistance vary markedly not only between 2-digit manufacturing industry groups but also between 4-digit industries *within* each 2-digit group, the latter being indicated by the standard deviations shown in Table 6.5.

To explain this wide dispersion in assistance rates within the manufacturing sector, it is necessary to consider also another major determinant of protection, namely, the role of groups who can benefit from that form of government intervention. Producer groups in each industry can be viewed as the demanders of protection for their industry, and political leaders as the suppliers. It seems reasonable to assume, following Downs (1957),[25] Buchanan and Tullock (1962) and Breton (1974), that political leaders tend to adopt policies so as to maximise their chances of remaining in office, while groups who expect to gain (lose) from a particular policy seek (oppose) its adoption by investing in lobbying and propagandising up to the point where they perceive that expected net benefits from further expenditure are zero. Thus the supply curve in this political market represents the marginal political cost of providing an extra unit of protection to an industry, in terms of reduced political support from groups opposed to such a policy change, and the demand curve represents the marginal political return to political leaders in terms of political support from the groups seeking that protection. With this simple political market in mind, the task of explaining interindustry differences in protection requires examining the differences in factors affecting the demand and supply curves for individual industries. Changes in these factors would then be expected to explain changes to the dispersion of protection rates.

Using this economic theory of politics, Anderson (1980)[26] argues that the effective rate of protection to an industry is likely to be higher, *ceteris paribus*,

- the lower the average wage per employee in the industry, AWPE
- the less an industry is growing as an employer of workers, DNOE
- the less the share of production exported, EXP

Table 6.6 Determinants of effective rates of protection to Australia's 4-digit manufacturing industries, 1968–1969

AWPE	17.0 (3.62)[a]
DNOE	5.69 (.59)
EXP	−.650 (4.15)[a]
IMP	−.254 (2.03)[b]
LI	45.1 (1.85)[b]
NOE	.293 (.95)
NOF	−14.2 (1.72)[b]
TC	−15.6 (2.10)[b]
VASO	−75.8 (2.82)[a]
Constant	125
R^2	.40
F-value	8.72
No. of observations	130

Note:
a t-values shown in parentheses, with significance levels shown as .01 = a, .05 = b.
Source: Anderson (1980).

- the greater the share of imports in domestic sales, IMP
- the more labour-intensive the industry, LI
- the larger the number of employees in the industry, NOE
- the smaller the number of firms in the industry, NOF
- the lower the natural protection provided by transport cost, TC
- the smaller the industry's value-added share of output, VASO

These hypotheses are indeed supported by the Australian data. As shown in the regression results in Table 6.6 for 1968–69, for example, the negative and significant coefficients for AWPE and VASO and the significantly positive coefficients for LI confirm that it is the labour-intensive, low-wage industries with low value-added shares of output that were most highly protected in Australia in 1968–69. Secondly, industries with fewer firms (NOF) and perhaps having larger numbers of employees (NOE) tend to be assisted more. Thirdly, the significant negative coefficients for EXP and TC suggest that exporting and naturally protected manufacturing industries tend to be assisted less.

Only the negative coefficient for IMP and the positive coefficient for DNOE are inconsistent with *a priori* expectations. However, when 3-digit instead of 4-digit industry data are used in the regression analysis, DNOE is negative and significant, suggesting that declining industry sub-sectors are being protected more than average but that the extent of assistance for some 4-digit industries within such sub-sectors is so great as to remove the incentive to decline so that, as perhaps with IMP, the expected relationship is not observed.

Summary

It would appear that the climate of public opinion, together with the incentives for vested interests to lobby for favours, determined the average level and the dispersion of Australian manufacturing protection up to the 1970s. When the public could be convinced that tariffs were in the national interest, the political costs of providing protection were low. The arguments that the tariff would redistribute income away from wealthy landowners to workers and would encourage emigration from Britain to Australia were used to influence public opinion in favour of protection. So too were the arguments that manufacturing boosted national defence and diversified Australia's industrial base. Public sympathy also ran high for industries which sprang up to help the war effort and then later faced increased import competition after postwar reconstruction and during the 1929–33 depression. All these factors kept down the political cost of introducing protection policies to a growing number of manufacturing industries during the first few decades of Federation. Tariff protection rather than more efficient forms of assistance such as direct subsidies were used presumably because they raised rather than depleted government revenue and were less overt in that they were not subject to annual budget scrutiny.

Once established, a protection policy has considerable value to the small number of groups benefiting from it. These groups therefore have an incentive to spend part of their benefit in lobbying for the retention of that policy. Groups adversely affected by it, on the other hand, are large in number and so have more difficulty organising to lobby against the policy. In addition, the loss to each of the latter groups is small relative to the gain to each protected group. And since there is considerable inertia in the political system anyway, in the sense that governments have time to deal only with what they perceive as the issues of highest priority, protection policies once in place tend to remain in place until the pressures for change exceed some threshold level.

Australia's protectionist stance might have been eroded during the 1950s and 1960s as other industrial countries wound down their tariffs following the Dillon and Kennedy rounds of multilateral trade negotiations. But this was not to be. Australia justified its high protection stance to the international community by arguing that since agricultural protection was excluded from the negotiations, Australia would receive no *quid pro quo* for a reduction in its manufacturing protection. Domestically there was little opposition to the stance, not least because manufacturing and primary production were expanding rapidly. In any case, the government argued that the stance was necessary during the 1950s for balance of payments reasons. Then in the 1960s, the Country Party found it convenient to continue propagating the apparent virtues of manufacturing protection (along with assistance to the least-competitive agricultural industries), in order to boost Party funds and thereby retain its influence in parliament despite the declining importance of rural votes.

By the mid-1960s, however, more people were recognising that Australia's postwar economic growth performance had been much weaker than that of most other industrial countries, and that protection policies were partly to blame. In fact, Australia's per capita income growth had been relatively slow for much of the century, averaging little more than half that of other advanced economies.[27] It would be unreasonable to conclude that protection was the only cause of this relatively poor performance. The exceptionally high levels of immigration from lower-income countries for much of the period may have constrained growth in *per capita* income. Moreover, domestic market regulations other than protection may be more distortive in Australia than elsewhere.[28] However, the consensus among economists that protection must have been one of the main contributors has received increasingly wider acceptance among Australian leaders during the past two decades.

Reasons for protection cuts in the 1970s

The effect of the growing awareness of the costs of protection has been to increase the political cost of maintaining or increasing existing levels of protection. Thus in 1971 the government approved a systematic review of the whole tariff structure. This review generated a steady stream of recommendations for 7 years from 1974 for reductions in protection to particular manufacturing industries. Most recommendations were accepted at least partially by both Labor and conservative governments. The industries affected accounted for about 10 percent of manufacturing value added, and their average effective rate of protection was halved.

More importantly, the changing climate of opinion concerning the national interest in protection made it easier for the Labor government in 1973 to reduce all tariffs by 25 percent as a way of reducing the inflationary pressures it inherited on taking office in late 1972. It was not difficult politically for the returned conservative government to add some further tariff reductions in January 1977, two months after a devaluation, in order to participate in the Tokyo Round of multilateral trade negotiations.

As discussed earlier, these changes ensured that the rate of protection to most manufacturing industries fell substantially during the 1970s. It is true that protection for textiles, clothing, footwear, and motor vehicles and parts increased enormously during the decade, but what is remarkable about that series of policy changes is not that protection for clothes and cars increased in response to declining international competitiveness, because this occurred also in North America and Western Europe; it is that other manufacturing protection was reduced, and by so much. The latter is remarkable not only because it signifies a major turnaround in Australia's policy stance but also because it occurred at a time when manufacturing output and employment were declining absolutely and when the rate of overall unemployment was high and rising. That is, it occurred at a time when one might expect to see protection being *increased*. That the latter did not occur has been attributed largely to the changed climate of opinion concerning the national interest in protection.[29] The Australian economics profession became increasingly outspoken against protectionism during the 1960s, and by the early 1970s was advocating trade liberalisation with near unanimity and with increasing technical sophistication. A similar change occured within the bureaucracy and in the financial press. These views were reflected in numerous official reports commissioned by the government, and were adopted by an increasing number of politicians. A leading role was played by the Industries Assistance Commission (formerly the Tariff Board). Private interest groups also responded to the changing trade policy atmosphere. In particular, the National Farmers' Federation made liberalisation of manufactured imports a major part of its political lobbying and public education programs.

The main reasons the textiles, clothing and footwear industries and the motor vehicle industry were singled out for protection increases in the latter 1970s and 1980s include the following: these are industries which had a well-established lobbying capacity before 1973, because for decades they had been highly dependent on protection; both groups of industries employed large numbers of workers (representing a total of 20 per cent of the manufacturing workforce in 1973); the motor vehicle industry is particularly important to the economies of South Australia and Victoria, and so also receives strong support from the governments of these states when seeking federal government protection; these industries enjoyed the highest tariff protection in 1973 and so were affected most by the 25 percent tariff cut; and the textiles, clothing and footwear industries, being very labour intensive, were the industries affected most in the mid-1970s by the increases in real wages and in import competition from industrialising Asia. Thus the expected net benefit from lobbing for protection increases was much higher for these than for other manufacturing industries.

Prospects for future protection cuts

The above analysis suggests there are a number of possibilities for improving the prospects for trade liberalisation, especially as the economy moves out of recession. There is scope for reducing the excessive influence of protected interest groups by requiring the disclosure of political campaign contributions, by limiting the extent of those contributions, and by providing more public funding of election campaigns to reduce the reliance of political parties on vested interests.

General measures can also be taken to decrease Governments' willingness to provide protection. One is to provide more information on the net benefits of trade liberalisation. While there seems to be widespread acceptance of the basic point that protection reduces average incomes, a number of other affects of protection are still not widely appreciated in Australia. One such effect is that the present structure of Australia's protection is equivalent to a regressive income tax.[30] This is largely because clothing and footwear are so highly protected, and these goods constitute a relatively large share of household expenditure of low-income earners. Another example is the misunderstanding of the likely structure of the Australian economy in the absence of protection. Many people imagine in a free-trading Australia that people would work either on farms or mine sites or would be unemployed. Economic analysis shows clearly that with appropriate aggregate demand policies and flexible prices, much of the change in jobs would be to the service sector.[31] In addition, however, it is likely that the more competitive manufacturing industries would expand,[32] particularly those able to supply expanding markets in Asia. This latter point in particular needs to be stressed more, as there would be less opposition to protection cuts if it were understood that the manufacturing sector as a whole may not be reduced much by trade liberalisation.

A further consideration along these lines relates to the effects of protection on different states. The resource-exporting states of Western Australia and Queensland plus the Northern Territory would certainly gain from reduced import barriers, at the expense of the highly protected states, particularly Victoria.[33] To reduce the latter's opposition to liberalisation, the former might agree to allow a greater share of federal grants to go to the protected states to allow investment in the restructuring of their economies. This would simply require that the resource-exporting, lightly populated states, which currently enjoy disproportionately high federal grants, accept closer to the same grant per capita as the more densely populated south-eastern states.

Finally, consideration might be given to ways of bringing export interests more directly into conflict with protected interest groups. This was in fact done in the negotiations leading to the 1957 trade agreement between Australia and Japan. While there is now no obvious opportunity for productive negotiations along these lines with a single country, it may be possible to achieve the same result in the context of regional trade negotiations amongst Pacific countries. In return for opening up our market for manufactured goods, Australian exporters would seek improved access to Asian markets for foodstuffs and processed minerals. Given the strong intra-regional trade bias and potentially strong trade complementarity between countries of the region, concessions could be given on a non-discriminatory, most-favoured-nation basis consistent with the GATT and yet most of the benefits would be reaped within the region.[34]

Appendix

Table 6.A1 Effective rates of assistance to 4—digit Australian manufacturing industries, 1979–80 and 1981–82[a] (per cent)

ASIC		*1979–80*	*1981–82*
21	*Food, beverages & tobacco*	11	8
2115	Meat, n.e.c.	5	6
2116	Poultry	–3	16
2117	Bacon, ham & smallgoods	6	–2
2121	Liquid milk & cream	–1	0
2122	Butter	8	2
2123	Cheese	4	2
2124	Ice cream & frozen confections	13	14
2125	Milk products, n.e.c.	12	–3
2131	Fruit products	25	22
2132	Vegetable products	7	7
2140	Margarine, oils & fats, n.e.c.	22	15
2151	Flour mill products	51	–31
2152	Starch, gluten & starch sugars	30	5
2153	Cereal foods & baking mixes	0	–4
2161	Bread	10	–7
2162	Cakes & pastries	–2	6
2163	Biscuits	12	8
2173	Confectionery & cocoa products	38	32
2174	Processed seafoods	7	5
2175	Prepared animal & bird foods	6	–1

Table 6.A1 Continued

ASIC		1979–80	1981–82
2176	Raw sugar & food products n.e.c.	3	3
2185	Soft drinks, cordials & syrups	5	4
2186	Beer	42	50
2187	Malt	7	4
2188	Wine & brandy	101	104
2189	Alcoholic beverages, n.e.c.	17	7
2190	Tobacco products	−2	−2
23	*Textiles*	55	54
2341	Cotton ginning	118	53
2342	Wool scouring & top making	28	27
2343	Manmade fibres & yarns	145	133
2344	Manmade fibre broadwoven fabrics	148	192
2345	Cotton yarns & broadwoven fabrics	50	45
2346	Worsted yarns & broadwoven fabrics	102	64
2347	Woolen yarns & broadwoven fabrics	75	52
2348	Narrow woven & elastic fabrics	61	57
2349	Textile finishing	152	92
2351	Household textile	73	67
2352	Textile floor coverings	14	48
2353	Felt & felt products	28	25
2354	Canvas & associated products, n.e.c.	24	26
2355	Rope, cordage & twine	30	22
2356	Textile products, n.e.c.	4	5
24	*Clothing & footwear*	137	204
2441	Hosiery	70	100
2442	Cardigans & pullovers	128	171
2443	Knitted goods, n.e.c.	223	215
2451	Men's trousers, shorts & work clothing	127	212
2452	Men's suits, coats & waterproof clothing	105	189
2453	Women's outerwear, n.e.c.	152	249
2454	Foundation garments	140	225
2455	Underwear and infants clothing	173	222
2456	Headwear and clothing, n.e.c.	61	75
2460	Footwear	143	229
25	*Wood, wood products & furniture*	14	13
2531	Log sawmilling	4	4
2532	Resawn and dressed timber	13	13
2533	Veneers & manufactured boards	31	30
2534	Wooden doors	13	14
2535	Wooden structural fitting & joinery, n.e.c.	12	15
2536	Wooden containers	17	17
2537	Hardwood woodchips	−1	0
2538	Wood products, n.e.c.	10	9
2541	Furniture excl. sheet metal	24	20
2542	Mattresses excl. rubber	−6	−7
26	*Paper & paper products, printing*	27	30
2631	Pulp, paper and paperboard	9	8
2632	Paper bags incl. textile bags	52	55
2633	Solid fibreboard containers	34	39
2634	Corrugated fibreboard containers	35	36
2635	Paper products, n.e.c.	20	29
2641	Publishing	−4	−4
2642	Printing & publishing	24	24
2643	Paper stationery	32	43
2644	Printing & bookbinding	41	42

2645	Printing trade services, n.e.c.	24	33
27	*Chemical, petroleum & coal products*	17	13
2751	Chemical fertilisers	0	0
2753	Synthetic resins & rubber	34	33
2754	Organic industrial chemicals, n.e.c.	24	20
2755	Inorganic industrial chemicals, n.e.c.	11	7
2762	Paints	13	13
2763	Pharmaceutical & veterinary products	13	−1
2764	Pesticides	38	37
2765	Soap & other detergents	18	18
2766	Cosmetics & toilet preparations	20	20
2767	Inks	50	50
2768	Chemical products, n.e.c.	16	168
2770	Petroleum refining	0	0
2780	Petroleum & coal products, n.e.c.	23	23
28	*Non-metallic mineral products*	5	5
2850	Glass & glass products	5	5
2861	Clay bricks	1	1
2862	Refractories	4	7
2863	Ceramic tiles and pipes	10	12
2864	Ceramic goods, n.e.c.	28	22
2871	Cement	0	1
2873	Concrete pipes & box culverts		
2872	Ready mixed concrete	0	0
2874	Concrete products, n.e.c.	−3	−3
2881	Plaster products & expanded minerals	14	14
2883	Glass wool and mineral wool products		
2882	Stone products	16	14
2884	Non-metallic mineral products, n.e.c.	18	16
29	*Basic metal products*	12	14
2941	Iron & steel basic products	10	14
2942	Iron castings	24	25
2943	Steel casting	25	26
2944	Iron & steel forging	24	24
2945	Steel pipes & tubes	14	18
2951	Copper smelting, refining	11	9
2953	Alumina	3	3
2954	Aluminum smelting		
2956	Nonferrous metals, n.e.c., smelting, refining	6	6
2957	Secondary recovery & alloying of nonferrous metals, n.e.c.	9	5
2961	Aluminum rolling, drawing, extruding	43	40
2962	Nonferrous metals, n.e.c., rolling, drawing, extruding	7	6
2963	Nonferrous metal casting	29	29
31	*Fabricated metal products*	33	34
3141	Fabricated structural steel	35	29
3142	Architectural aluminum products	26	28
3143	Architectural metal products, n.e.c.	40	39
3151	Metal containers	78	87
3152	Sheet metal furniture	40	43
3153	Sheet metal products, n.e.c.	40	43
3161	Cutlery & hand tools, n.e.c.	22	23
3162	Springs & wire products	29	31
3163	Nuts, bolts, screws & rivets	29	31
3164	Metal coating & finishing	22	22
3165	Nonferrous steam, gas & water fittings	14	14
3166	Boiler & plate work	25	27

Table 6.A1 Continued

ASIC		*1979–80*	*1981–82*
3167	Metal blinds & awnings	26	19
3168	Fabricated metal products, n.e.c.	19	21
32	*Transport equipment*	74	79
3231	Motor vehicles	149	158
3232	Truck & bus bodies, trailers & caravans	30	34
3233	Motor vehicle instruments & electrical equipment, n.e.c.	98	106
3234	Motor vehicle parts, n.e.c.	110	119
3241	Ships	12	17
3242	Boats	31	31
3243	Railway rolling stock & locomotives	33	45
3244	Aircraft building & repair	−4	−5
3245	Transport equipment, n.e.c.	19	16
33	*Other machinery & equipment*	22	22
3341	Photographic goods, optical instruments	5	6
3342	Photographic film processing	14	14
3343	Measuring, professional & scientific equipment, n.e.c.	1	1
3351	Radio & TV receivers, audio equipment	46	39
3352	Electronic equipment, n.e.c.	21	27
3353	Refrigerators & household appliances	40	35
3354	Water heating systems	27	34
3355	Electric & telephone cable & wire	5	5
3356	Batteries	42	42
3357	Electrical machinery & equipment, n.e.c.	10	14
3361	Agricultural machinery	20	19
3362	Construction machinery	31	32
3363	Materials handling equipment	46	42
3364	Wood & metal working machinery	5	7
3365	Pumps & compressors	14	16
3366	Commercial space heating & cooling equipment	28	27
3367	Dies, saw blades & machine tool accessories	34	35
3368	Food processing machinery	21	22
3369	Industrial machinery & equipment, n.e.c.	17	19
34	*Miscellaneous manufacturing*	26	27
3451	Leather tanning & fur dressing	32	39
3452	Leather & substitute goods, n.e.c.	42	34
3461	Rubber tyres, tubes, belts, hoses, sheets	57	42
3462	Rubber products, n.e.c.	37	31
3471	Flexible packaging & abrasive papers	28	37
3472	Rigid plastic sheeting	21	21
3473	Hard surface floor coverings, n.e.c.	36	36
3474	Plastic products, n.e.c.	14	18
3481	Ophthalmic articles	19	19
3482	Jewellery & silverware	45	44
3483	Brooms & brushes	21	18
3484	Signs & advertising displays	24	26
3485	Sporting equipment	25	19
3496	Writing & marking equipment	19	17
3487	Manufacturing, n.e.c.	42	42
	Total manufacturing	24	26

Notes:
ASIC Australian Standard Industrial Classification (1978 edition).
n.e.c. Not elsewhere classified.
n.a. Not available
a The forms of assistance covered by this table include tariffs, quantitative restrictions on imports, production and export subsidies, and special pricing schemes for sugar and petroleum products. Forms of assistance not taken into account include government purchasing practices and assistance from state governments. The pattern of production in 1977–78 was used for the estimates.
Source: Industries Assistance Commission (1980a, 1983).

Endnotes

* The authors are grateful to Christopher Findlay for helpful comments on an earlier draft of this chapter.

1 This and subsequent sections draw on Corden, W.M. (1963), 'The Tariff', in A. Hunter (ed.), *The Economics of Australian Industry*, Melbourne: Melbourne University Press; Anderson, K. and Garnaut R. (forthcoming), *Australian Protection and Trade with Developing Countries*, Monograph to be published by the World Bank, Chapters 3 and 4; and Anderson, K. (1986), 'Tariffs and the Manufacturing Sector', in R. Maddock and I. McLean (eds.), *The Australian Economy Since 1900: Performance and Policies*. Cambridge: Cambridge University Press, (forthcoming).

2 Maddock, R. and McLean, I. (1984), 'Supply-side Shock: The Case of Australian Gold', *Journal of Economic History*, 44(4), pp. 1047–67.

3 Crawford, J.G. (1934), 'Tariff Level Indices', *Economic Record*, 10, pp. 213–21.

4 Carmody, A.T. (1952), 'The Level of the Australian Tariff: A Study in Method'. *Yorkshire Bulletin of Economic and Social Research*, 4(1), pp. 51–65.

5 Moffatt, G.G. (1970), *Import Control and Industrialisation: A Study of the Australian Experience*. Melbourne: Melbourne University Press.

6 Arndt, H.W. (1965), 'Australia—Developed, Developing or Midway?', *Economic Record*, 41(95), pp. 318–40; and Snape, R.H. (1984), 'Australia's Relations with GATT', *Economic Record*, 60(168), pp. 16–27.

7 Anderson, K. and Garnaut, R. (forthcoming).

8 Government assistance to other tradeable sectors has not been absent, but has always been relatively minor compared with the tariff assistance afforded the manufacturing sector. Even during the 1920s and 1930s when some agricultural industries were heavily protected, the (weighted) average rate of assistance to agriculture as a whole was still low compared with assistance to manufacturers.

9 Kuznets, S.S. (1971), *Economic Growth of Nations: Tota Output and Production Structure*. Cambridge, Mass.: Harvard University Press, Tables 22 and 39.

10 World Bank, (1980), *World Tables (1980 Edition)*, New York, Oxford University Press.

11 For further details, see Anderson, K. (1986).

12 It has been argued in the past that Australia may have been able to influence the price of wool. In the 1920s wheat was also considered by some authors to have a low export demand elasticity: Brigden, J.B. (1925), 'The Australian Tariff and the Standard of Living', *Economic Record*, 1(1), pp. 29–46; Brigden, J.B. et al. (1929), *The Australian Tariff: An Economic Enquiry*, Melbourne: Melbourne University Press; Giblin, L.F. (1936), 'Some Economic Effects of the Australian Tariff', The Joseph Fisher Lecture in Commerce, University of Adelaide, 23 June, 1936. Others, however, strongly disagreed that Australia would have monopoly power even in wool and wheat: Benham, F.C. (1926), 'The Australian Tariff and the Standard of Living: A Reply', *Economic Record*, 2, pp. 20–42; Viner, J. (1929), 'The Australian Tariff', *Economic Record*, 5(9), pp. 306–15. See Reitsma, A.J. (1960), *Trade Protection in Australia*, Brisbane: University of Queensland Press, Chapter 6, for a detailed discussion of the debate on the terms of trade argument. With strong competition from synthetic and other natural fibres, and with the growth in the wool industry in a number of other countries, it is likely that any monopoly power in wool which Australia may have enjoyed in the past has now virtually disappeared. See Freebairn, J.W. (1978), 'Projections of Australia's World Trade Opportunities: Mid and Late Nineteen Eighties', IMPACT Working Paper No. 1–07, Melbourne: Industries Assistance Commission, for a recent survey of estimates of Australia's export demand elasticities. In any case, the

more efficient policy if any such monopoly power existed would be a tax on exports of that product rather than taxes on every item imported. This is because the latter approach not only would involve higher tax collection costs but would, through the tariff's effect on the exchange rate, provide disincentives to exporters of other export products besides wool.

13 Reitsma, A.J. (1960), Chapter 7. This school of thought was popularized by List, F. (1885), *The National System of Political Economy*, London: Longmans, Green (translated from the German text first published in 1841). Marshall, A. (1890), *Principles of Economics*, London: Macmillan, used the presumed presence of these supply conditions in industry and in agriculture to argue that protection was less costly in the US, then mainly an exporter of primary products, than in Britain, an exporter of manufactures. For a discussion of why the infant industry argument should not be taken as a serious justification for import protection, see Baldwin, R.E. (1969), 'The Case Against Infant Industry Protection', *Journal of Political Economy*, 77(3), pp. 295–305; and Corden, W.M. (1974), *Trade Policy and Economic Welfare*, Oxford: Clarendon Press, Chapter 9.

14 Brigden, J.B. (1925); Brigden, J.B. et al., (1929); and Reddaway, W.B. (1937), 'Some effects of the Australian Tariff', *Economic Record*, 13, pp. 22–30 and 249–50.

15 In the process moderate tariff levels could raise tariff revenue for the government, so reducing the need for other (and perhaps more administratively expensive) forms of taxation, although this point was not included in the argument at the time. Tariffs also benefited industries, of course. Indeed, an Arbitration Commission was established in part to ensure that workers wage increases did not absorb all the profit increases due to protection. On the perceived importance of "peopling" Australia with European settlers, see Borrie, W.D. (1963), 'The Peopling of Australia', in H.W. Arndt and W.M. Corden (eds), *The Australian Economy: A Volume of Readings*, Melbourne: Cheshire.

16 Supporters of the position taken by Brigden (1925); Brigden, J.B. (1927a), 'The Australian Tariff and the Standard of Living: A Rejoiner', *Economic Record*, 3, pp. 102–16 and (1927b), 'Comment on Mr. Benhan's Restatement', *Economic Record*, 3, pp. 249–51, included Giblin, L.F. (1927), 'The Australian Tariff and the Standard of Living: A Note on Mr. Benham's Statistics', *Economic Record*, 3, pp. 148–56; Giblin, L.F. (1936) and Copland D.B. (1931), 'A Neglected Phase of Tariff Controversy', *Quarterly Journal of Economics*, 45(1), pp. 289–308; and Copland, D.B. (1935), 'Notes on Tariff Theory with Special Reference to the Australian Tariff', *Economic Record*, 11 (Suppl.), pp. 33–9. Both were members of the Brigden Committee that produced the 1929 tariff report. Critics included Benham, F.C. (1926); Benham, F.C. (1927), 'The Australian Tariff and the Standard of Living: A Restatement', *Economic Record*, 3, pp. 239–48; and Viner, J. (1929). An account of the debate is given in Cain, N. (1973), 'Political Economy and the Tariff: Australia in the 1920s', *Australian Economic Papers*, 12, pp. 1–20.

17 Viner, J. (1929).

18 Stolper, W.F. and P.A. Samuelson (1941), 'Protection and Real Wages', *Review of Economic Studies*, 9(1), pp. 58–73. The Australian situation had been discussed at that time in a number of theoretical papers published in the *Quarterly Journal of Economics* by Copland, D.B. (1931); Anderson, Karl (1938), 'Protection and the Historical Situation: Australia', 53(4), pp. 86–104; and Anderson, Karl (1939), 'The Australian Case for Protection Re-examined: Comment', 54(4), pp. 149–51; Samuelson, M. (1939), 'The Australian Case for Protection Re-examined', 54(4), pp. 143–49. Interest in the history of thought concerning the so-called Australian case for protection was rekindled recently in an exchange between Paul Samuelson (1981a), 'Summing Up on the Australian Case for Protection', *Quarterly Journal of Economics*, 96(1), pp. 147–60; (1981b), 'Justice

to the Australians', 96(1), pp. 169–70; and Manger G.J. (1981a), 'The Australian Case for Protection Reconsidered', *Australian Economic Papers*, 20(37), pp. 193–204; (1981b), 'Summary Upon the Australian Case for Protection: Comment', *Quarterly Journal of Economics*, 96(1), pp. 161–7.

19 Lloyd, P.J. (1982), 'Protection for Whom?', *Economic Papers*, 1(2), pp. 1–16.

20 Changes to Australia's tax laws also would ease the burden of adjustment. Australians have been encouraged by the tax provisions to invest much of their long-term savings in superannuation and owner-occupied housing. Superannuation is often not portable, and there are large taxes and other transactions costs involved in buying and selling a house. Also, the asset value of the home is tied to housing values in a particular area. If a major industry in that area declines, so too does the wealth of home-owning workers.

21 Corden, W.M. (1984), 'Booming Sector and Dutch Disease Economics: A Survey', *Oxford Economic Papers*, forthcoming.

22 Corden, W.M. (1981), 'Exchange Rate Protection', in R.N. Cooper (ed.), *The International Monetary System Under Flexible Exchange Rates*, Cambridge, Mass.: Ballinger.

23 Johnson, H.G. (1965), 'An Economic Theory of Protectionism, Tariff Bargaining and the Formation of Customs Unions', *Journal of Political Economy*, 73(3), pp. 256–83.

24 Anderson, K. (1986), in R. Maddock and I. McLean.

25 Downs, A. (1957), *An Economic Theory of Democracy*, New York: Harper and Row; Buchanan, J. and G. Tullock (1962)1, *The Calculus of Consent*, Ann Arbor: University of Michigan Press; and Breton, A. (1974), *The Economic Theory of Representative Government*, Chicago: Aldine.

26 Anderson, K. (1980), 'The Political Market for Government Assistance to Australian Manufacturing Industries', *Economic Record*, 56(153), pp. 132–44.

27 Maddision, A. (1982), *Phases of Economic Development*, London: Oxford University Press.

28 Pincus, J.J. and G.A. Witchers (1983), 'Economics of Regulation', in F.H. Gruen (ed.), *Surveys of Australian Economics*, Vol. 3, Sydney: George Allen and Unwin.

29 Australia's manufacturing sector was under strong pressure to decline in the 1970s because of the mining export boom (induced by both domestic supply discoveries and increased export prices), large institutionalised increases in real wages, particularly for female and unskilled workers, and increased import competition from manufacturers in Asia. These factors, together with the changes in protection levels, were the main contributions to the decline in the manufacturing sector's shares of GDP and employment from around 27 per cent in 1970 to around 20 per cent in the early 1980s: Anderson and Garnaut, (forthcoming), Chapter 6.

30 Industries Assistance Commission, (1980b), *Tariffs as Taxes*, Canberra: Australian Government Publishing Service.

31 Corden, W.M. (1979), 'Tell Us Where the Extra Jobs Will Be', *Bank of New South Wales Review*, 30 (October), pp. 1–5.

32 Dixon, P.B., B.R. Parmenter and A.A. Powell (1983), 'Trade Liberalization and Labour Market Disruption', General Paper No. G-46, IMPACT Project, University of Melbourne.

33 Industries Assistance Commission, (1981), *The Regional Implications of Economic Change*, Canberra: Australian Government Publishing Service.

34 Garnaut, R., 'Australian Trade Policy and Western Pacific Economic Growth', *Economic Papers 68*, pp. 14–30. Anderson, K. et al. (1985), 'Pacific Economic Growth and Prospects for Australian Trade', Research Paper No. 122, Canberra: Australia-Japan Research Centre.

7 The effect of manufacturing sector protection on ASEAN and Australia: a general equilibrium analysis

GLENN W. HARRISON AND E.E. RUTSTRÖM*
May 1985

Overview

THIS paper employs a numerical general equilibrium model of the Pacific Basin to study the effects of various trade liberalisation strategies on ASEAN and Australia. We focus on the alternatives to participation in multilateral tariff liberalisation schemes, such as the Kennedy and Tokyo Rounds concluded under the auspices of the GATT. The tariff reductions negotiated in these Rounds concentrated on manufacturing trade, and we pursue that emphasis in our simulation of alternative policies.

A major conclusion of our study is that the economic benefits from ASEAN and Australian participation in GATT-type multilateral tariff liberalisation schemes, although positive and robust to uncertainty about the parameters used in our model, are not particularly large. Moreover, these benefits to ASEAN and Australia are small in comparison to the benefits from liberalisation of non-tariff barriers. The political difficulty of multilateral liberalisation is not considered here. However we strongly believe that one important role for economic analysis and quantitative modelling is to identify areas in which existing political constraints should be relaxed.

Section 2 introduces the general equilibrium (GE) model, which is described in detail in Harrison (1984a).[1] It is firmly neoclassical and static in structure, and is empirically calibrated to the global economy of 1975. Policy simulations are reported in Section 3. Three sets of policies are considered: (i) global liberalisation in tariffs and non-

* G.W. Harrison is affiliated with the Department of Economics, University of Western Ontario and University of Arizona (Visiting); E.E. Rutström is affiliated with the Stockholm School of Economics.

tariff barriers (NTBs); (ii) global liberalisation of trade distortions in manufacturing sectors; and (iii) regional liberalisation of tariffs and NTBs by ASEAN nations alone and by ASEAN and Australia. Related policy applications appear in Harrison and Kimbell (1985), and Harrison (1984a, 1985a).[2]

We argue throughout this paper that GE methods of analysis, whether theoretical or numerical, are of some value in the policy process but we do not want to appear dogmatic on the possible use of alternative methods of analysis or models. Indeed, we suggest in Section 5 that significant reformulations of existing numerical GE models are necessary to address properly many trade policy issues of importance in ASEAN and Australia. Moreover, the alternatives to GE analysis are not well defined since many GE features may be implicit. An important example is the use of input-output modelling assumptions underlying 'effective protection' calculations such as those presented elsewhere in this volume. We now consider in detail the relationship between those calculations and our own. In Section 4 we directly compare the results of effective protection calculations and results from our GE model concluding that the former can be unreliable indictors of resource pulls due to protection.

Effective protection calculations

The development of effective protective rates in empirical trade analysis has been an attempt to design a tool to forecast effects of protectionist policies on resource allocation in a more complex setting than in the simple world without intermediate goods where nominal tariff rates perform this task adequately.

Corden (1966)[3] uses the input-output structure of the economy to derive a formula for ERP which is simply a ratio of value added under protection to value added in free-trade. His basic conclusion is that, after ordering economic activities according to the rate of effective protection (ERP) it is possible to determine the direction of sectoral resource reallocation: there will be a movement of resources out of activities with low (effective) protection into activities with high (effective) protection. As Corden is assuming fixed input-output coefficients he cannot tell the net effects on the activities in between these two extremes. More information on production substitution elasticities is needed for this to be possible. This implies the need for a second step after the calculation of ERP. The ERP is designed to indicate the direction of resource pull of a specific tariff scheme, and thus should not already incorporate any of the reallocation effects it is supposed to indicate. The input-output coefficients to be used in the ERP calculations should ideally be the coefficients applicable under free trade. The second step in forecasting would then be, following the computation of changes in relative net benefits of the effective protection scheme, to use estimated production substitution elasticities in an economy-wide analysis of resource shifts. At this stage the procedure is approximating a GE framework; indeed we employ variable input-output coefficients in the GE model presented in Section 2.

Another complication with ERPs is substitution between traded and nontraded final goods that will change the *real* exchange rate and thereby provide a uniform rate of effective (positive or negative) protection for all tradeables. Incorporation of this substitution requires introduction of the demand side of the economy into the calculations, which is another step towards an explicit GE analysis.

A third complication is to take the effects of foreign tariffs into account. At the final goods level an import tariff imposed by a trading partner acts as equivalent to an export tax. A scale of combined effective rates could therefore be constructed to analyse the resource movements under a multilateral tariff reduction, although this is rarely done in ERP analyses. A further complication of foreign tariffs is included if one allows for trade in intermediate goods. When calculating changes in value added for an industry as a measure of ERP the cost of intermediate goods is subtracted, and these could include not only protection of domestically produced intermediates but also protection in the country of production of imported intermediates. This is especially important when discussing multilateral trade-liberalisation schemes. This is not to our knowledge accounted for in existing ERP analyses, but is automatically taken into account in a multi-country GE framework.

One important difficulty encountered when empirically calculating ERPs is that the free trade input-output coefficients are, in general, not known. The substitution effects in production (due to the existing tariff distortions) are already incorporated in the ERP estimate and the structure of intermediate imports from a given trade partner is also assumed to be constant because of fixed input-output coefficients in trade in intermediates. Any resource reallocation consideration of a multilateral trade liberalisation scheme will be based on the trade pattern applying in the (distorted) situation with barriers to trade. Again, a GE model like ours, especially with some degree of substitutability in intermediate trade, can easily take this into account when considering the comparative static effects of a change in policy.

Bhagwati and Srinivasan (1973)[4] raise the question of sufficiency conditions for ERP theory to perform the same forecasting role in a world with traded inputs as nominal tariffs perform in traditional models with no traded inputs. Under certain conditions they are able to devise a 'price' of value added to rank production activities so that reallocation of 'quantities' of value added can be predicted. They find that this is possible only under the restriction of 'separable production functions' when there are more than two final goods. The GE model presented below assumes such production functions, albeit not in the especially restrictive class of Leontief production functions typically adopted in ERP models. A further result is that the pattern of 'prices' of value added cannot in general be computed solely from knowledge of the tariff structure; information on production functions is also needed, especially information on how the share of traded inputs on output prices. The problem is not as severe in Cobb-Douglas production functions, where these shares are constant, as it is in the general CES production function. We conclude that this effort and amount of information that is required in order to compute an ERP with good predictive power implies there is no real reason for preferring ERP over GE computations.

One important feature of the GE model presented in the next section is the use of a technique for allowing intermediate input substitutability that does not involve significant increases in computational expense.[5] This feature is particularly important for international trade models in the face of the empirical significance of trade in intermediates and the evidence that import volumes respond to price changes.

2 A Pacific Basin general equilibrium model

We employed three broad criteria in specifying the model. The first was that it be *understandable*, in the sense of having a (neoclassical) structure that would be familiar to all economists despite great sectoral detail and a large number of trading regions. The second was that it be *readily operational* with existing data sources. The third requirement was that it be *repeatedly soluble* for the purposes of systematic sensitivity analysis. A number of additional requirements, specific to a diverse range of intended policy applications, are of secondary importance but nonetheless have influenced the chosen specification.

Trading regions

The model identifies eleven trading regions, listed in Table 7.1, along with several aggregate statistics, plus a residual 'Rest of World'. These regions represent a diverse range of industrialisation, size, and 'openness' to international trade. In terms of geographic coverage, our model subsumes the three region (US, Japan, EEC) model

Table 7.1 Trading regions

Region	GDP	Exports	Imports	Population	GDP per capita
Indonesia	30.5	6.9	5.5	135.2	226
Malaysia	9.3	3.8	3.5	11.9	781
Philippines	15.8	2.3	3.5	42.1	375
Singapore	5.6	5.1	7.5	2.3	2435
Thailand	14.6	2.2	2.8	41.9	348
Australia	87.3	11.7	9.5	13.8	6326
Korea	20.6	5.0	6.7	35.3	583
Japan	501.9	54.7	49.7	111.6	4497
USA	1518.3	107.1	98.1	213.6	7108
Canada	165.2	33.9	34.3	22.7	7277
EEC	1373.2	146.3	148.0	258.0	5323

Notes:
GDP, Exports (fob) and Imports (fob) are measured in billions of US dollars in 1975, and were obtained from lines 99b, 77aad and 77abd, respectively, of the *International Financial Statistics* of the IMF (period average exchange rates used). Population is measured in millions, and is obtained from line 99z of the *IFS*. GDP per capita is measured in US dollars. Note that the Exports and imports listed here are not the model-equivalent values. The model also includes a twelfth region, a residual 'Rest of World'.

presented in Whalley[6] (1980a, 1980b, 1982, 1985) and Brown and Whalley (1980),[7] while providing certain country specific detail for less developed countries abstracted from the seven region model presented in Whalley.[8]

Commodities and production structure

We adopt two levels of commodity aggregation in Table 7.2 because of the relative ease of computation and interpretation of results with the aggregated model. In addition to the commodities listed, each household in each trading region allocated income to 'savings', which are in turn allocated to the purchase of a bundle of investment goods (primarily, but not exclusively, in the household's own region).

Each of the commodities listed in Table 7.2 is assumed to be produced in each region and is, in principle, tradeable. Each commodity is distinguished by producing region, implying that our model is the familiar 'Armington assumption' distinguishing products by point of production. Each sector uses intermediate inputs from its own region and from all other regions, as well as primary factors (labour and capital). Although it is useful to visualise the use of intermediate inputs in the form of a complete multiregional (international) input-output table, an important feature of the present model is that the implied technical coefficients are not fixed with respect to relative input prices. That is, we do not employ a Leonief technology in the use of intermediate inputs, but assume instead a Cobb-Douglas technology. In this respect our analysis of tariff incidence represents a

Table 7.2 Commodities considered

Twenty sector aggregation	Seven sector aggregation
1. Agriculture, forestry & fishing	1. Agriculture, forestry & fishing (1)
2. Minerals & extractive ores	2. Mining & Quarrying (2,3)
3. Energy products	3. Manufacturing (4/15)
4. Food, beverage & tobacco	4. Utilities (16)
5. Textiles, clothing, footwear & leather	5. Construction (17)
6. Lumber & wooden products	6. Trade & transport (18)
7. Pulp, paper & printing	7. Services (19, 20)
8. Chemicals	
9. Rubber & plastic products	
10. Nonmetallic mineral products	
11. Basic & fabricated metal products	
12. Industrial machinery	
13. Electrical & other machinery & equipment	
14. Motor vehicles	
15. Other transport equipment	
16. Electricity, gas & water supply	
17. Construction	
18. Trade, transport & communications	
19. Services	
20. Public administration & community services	

Note:
The bracketed numbers for the seven sector aggregation indicate the aggregated sectors from the twenty sector list.

direct generalisation of ERP calculations that do not allow variable technical coefficients. Thus intermediate inputs are substitutable and, as a composite, are substitutable with a composite of primary inputs.

The two primary factors employed in each sector are characterised by a CES technology. In general, the Heckscher-Ohlin (HO) factor mobility assumptions are adopted: each factor is free to move within the sectors of each region but not between regions.

There are two major data sources required to calibrate this production structure: a complete multiregional IO table for the regions listed in Table 7.1 (including sectoral value added data), and estimates of the elasticities of substitution between labour and capital. The Institute of Developing Economies and Harrison[9] describe the construction of the IO table, and Harrison and Kimbell[10] describe the data sources for our elasticity estimates. The relative availability of national IO data for 1975 determined the dating of our model. We regard this dating as perhaps the most serious weakness of our database, and hence of the empirical GE model used here.

Demand Structure

The demand pattern in each trading region is represented by a single private unit or 'household' and a single public (or government) unit.

Thus the model identifies 24 units in all. A key feature of our model is the explicit formulation of demand systems for each household which contrasts sharply with the assumptions implicit in the use of ERP.

Private households maximise a nested utility function with three levels. The 'top' level is a Klein-Rubin utility function,[11] leading to an Extended Linear Expenditure System (ELES) defined in principle over eight commodity groupings (Food, Clothing, Housing, Durables, Personal Care, Transportation, Recreation, and Other Services) and savings. The 'middle' level is a CES function defined over the commodities within each of these eight commodity groupings. Thus 'Agriculture, forestry and fishing' and 'Food, beverage and tobacco' from Table 7.2 are combined in a CES function to form the composite grouping 'Food'.[12] Finally, the 'bottom' level is a CES function defined over each of the commodities listed in Table 7.2 differentiated by origin.

The consumption problem of the private household may therefore be viewed in three stages. Given the income to be allocated to consumption (i.e., non-savings), the allocation of expenditure to the eight commodity groupings is decided. Then, conditional on the expenditure of each group, the allocation to each of the (varying number of) commodities within each group is decided. Finally, the household decides between alternative sources of each commodity given the expenditure allocated to that commodity. Specific functional forms aside, this type of utility nesting structure is common to recent international trade GE models. Harrison and Kimbell discuss the calibration of the utility function of each household, relying heavily on elasticity estimates reported in Lluch, Powell and Williams.[13]

Household savings are allocated entirely to the purchase of Cobb-Douglas composite of commodities from all regions for the purpose of capital formation. These expenditures refer to purchases of real goods and services.

Public units in each region spend their revenues on various own-region and foreign commodities. A Cobb-Douglas utility function is used for these units, and is calibrated using expenditure shares by each government (for current consumption purposes *and* capital formation).

Private household income is generated from the sale of their factor endowment to own-region industries and from transfers received from their government. Each government receives revenue from the taxes, tariffs and non-tariff barriers that it levies on own-region and foreign economic activity; these policy instruments are discussed in more detail in the next section. In principle the model allows for intergovernment transfers, in the form of (untied) aid; in practice we have been unable to obtain adequate data to include these transfers in the present model.

Although each and every unit has a 'balanced budget' in equilibrium, there is no explicit or implicit presumption in the model that *bilateral* trade flows between any *two* regions balance.

Tariffs and non-tariff barriers

In principle the model incorporates a wide range of taxes, tariffs and non-tariff barriers (NTBs) differentiated by commodity, region and stage of production of (legal) incidence, and taxing government. In practice, however, data limitations severely circumscribe and detail, coverage and accuracy of our 'model equivalent' estimates.[14]

The basic source of data of tariffs was the international input-output table presented in Institute of Developing Economies, and Harrison.[15] We draw a distinction between tariffs levied on imports of intermediate inputs and those levied on imports of final goods. Recall that we allow a given sector to import intermediates from *all* other sectors in all regions, thus the total import duties paid by this sector reflects the various tariff rates applicable to the range of intermediates it imports, *weighted* by the expenditure on each imported intermediate. The *ad valorem* tariff implied by this procedure need bear no similarity to the posted tariff on imports of the commodity of the sector in question.[16] Moreover, the same tariff rate applies to all intermediate input imports of the given sector. The implied tariff rates on final demand imports therefore bear a direct similarity to posted rates. Although our model-equivalent tariff rates on intermediate input imports do not correspond directly to posted rates, they can be reconciled satisfactorily with the 1976 rates used by Whalley[17] for the US, EEC and Japan.

The available data on *ad valorem equivalents* of NTBs are notoriously poor. We rely heavily on the aggregative estimates listed in Whalley[18] and the detailed estimates employed in Whalley and Yeats[19] (1979). We adopt the measures of the *ad valorem* equivalent of agricultural NTBs in Japan and Korea reported in Saxon and Anderson, and Anderson,[20] respectively. The measures adopted for Australia, Canada, Indonesia, Malaysia, the Philippines, Singapore and Thailand were derived from Tyers and Chisholm.[21] Government procurement practices are approximated by a 50 percent tariff applied to imports of each public household in each region, following Whalley.[22]

Table 7.3 presents some aggregative measures of the *ad valorem*

Table 7.3 ***Ad valorem* tariff and NTB tariff equivalents**

Whalley model

Policy instrument	Commodity	Region: EEC	US	Japan	Other developed	OPEC	NIC	LDC
Tariff	Agriculture & food	.042	.063	.233	.048	.0	.500	.080
	Minerals & extractive ores	.009	.040	.0	.036	.0	.230	.350
	Energy	.037	.035	.077	.059	.0	.230	.350
	Nonmechanical manufacturing	.085	.079	.068	.047	.15	.860	.960
	Machinery & transport	.082	.045	.084	.057	.15	.860	.800
	Services and non-traded	.0	.0	.0	.0	.0	.0	.0
NTB	Agriculture & food	.332	.438	.721	.497	.0	.125	.202
	Minerals & extractive ores	.484	.0	.460	.315	.0	.058	.088
	Energy	.282	.056	.377	.239	.0	.058	.088
	Nonmechnical manufacturing	.079	.075	.081	.078	.0	.215	.240
	Machinery & transport	.017	.0	.018	.012	.0	.200	.200
	Services & non-traded	.0	.0	.0	.0	.0	.0	.0

Harrison model

Policy instrument	Commodity	Region: Indonesia	Malaysia	Philippines	Singapore	Thailand	Australia	Korea	Japan	USA	Canada	EEC
Tariff	Agriculture, forestry & fishing	.24	.34	.63	.07	.09	.04	.16	.71	.06	.05	.042
	Mining & quarrying	.02	.01	.08	.0	.01	.0	.23	.0	.04	.01	.01
	Manufacturing	.61	.13	.39	.05	.20	.24	.85	.07	.06	.05	.08
	Utilities	.0	.0	.0	.0	.0	.0	.0	.0	.0	.0	.0
	Construction	.0	.0	.0	.0	.0	.0	.0	.0	.0	.0	.0
	Trade & transport	.24	.04	.20	.03	.09	.03	.02	.06	.0	.02	.05
	Services	.0	.0	.0	.0	.0	.0	.0	.0	.0	.0	.0
NTB	Agriculture, forestry & fishing	.35	.09	.03	.01	.21	.09	.92	1.34	.42	.51	.33
	Mining & quarrying	.0	.01	.01	.0	.02	.0	.06	.50	.0	.32	.37
	Manufacturing	.35	.19	.21	.02	.14	.17	.22	.08	.04	.07	.06
	Utilities	.0	.0	.0	.0	.0	.0	.0	.0	.0	.0	.0
	Construction	.0	.0	.0	.0	.0	.0	.0	.0	.0	.0	.0
	Trade & transport	.19	.08	.12	.0	.16	.20	.10	.01	.03	.02	.02
	Services	.0	.0	.0	.0	.0	.0	.0	.0	.0	.0	.0

Notes: Whalley estimates from Whalley (1984, Appendix D, Table D1); Harrison estimates from sources described in text.

equivalents of tariffs and NTBs in the present model, as well as the roughly comparable values used by Whalley[23] in his seven region model. Note that the NTB tariff equivalents are expressed in terms of the *incremental* protection they afford in addition to that provided by tariffs. Although the estimates are similar, several differences should be noted. Our manufacturing sector tariff rates for ASEAN are significantly lower than those used by Whalley for newly industrialising countries (NIC) and LDCs. In part this is 'compensated' by our rather higher NTB rates for manufacturing. However, Whalley draws his tariff estimates from Balassa,[24] reflecting a *simple* average of tariff rates in the early 1960s for Brazil, Chile, Malaysia and Mexico (the NIC estimates) and Pakistan and the Philippines (the LDC estimates). Our manufacturing sector tariff estimates are country specific, and are based on nominal tariff estimates for individual ASEAN countries reported elsewhere in this volume. The NTB estimates for Agriculture, forestry and fishing sectors in the present model are also significantly higher for Japan and Korea than those used by Whalley.

Domestic tax systems are often viewed as NTBs,[25] and their discriminatory features are included in the model. Asher and Booth[26] provide an excellent description of the comparative role of indirect taxes in ASEAN countries; our model reflects many of their estimates of sales, excise and foreign trade taxes.

Our model incorporates information on changes in foreign tariffs and NTBs when evaluating the impact of multilateral trade liberalisation policies. Again, this treatment contrasts sharply with that implicit in ERP analyses.

Solution procedures

A benchmark equilibrium solution for 1975 was obtained by solving 'backwards' for certain parameter values in the usual fashion. Apart from the treatment of intermediate input substitutability these procedures were standard to the literature. Mansur and Whalley, and Piggott and Whalley[27] provide general discussions of these procedures, and Kimbell and Harrison[28] discuss the calibration of models with immobile factors (i.e., our Heckscher/Ohlin assumption). Given some counterfactual policy change, we solve the model for a new equilibrium using the Factor Price Revision Rule introduced by Kimbell and Harrison.[29]

Sensitivity analysis

The policy relevance of numerical GE models, and their avowedly 'empirical' nature, render them open to casual criticism. Most economists are familiar with their underlying neoclassical structure; we are *not* therefore concerned to defend them here from criticisms based on rejection of that structure (see Section 5 for several structural criticisms that we consider important). On the other hand, criticism of the particular *empirical calibration* adopted can lead to an uninformed debate. The question is, How is one to identify the robustness of the results for some particular policy decision? Our response to this important question is to undertake a systematic sensitivity analysis of our policy simulations in Section 3. We undertake sensitivity analysis for each policy simulation with respect to three sets of elasticities: the elasticities of substitution between primary factors, the import demand elasticities, and the own price demand elasticities. Harrison (1985a, 1985b), and Harrison and Kimbell (1985) discuss the measures of uncertainty employed for these elasticities.[30]

3 Welfare and terms of trade (TOT) effects

This section reports the welfare incidence of a series of hypothetical trade liberalisation strategies. The welfare change for the private household in each region is measured by the Hicksian equivalent variation between the benchmark equilibrium (the GE solution that replicates the observed data in the base year, 1975) and the various counterfactual equilibria (the GE solution with some policy change that does not, in general, correspond to any observed historical episode). This measure is then expressed as a percentage of GDP in the base year in order to allow comparisons between regions of such diverse size.

Table 7.4 presents the welfare impacts of three separate global policies: one that eliminates tariffs in all regions, another that eliminates NTBs in all regions, and a third that eliminates tariffs and NTBs simultaneously in all regions. The 'Point estimate solution' column reports the impacts of the policy when all parameters are set equal to their respective point estimates. The remaining three columns report summary statistics for the set of counterfactual equilibria implied by our sensitivity analysis. The mean welfare impact is the average change with the prior probability density functions derived from the econometric sources of the point estimates being used to weight the results. The standard deviation of the welfare impact is similarly computed from the pdf of welfare impacts.[31] The final column reports the probability of welfare gain, obtained by numerically evaluating the (proper) pdf of welfare gains. This column provides a useful measure of the confidence one can attach to *qualitative* inferences from the model.

Table 7.4 Welfare effects of global trade liberalisation, 1975

Policies	Affected region	Point estimate solution*	Mean	Standard deviation	Probability of welfare gain
Tariffs	Indonesia	−0.56	−0.70	0.03	0.28
	Malaysia	1.93	2.65	0.04	0.94
	Philippines	−0.80	−0.98	0.04	0.16
	Singapore	1.65	2.01	0.02	0.95
	Thailand	−1.94	−1.75	0.08	0.18
	Australia	0.03	0.49	0.08	0.91
	Korea	1.34	2.34	0.07	0.98
	Japan	1.11	1.79	0.03	0.95
	USA	0.21	0.24	0.02	0.87
	Canada	0.26	0.41	0.04	0.87
	EEC	0.41	0.61	0.06	0.94
NTB	Indonesia	1.67	1.78	0.23	0.90
	Malaysia	2.41	2.36	0.39	0.79
	Philippines	2.75	2.82	0.31	0.84
	Singapore	2.87	3.01	0.41	0.96
	Thailand	4.31	4.81	0.50	0.98
	Australia	1.26	1.43	0.19	0.93
	Korea	3.59	3.73	0.46	0.76
	Japan	0.92	1.05	0.09	0.85
	USA	−0.44	−0.38	0.07	0.31
	Canada	1.82	1.96	0.13	0.96
	EEC	0.37	0.41	0.08	0.77
Tariffs & NTBs	Indonesia	1.30	1.09	0.20	0.78
	Malaysia	4.22	4.81	0.27	0.73
	Philippines	2.00	1.97	0.19	0.82
	Singapore	3.51	3.90	0.37	0.90
	Thailand	2.16	3.07	0.34	0.76
	Australia	1.31	2.04	0.11	0.90
	Korea	4.74	4.89	0.26	0.69
	Japan	2.01	2.56	0.07	0.91
	USA	−0.30	−0.19	0.06	0.40
	Canada	1.56	1.55	0.09	0.91
	EEC	0.87	1.26	0.09	0.85

* Point estimate solution' refers to the comparison of the model's benchmark equilibrium solution and counterfactual equilibrium solution when both solutions adopt the point estimate for all parameters. Values in the Table are percentages of GDP in the affected region.

NTB vs tariff liberalisation

Several features of the results in Table 7.4 are noteworthy. First, it is apparent that NTB liberalisation confers greater benefits than tariff liberalisation to each nation except the USA. Note that the benchmark year tariffs are those applying prior to the Tokyo Round multilateral tariff reductions, suggesting that our conclusion about the relative importance of liberalising NTBs is of even greater relevance for current circumstances. This conclusion is echoed in results obtained in the Whalley (1984b) *four region* GE trade policy model. Whalley (1980a)[32] concludes that:

> 'Abolition of tariff barriers alone would increase world trade by (1973) US$10 billion. While abolishing non-tariff barriers would increase trade by US$20 billion. The two sets of policies thus have a compounding effect on the level of world trade. The effects of these two areas of policy appear unequally balanced and suggests that more stress should be placed on non-tariff barriers in the current Tokyo Round negotiations'.

The simulations reported in Brown and Whalley[33] indicate that the global welfare gains from (multilateral) NTB liberalisation are roughly double those from tariff liberalisation. Whalley[34] concludes that '. . . under some assumptions the changes in the NTB codes in the (Tokyo Round) Agreement may be more significant than the tariff cuts', although quantification of the negotiated NTB reduction is particularly speculative in this case.

Table 7.5 presents the welfare impact of global trade liberalisation in the *seven region* model of Whalley (1984b). For comparison with our results in Table 7.4 we report the Hicksian equivalent variation as a percentage of base case GNP (shown in Table 7.5). In this model NTBs are only slightly more important than tariffs. Indeed, when NICs and LDCs jointly abolish tariffs and NTBs, tariffs have a significantly greater impact on welfare. Moreover, NICs and LDCs each lose substantially from unilateral trade liberalisation. Two reasons for these results in the Whalley model are: (i) the extremely high tariff rates assumed for NICs and LDCs, especially in manufacturing sectors (recall Table 7.3), and (ii) the market power enjoyed by NICs and LDCs as trading blocs. Neither of these features, are present in our model.

Uniformity effects

The results in Table 7.4 are qualitatively identical to, but proportionately larger than, those reported in Harrison, and Harrision and Kimbell[35] for *fifty percent* tariff and NTB reductions. It is not too surprising that a one hundred percent reduction in a set of distortions (such as shown in Table 7.4) generates greater welfare impacts, but what is notable is that the welfare losses incurred by some countries are less severe and the welfare gains incurred by *some* countries are more than double those reported in the other studies. These findings are easily explained—*removing* all distortions of a certain type not only reduces the size of the distortion, but it 'equalises' it (by definition, at zero) across all sectors and all regions. The latter effect is present in the results shown in Table 7.4. It is reminiscent of the emphasis that is often (correctly) placed on the variance of tariffs rather than their absolute level.[36] Corden examines this 'uniformity

Table 7.5 Welfare impact of trade liberalisation in Whalley model

Region	1977 GNP ($bill.)	Multilateral abolition by NIC AND LDC of . . .			Multilateral abolition by all regions of . . .		
		Tariffs	NTBs	Tariffs & NTBs	Tariffs	NTBs	Tariffs & NTBs
EEC	1909.0	1.06	0.38	2.00	1.27	0.24	1.86
US	2202.5	0.32	0.09	0.59	0.17	0.21	0.44
Japan	806.5	1.47	0.51	2.65	1.72	0.26	2.58
Other developed	2351.0	0.11	0.02	0.20	−0.10	0.05	−0.08
OPEC	256.2	2.26	0.27	2.73	3.79	0.78	3.24
NIC	677.3	−2.97	−0.43	−4.69	−2.52	0.78	−3.10
LDC	879.7	−1.12	−0.21	−3.22	−0.70	1.06	−2.13

Source: Derived from simulation results in Whalley (1984b, Table 10.2) and GNP values in Whalley (1984b; Table 1.2).

effect' in some detail in the context of customs union formation. In essence, it reflects the beneficial effects of an *equalisation* of a distortion offsetting the detrimental effects of any increase in the (average) absolute level of the distortion. In our simulations we are equalising the distortion *and decreasing* its (average) absolute level, each resulting in efficiency gains. The point is that we are reaping greater gains by setting the distortion uniformly (at zero) across sectors and regions than we would if we had merely lowered the absolute level of the existing (non-uniform) set of distortions to a point where the *average* distortion is zero. The welfare losses in Table 7.4 for Thailand, the Philippines and Indonesia for the *tariff* liberalisation may be due in part to this effect: in these cases a removal of tariffs, leaving NTBs in place, generally *increases* the sectoral variance of trade distortions despite decreasing their absolute level.

Symmetry

To the extent that the results for tariffs and NTBs in Table 7.4 are additive and symmetric, they support a widespread and cynical view of the last few decades of multilateral trade 'liberalisation'. Hamilton and Whalley,[37] for example, note that:

> 'Those who are pessimistic about the GATT process argue that successive GATT rounds have not produced substantive progress in limiting and reducing non-tariff barriers. They also argue that non-tariff barriers are increasingly being used as a way of offsetting reductions in protection produced by negotiated tariff cuts. A cynical view of trade liberalisation sometimes expressed is that countries participate in GATT tariff cuts hoping that partners will take the tariff cut seriously, with the reduction in own protection subsequently offset by an erection of non-tariff barriers.

This view is supported in the present model. The results in Table 7.4 are approximately additive in the sense that removing tariffs *and* NTBs together results in a pattern of welfare impacts close to those implied by adding the impacts shown for each policy separately. The joint removal of tariffs and NTBs does lead to slightly larger welfare effects than are implied by adding their separate effects (for the 'uniformity' reasons discussed above). The results for NTBs are also approximately symmetric in the sense that an *across the board* increase leads to welfare impacts that are roughly equal in size and opposite in sign to those impacts resulting from a decrease in NTBs. These approximations are much more exact if one is looking at *partial* reductions (eg. fifty percent) or increases in the particular policies rather than their complete elimination (again, due to the 'uniformity' effect).

Proportional welfare gains

The welfare impacts may seem 'small' when one considers the radical policy changes involved (viz., complete elimination of certain trade barriers for all countries). However, it should be noted that these impacts represent *annual* gains or losses that could accrue over many years to represent 'large' impacts. Moreover, the welfare gains for Malaysia (2.65%), Singapore (2.01%), Korea (2.34%) and Japan (1.79%) due to *tariff* liberalisation are hardly small. Similarly, the gains to ASEAN nations (especially Thailand) from NTB liberalisation are quite large in relation to previous estimates. Also note that many of the stated welfare impacts are extremely robust to uncertainty about the parameters of the model (i.e., high values for probability of welfare gain), even if the mean impact is deemed 'small' by some standard. In Section 5 we discuss several extensions and modifications of our model that are likely to significantly increase the welfare effects reported here.

Liberalisation of trade in manufactures

The results in Table 7.6 demonstrate how the overall welfare gains reported in Table 7.4 are reduced when one only considers the multilateral liberalisation of protection (tariffs *and* NTBs) directed at manufacturing sectors. Recall from Table 7.2 that we define 'manufacturing' as sectors 4 through 15. This implies that we *are* reducing protection, *inter alia*, on the 'textiles, clothing, footwear and leather'. This sector has, of course, been one of the holiest of sacred cows in recent GATT ordained multilateral trade negotiations, even though it has been desecrated on a bilateral basis with the distortions of the Multi-Fibre Agreement.

Table 7.6 Welfare impact of global manufacturing trade liberalisation, 1975

Policies	Impacted region	Point estimate solution	Mean	Standard deviation	Probability of welfare Gain
Tariffs & NTB	Indonesia	1.30	1.26	0.19	0.91
	Malaysia	1.89	1.72	0.28	0.82
	Philippines	1.72	1.65	0.30	0.86
	Singapore	2.41	2.44	0.34	0.93
	Thailand	1.01	1.04	0.21	0.71
	Australia	0.85	0.84	0.12	0.94
	Korea	2.80	2.85	0.31	0.75
	Japan	0.94	0.93	0.12	0.87
	USA	−0.31	−0.33	0.10	0.28
	Canada	0.91	0.91	0.10	0.97
	EEC	0.14	0.16	0.06	0.73

Note:
These results are generated by the 'best approximate' aggregate model, described in Harrison (1985a) and Harrison, G.W. and Manning, R. (1984), 'Best Approximate Aggregation of Input-Output Systems', Working Paper No. 8401, Centre for the Study of International Economic Relations, Department of Economics, University of Western Ontario.

The importance of the results in Table 7.6, when compared with those in Table 7.4, is the focus they place on agricultural protection by NTBs. Consider, for example, the large gains for Japan and Korea in Table 7.4 and their smaller gains in Table 7.6. These results for Japan and Korea reflect the efficiency gains in production from removal of distortions favouring agricultural and food products as opposed to the significant decline in national income due to changes in the domestic relative price. In each case the former effect dominates. Similar remarks also apply to Thailand, where there is also a significant worsening of the domestic relative price.

Terms of trade effects

The reliability of welfare gains from trade liberalisation in neoclassical models depends primarily on the TOT effects that result. If a region is treated as a 'small country' from the perspective of trade theory this implies that there are no TOT effects. As a consequence, trade liberalisation must encourage a more efficient allocation of resources and hence a greater level of welfare. To the extent that the TOT may move against a trade liberalising country with market power, there may be a net decrease in welfare. The empirical issue is therefore whether ASEAN nations experience significant TOT effects.

We measure TOT changes using the net barter TOT: an improvement occurs when the weighted price of a country's exports increases relative to the weighted price of a country's imports. This measure is also adopted by Whalley,[38] who further employs base case domestic *production* levels to construct the relevant indices for a given country. He chooses production weights, rather than consumption weights, to avoid capturing the effects of direct tariff changes. This choice may appear to be of subsidiary technical significance, but is substantively important. In the Whalley models domestic tariffs and NTBs do not directly increase the prices that domestic producers must pay for intermediate input purchases: their 'legal' incidence is solely on final consumers. In our model the legal incidence of import tariffs and NTBs depends on the (empirical) share of those imports purchased by producers and consumers for intermediate and final consumption, respectively. Producers import significant quantities of manufactured goods for intermediate purposes, typically between 30 and 60 percent of all imports of a given commodity *whatever* the accounting conventions. The crucial implication of this feature of our model for TOT measurement is that domestic production (as well as domestic consumption) will also capture the effects of tariff/NTB changes. We therefore adopt base case import and export quantities as weights to evaluate TOT effects. Although this is consistent with accepted practice in the literature,[39] it is important for the interpretation of our results.

Table 7.7 presents the TOT effects of all of the trade liberalisation schemes considered here that abolish tariffs and NTBs. These effects are shown in percentage form, and correspond to the mean TOT change when we allow for parameter uncertainty. The large trading regions experience significant TOT changes when all global trade is liberalised, with Japan and the EEC experiencing large gains. A notable result is the deterioration of the US terms of trade, reflecting the market power that the US foregoes when it agrees to a multilateral liberalisation. Whalley[40] reports TOT gains of 12.7 percent, 1.5 percent and 23.1 percent for the EEC, US and Japan, respectively, for the same policy. Our TOT effects for EEC and Japan are similar to these and a little larger (Whalley reports a TOT loss of 3.9 percent for the US when it unilaterally abolishes its tariffs). These TOT gains and losses deteriorate as we consider less comprehensive trade liberalisation schemes in terms of sectoral and geographic coverage.

The ASEAN nations experience a noticeable deterioration of the TOT effects as we move from liberalising all global trade to liberalising only manufacturing trade. Note that the TOT effects for ASEAN nations are generally quite small in absolute terms. However, the implications of these for domestic production structure can be significant if the domestic 'production possibilities curve' is relatively flat. This curve reflects the domestic opportunity cost of producing one bundle of goods rather than another assuming that each is produced efficiently. A generic feature of our model is that this curve

Table 7.7 Terms of trade effects of alternative trade liberalisation schemes, 1975

Affected region	All global trade (Table 7.4)	Global manufacturing (Table 7.6)	ASEAN manufacturing (Table 7.8)	ASEAN and Australia manufacturing (Table 7.9)
Indonesia	0.04	−0.06	0.01	0.01
Malaysia	0.12	−1.10	0.03	0.02
Philippines	0.03	−0.01	0.02	0.02
Singapore	0.01	0.05	0.02	0.01
Thailand	0.01	−0.02	−0.01	0.00
Australia	4.12	−0.06	−0.02	0.03
Korea	8.60	2.46	0.00	0.00
Japan	29.11	8.75	0.00	0.00
USA	−1.46	−0.86	0.00	0.00
Canada	4.98	1.00	0.00	0.00
EEC	22.31	6.04	0.00	0.00

Notes:
Estimates shown are mean percentage TOT effects, corresponding to the mean welfare effect shown in the indicated Tables; all policies consider liberalisation to tariffs and NTBs.

is quite flat. The source of this feature is the importance in the model of domestic and international trade in intermediate goods. Even though individual sectors may have widely differing *direct* primary factor intensities for given factor prices, the *embodied* primary factor intensities are relatively alike across sectors. To the extent that the production possibility curve reflects embodied rather than direct factor intensities, it will therefore be relatively less concave to the origin (ie, it will be flatter). Thus 'small' TOT effects can still have 'large' effects on the efficient structure of domestic production and hence on welfare (given the underlying set of production possibilities). This point is important when comparing the TOT effects in Table 7.7 with the welfare effects in other tables for ASEAN and Australia.

Welfare effects in ASEAN

Comparing the welfare effects on ASEAN nations in Tables 7.4 and 7.6, we note that virtually each nation does not gain as much in welfare terms from a global trade liberalisation that is concentrated solely on manufacturing sectors. Indonesia is an apparent exception to this intuitive result: her mean welfare gain *increases* from 1.09 percent in the across the board liberalisation (Table 7.4) to 1.26 percent in the manufacturing sector only liberalisation (Table 7.6). However this counter-intuitive result can also be seen to be statistically insignificant since the standard deviation in each case is around 0.2 percent. Thus the reported increase of 0.17 percent in Indonesian welfare is less than one standard deviation from either reported mean. We conclude that there is no significant difference between the results for Indonesia, and that welfare gains to all other ASEAN nations are significantly reduced by the more restrictive liberalisation.

Liberalisation in ASEAN and Australia

Table 7.8 reports the welfare impacts of liberalisation of manufacturing sector tariffs and NTBs in all ASEAN nations and also in ASEAN *and* Australia. Not surprisingly, the welfare impacts on the larger trading regions (Japan, USA and the EEC) are trivial. Korea and Canada are, however, adversely affected along the lines of familiar 'trade diversion' arguments from the customs union literature, even though the liberalisations in Table 7.8 are not designed to be discriminatory.

Compare the results for the two regional liberalisation strategies in Table 7.8 with those for global liberalisation in Table 7.6. Indonesia and the Philippines clearly do not gain as much in the regional liberalisation exercise, due to the relatively small share of their trade with ASEAN ('small' in absolute terms and in comparison with other ASEAN countries). It is interesting that Australia loses from an intra-ASEAN trade liberalisation and gains significantly if it also liberalises. Australia's gain appears to be at the expense of Singapore and largely involves trade in manufactured intermediates rather than final consumer goods. Apart from Singapore, each ASEAN nation gains (or does not lose significantly) from a broadening of the liberalisation to include Australia.

One final point to note about the regional liberalisation policies is that we would not expect any across the board retaliatory action from any of the other trading regions given the trivial effects on their *welfare*. On the other hand, it is possible that certain sectors (e.g., fishing and agriculture in Korea and Japan) might be affected more seriously, eliciting some sector specific response. Further, our results do not address the likelihood of a retaliatory response to any (geographically) *discriminatory* regional liberalisation scheme such as would be involved in the formation of a customs union.

4 Comparisons with effective protection results

In Section 1 we reviewed the familiar arguments against the use of ERPs to predict the pattern of primary resource pulls due to trade distortions. We argued that the various modifications and extensions to basic ERP calculations that are required to produce acceptable predictions virtually lead one to construct a 'de facto' applied GE

Table 7.8 Welfare effects on regional manufacturing trade liberalisation, 1975

Policies	Liberalising nations	Affected region	Point estimate solution	Mean	Standard deviation	Probability of welfare gains
Tariffs and NTB	ASEAN nations	Indonesia	0.7	0.7	0.004	0.94
		Malaysia	0.9	1.1	0.01	0.91
		Philippines	0.2	0.2	0.01	0.80
		Singapore	1.0	1.1	0.01	0.87
		Thailand	0.9	0.8	0.005	0.91
		Australia	−0.2	−0.2	0.001	0.31
		Korea	−0.01	−0.004	0.0	0.15
		Japan	−0.003	−0.002	0.0	0.12
		USA	−0.001	0.0	0.0	0.29
		Canada	−0.03	−0.02	0.001	0.26
		EEC	0.0	0.0	0.0	0.31
Tariffs and NTB	ASEAN and Australia	Indonesia	0.7	0.7	0.003	0.95
		Malaysia	1.0	1.0	0.01	0.93
		Philippines	0.3	0.03	0.01	0.84
		Singapore	0.7	0.7	0.01	0.81
		Thailand	0.6	0.6	0.004	0.90
		Australia	0.6	0.5	0.01	0.90
		Korea	−0.02	−0.002	0.0	0.13
		Japan	−0.002	−0.003	0.0	0.10
		USA	−0.002	−0.001	0.0	0.26
		Canada	−0.07	−0.02	0.002	0.13
		EEC	0.0	0.0	0.0	0.32

Notes:
A value of '0.0' indicates that the computed value was not significant to four decimal places (i.e. it is less than 0.00005 in absolute value). All computations were undertaken at much higher levels of accuracy.

model of the type discussed in the previous two sections. For this reason alone one might prefer to use a GE model to predict resource pulls. This is quite apart from the inability of ERP analysis to generate any measures of the welfare effects of trade distortions, which is a more common measure of their incidence.

In this section we provide a constructive and illustrative demonstration of the potential difficulties of using ERP calculations. We employ the data base assembled for our GE model in a standard, unadorned, ERP exercise and compare the results with those obtained using the full GE model. This comparison deliberately employs the same data base, in order to focus on the role of model structure on the results. Harrison presents further ERP calculations for ASEAN with this data base. Evans and later, Meltzer[41] report similar comparative exercises with multisectoral Johansen type models for a single country. They reach essentially the same qualitative conclusions on the comparative reliability of the standard ERP methodology.

The policy exercise we will consider is the complete multilateral abolition of global tariffs and NTBs, the GE results of which were

Table 7.9 Comparison of ERP and GE calculations of resource pulls of protection in ASEAN and Australia

	ASEAN			Australia		
Sector	Basic ERP	Modified ERP	GE	Basic ERP	Modified ERP	GE
Agriculture, forestry & fishing	15.7	0.3	−13.5	6.9	12.8	−4.9
Mining & quarrying	−10.1	−9.3	−6.4	−78.1	−12.0	−15.2
Manufacturing	69.8	46.3	5.9	49.5	23.1	7.4
Utilities	1.2	−0.9	0.5	4.3	−0.1	1.8
Construction	−11.0	−8.3	−1.1	−4.9	−7.1	−0.9
Trade & transport	−1.5	−4.2	1.2	0.1	0.9	0.5
Services	−5.7	−0.8	−3.2	−4.6	−0.9	−1.8

Note:
GE results reported here refer to the point estimate solution (as are the free trade coefficients and prices used in the modified ERP calculations) for global trade liberalisation.

presented in Table 7.4. This is a rather adventurous policy to consider with an ERP calculation and it involves removal of the trade distortions of countries other than the one whose resource pulls are being considered. This is no major problem here given the availability of a complete international IO table detailing trade in intermediates between each country. The second adventurous aspect is that if ever a liberalisation policy were likely to change the base case input-output coefficients, this is the policy to do it. We may overcome this problem here by using the free trade input-output coefficients determined endogenously as a by-product of the counterfactual simulation of the GE model in a 'modified ERP' calculation. Clearly this begs the question of how an ERP analyst would arrive at these free trade coefficients without a GE model, but it does give our 'straw man' ERP calculation its best shot at approximating the GE calculation.[42]

Table 7.9 presents the results of our comparative calculations for ASEAN as a whole and Australia. The results for ASEAN were obtained initially for each country and then summed with appropriate weights. Several features of these results are noteworthy. First, the ERP values (however calculated) are smaller in absolute and relative terms than those reported elsewhere in this volume for *unilateral* tariff reductions in *individual* nations. This difference is almost solely attributable to our inclusion of partner country tariff reductions in a multilateral liberalisation context *and* the explicit treatment of intermediate trade. Second, the modified ERP results are 'closer' to the GE results than to the ERP results. This suggests that allowing for substitutability in intermediate trade may be an important modification to basic ERP analyses. Third, the GE resource pulls are often significantly lower in absolute terms than either ERP estimate. This is largely attributable to the inclusion of supply side constraints on the expansion or contraction of certain sectors (ie, factor price effects): also demand considerations will also play some role here. Finally, we note that the ERP calculations are generally in the *wrong direction* when predicting the resource pull for agriculture, forestry and fishing. They also significantly *overstate* the protective effect of trade distortions on manufacturing (although here one does need a more disaggregated analysis). Moreover, the ERP rankings of resource pull only agree with the GE rankings with respect to one 'end-point' (or sector of extreme impact): the resource pull into manufacturing due to trade distortions. The other 'end point' sectors of ASEAN from the ERP calculations, construction or mining and quarrying, do not correspond to the sector which is most hurt by protection in the GE model (agriculture, forestry and fishing). It is important to note that we are implicitly presuming throughout that our GE calculations are better predictors of the 'true' resource pull, and we therefore judge the performance of the ERP calculations by how closely they approximate *our* results.

5 Alternative formulations

In this section we indicate several alternatives to the neoclassical GE model presented and applied in the previous three sections. It is our firm belief that numerical GE models are only beginning to reach the point at which they can be used to indicate policy relevant effects of trade policies. Given the 'ad hoc' nature or inadequacy of alternative quantitative techniques currently in use, we urge careful consideration of the issues noted below in future modelling work.

Rent seeking

Apart from the direct losses incurred by barriers to trade, such as tariffs and quotas, many developing countries incur an additional loss in terms of resource use in rent seeking under an import quota system. Under a tariff system the revenues are collected and, in principle, redistributed by the government to different sectors of the economy. Apart from any distortionary effects that might be created by particular redistribution schemes, there are no losses other than the above mentioned consumer and producer losses. It is solely these losses (the 'Harberger Triangles') that are captured in the welfare impacts reported in Section 3. Under an import licensing scheme the revenues from a restrictive trade policy actually accrue to the licensee *but could be totally lost* in competition to obtain the licence and hence the associated rents. This is known as the rent seeking process. One example would be when licences are allocated in proportion to firms' capacity. In this case investments in excess capacity would increase the probability of receiving the licence. Krueger[43] has shown that under the assumption that rent seeking activities are perfectly competitive, the value of the waste of resources devoted to rent seeking will equal the value of the rents of the import licences, that is the revenues from the restrictive trade policy. These are known colloquially as Krueger Rectangles. Hamilton, Mohammad and Whalley[44] have incorporated competitive rent seeking activities for developing countries in their global numerical GE model.

The formalisation of such wasteful rent seeking activity should ideally be based on true resource use, which of course could be different for different kinds of rent seeking. By assuming that virtually all rent seeking is performed as investments in excess capacity, the rent seeking activity would use the same factor proportions as the licensee's activities. Hamilton, Mohammad and Whalley approximate this behavioural assumption by reducing the factor endowments of the importing region proportionally, with the value of factor loss equalling the value of tickets. In the same paper it is shown that large rent seeking costs will have a significant impact on the results of simulating the model.

An important theme to emerge from these early numerical results is that Krueger Rectangles appear to be *much larger* than Harberger Triangles. In other words, the welfare losses from imposing quotas are much larger when we take into account the resource waste that is involved in seeking the rents accruing to these quotas than when we just measure the traditional efficiency loss due to resources being misallocated as a result of the quota induced distortion. In the traditional analysis factors are always employed in some gainful activity; this may not be in the *most* gainful activities possible. In the newer rent seeking analysis, however, resources are not used in *any* gainful employment. It is not surprising, therefore, that the latter effects of trade distortions often swamp the former traditional effects.

On the other hand, one should not casually infer that allowing for such rent seeking activities will always lead to greater welfare gains due to trade liberalisation. It is perfectly possible for *tariff* liberalisation to *reduce* welfare in the presence of quotas and rent seeking, by increasing the value of the rents accruing to the quota (and hence encouraging greater allocation of resources of rent seeking activities).

In any case, it is apparent that more scholarly attention should be devoted to estimating the extent and nature of rent seeking activities. Hamilton[45] illustrates the type of data collection and analysis that is needed for ASEAN countries.

Economies of scale

Pearson and Ingram[46] evaluate the potential gains from economic integration in Ghana and Ivory Coast, emphasising the importance of unrealised internal economies of scale under existing protection schemes in developing countries. Their approach closely follows Corden,[47] who incorporated economies of scale into customs union theory and suggested cost reduction and trade suppression effects in addition to the standard Vinerian trade creation and trade diversion effects. The effect is sometimes called 'fragmentation of production' in LDCs and considers a situation in which several firms produce similar products at scales of output less than the level of long run minimum average costs. Fragmentation of production could, for example, be expected under certain forms of import licensing, leading to rent seeking behaviour. Krueger[48] states that a pro rata allocation of licences in proportion to the applications for those licences would result in a larger than optimal number of firms operating on the downward sloping portion of their cost curves.

In the terminology of Harris[49] the intra-industry adjustments from changing trade barriers under imperfect competition would be quite strong and might shift the comparative advantage for a country to such an extent that traditional analysis of changing trade patterns might be incorrect. Harris also introduces non-competitive industries in a general equilibrium framework for Canada.[50] Two pricing hypotheses are used: the monopolistic competition hypothesis using a Lerner rule to set prices conditional on an elasticity of a perceived demand curve, and the Eastman-Stykolt hypothesis where domestic firms determine their prices from a collusive focal point, which is provided by the world price of the foreign competing good plus a markup of the domestic tariff. The pricing rule in the model is a linear combination of these two hypotheses with exogenously given weights. Harris concludes that introducing scale economies does change the results from a cut in tariffs—there will be significant efficiency gains from rationalisation of manufacturing industries. These welfare gains, however, are extremely sensitive to the pricing hypothesis adopted.[51]

Labour market distortions

Most developing countries are characterised by high urban unemployment rates and/or under utilisation of labour in the agricultural sector. It is therefore important to introduce in the general equilibrium framework a labour market formulation that allows for modifications of traditional neoclassical assumptions to take these phenomena into account.

A situation with under utilisation of labour is most often referred to as a 'surplus labour' economy and was first discussed by Lewis.[52] It has been developed further by Ranis and Fei (1961), Fei and Ranis (1964) and Dixit (1971),[53] *inter alia*. Lewis was mainly concerned with analysing the agricultural sector of the economy, while later authors have analysed the interactions between that sector and the industrial sector.

Urban unemployment was analysed in Harris and Todaro[54] by the introduction of a fixed wage in the urban industrial sector, with migration from the countryside depending on the relationship between the actual wage in agriculture and the expected wage in urban industry. The basic assumption is that there is full employment in rural agricultural areas and no imperfections in rural labour markets. Expectations of urban wages differ from the actual wages because of the probability of becoming unemployed after moving to the urban area. The Harris-Todaro analysis has been the basis for further studies of first best and second best policy alternatives for reaching full employment and/or maximum social welfare. Among the most notable contributions in this area are Bhagwati and Srinivasan and Blomqvist.[55] Blomqvist and McMahon[56] simulate different policy alternatives in a numerical general equilibrium model for Kenya with Harris-Todaro features.

6 Summary

A numerical general equilibrium model of the Pacific Basin is employed to study the welfare effects on ASEAN and Australia of alternative manufacturing trade liberalisation strategies. The model indicates that ASEAN and Australia would generally benefit from strategies that are more regional in focus as well as the GATT-type multilateral strategy. Moreover, these benefits are shown to be robust to the particular empirical calibration of the model. In other words, reasonable uncertainty about certain key elasticities used in the model does not appear to imply significant uncertainty about the direction of welfare change due to any particular policy. We also demonstrate that popular ERP calculations of the resource pull of trade liberalisation policies can generate seriously misleading conclusions (at least relative to a complete GE calculation). Finally, we argue that significant reformulations of the *qualitative* structure of existing general equilibrium models are necessary to address many trade policy issues of importance to ASEAN and Australia. These include an allowance for resource wasting rent seeking activities in the context of nontariff trade barriers, allowance for economies of scale in manufacturing sector production in small economies, and the implications of alternative formulations of labour markets in developing economies.

Endnotes

* We are grateful to the Reserve Bank of Australia, the ASEAN-Australia Economic Relations Research Project, and the Social Sciences and Humanities Research Council of Canada for support at various stages of this project. We are also grateful to Christopher Findlay, Augustine Tan, an anonymous referee, and participants at a workshop on Trade and Growth in Pacific Asia (Centre for Pacific Asia Studies, Stockholm University, April 1985) and seminars at the Asian Development Bank (Manila), the Center for Strategic and International Studies (Jakarta), the East-West Center (Honolulu), the Institute for South East Asian Studies (Singapore), the University of Malaya (Kuala Lumpur) and the University of the Philippines (Manila) for helpful comments.

1 Harrison, G.W. (1984a), 'Economic Interdependence Between ASEAN and Australia: A General Equilibrium Approach', Working Paper, ASEAN-Australia Economic Relations Research Project, Australian National University.

2 Harrison, G.W. and Kimbell, L.J. (1985), 'Economic Interdependence in the Pacific Basin: A General Equilibrium Approach', in J. Piggott and J. Whalley (eds), *New Developments in Applied General Equilibrium Analysis*, New York: Cambridge University Press; Harrison, G.W. (1984a); and Harrison, G.W. (1985a), 'A General Equilibrium Analysis of Tariff Reductions', in T.N. Srinivasan and J. Whalley (eds), *General Equilibrium Trade Policy Modelling*, Cambridge: MIT Press.

3 Corden, W.M. (1966), 'The Structure of a Tariff System and the Effective Protection Rate', *Journal of Politcal Economy*, vol. 74 (June), pp. 221–37.

4 Bhagwati, J.N. and Srinivasan, T.N. (1973a), 'The General Equilibrium Theory of Effective Protection and Resource Allocation', *Journal of International Economics*, vol. 3, August, pp. 259–282. See also Tan, A.H.H. (1970), 'Differential Tariffs, Negative Value-Added and the Theory of Effective Protection', *American Economic Review*, vol. 60, pp. 107–116.

5 Whalley, J. (1980b), 'Discriminatory Features of Domestic Factor Tax Systems in a Goods Mobile—Factors Immobile Trade Model: An Empirical General Equilibrium Approach', *Journal of Political Economy*, vol. 88, p. 1185, notes that significant computational expense involved in solving his 33 sector model when CES intermediate input substitutability is allowed. Indeed the additional expense forces him to aggregate his model to five sectors. Contrast Whalley, J. (1980a), 'General Equilibrium Analysis of US-EEC-Japanese Trade and Trade Distorting Policies', *Economic Applique*, vol. 33. The use of Cobb-Douglas intermediate input substitutability was apparently first developed by Boadway, R. and Treddenick, J. (1978), 'A General Equilibrium Computation of the Effects of the Canadian Tariff Structure', *Canadian Journal of Economics*, vol. 11, p. 430ff. Harrison, G.W. (1984a) provides a formal statement of this approach and its benchmark calibration.

6 Whalley, J. (1980a, 1980b); Whalley, J. (1982), 'An Evaluation of the Tokyo Round Trade Agreement Using General Equilibrium Computational Methods', *Journal of Policy Modelling*, vol. 3; Whalley, J. (1985b).

7 Brown, F. and Whalley, J. (1980), 'General Equilibrium Evaluations of Tariff-

Cutting Proposals in the Tokyo Round and Comparisons with More Extensive Liberalization of World Trade', *Economic Journal*, vol. 90, pp. 838–66.

8 Whalley, J. (1984a), 'The North-South Debate and the Terms of Trade: An Applied General Equilibrium Approach', *Review of Economics and Statistics*, vol. 66; and Whalley, J. (1984b), *Trade Liberalization Among Major World Trading Areas: A General Equilibrium Approach*, Cambridge: MIT Press.

9 Institute of Developing Economies (1982), *International Input-Output Table for ASEAN Countries: 1975*, IDE Statistical Data Series no. 39, Tokyo: Asian Economic Press; and Harrision, G.W. (1984b), 'Structural Interdependence Between ASEAN and Australia: An Input-Output Approach', Working Paper, ASEAN-Australian Economic Relations Research Project, Australian National University.

10 Harrison, G.W. and Kimbell, L.J. (1985).

11 Lluch, C. (1973), 'The Extended Linear Expenditure System', *European Economic Review*, vol. 4, pp. 21–32; and Howe, H.J. (1975), 'Development of the Extended Linear Expenditure System', *European Economic Review*, vol. 6, pp. 305–10, advance alternative interpretations of the formal household problem leading to the ELES; see also Lluch, C., Powell, A.A. and Williams, R.A. (1977), *Patterns in Household Demand and Saving*, New York: Oxford University Press, p. 14.

12 In several cases, given the level of commodity aggregation adopted, this level of the utility tree is redundant (e.g., the grouping 'clothing' includes only one commodity from Table 7.2, 'textiles, clothing, footwear and leather') or ambiguous (e.g., the commodity 'Services' in Table 7.2 is allocated to two groupings, 'Recreation' and 'Other services').

13 Harrison, G.W. and Kimbell, L.J. (1985); Lluch, C., Powell, A.A. and Williams, R.A. (1977).

14 Harrison, G.W. (1985a); and Harrison, G.W. (1985b), 'Sensitivity Analysis and Trade Policy Modelling', in T.N. Srinivasan and J. Whalley (eds), *General Equilibrium Trade Policy Modelling*, Cambridge: MIT Press, provides details on the data sources of the estimates used, as well as a description of the various taxes included in the model. Given the present focus on trade policy, we do not discuss the treatment of taxes in any detail.

15 Institute of Developing Economies (1982); Harrison, G.W. (1984b).

16 Unless, of course, all of those imports were directed solely to the corresponding domestic sector (i.e., the off diagonal *elements* of the off diagonal trade *blocks* in our international IO table are *all* zero). This is not the case in the present IO table.

17 Whalley, J. (1980b), Table 2.

18 Whalley, J. (1984b), Table D1.

19 Whalley, J. (1980a); and Yeats, A.J. (1979), *Trade Barriers Facing Developing Countries*, New York: St. Martin's Press.

20 Saxon, E. and Anderson, K. (1982), 'Japanese Agricultural Protection in Historical Perspective', Research Paper no. 92, Australia-Japan Research Centre, Research School of Pacific Studies, Australian National University; and Anderson, K. (1981), 'Northeast Asian Agricultural Protection in Historical and Comparative Perspective: The Case of South Korea', Research Paper no. 82, Australia-Japan Research Centre, Australian National University.

21 Tyers, R. and Chisholm, A. (1982), 'Agricultural Policies in Industrialized and Developing Countries and International Food Security', unpublished manuscript, Research School of Pacific Studies, Australian National University, Table A1.

22 Whalley, J. (1982).

23 Whalley, J. (1984b).

24 Balassa, B.A. (1971), *The Structure of Protection in Developing Countries*, Baltimore: Johns Hopkins University Press, Table 3.1.

25 See Lloyd, P.J. (1983), *Non-Tariff Distortions of Australian Trade*, Canberra: Australian National University, Chapter 7.

26 Asher, M.G. and Booth, A. (1983), *Indirect Taxation in ASEAN*, Singapore: Singapore University Press.

27 Mansur, A. and Whalley, J. (1984), 'Numerical Specification of Applied General Equilibrium Models: Estimation, Calibration and Data', in H.E. Scarf and J.B. Shoven (eds), *Applied General Equilibrium Analysis*, New York: Cambridge University Press, Section 3; and Piggott, J. and Whalley, J. (1985), *Economic Effects of U.K. Tax-Subsidy Policies: A General Equilibrium Appraisal*, New York: Cambridge University Press, Chapter 4.

28 Kimbell, L.J. and Harrison, G.W. (1984), 'General Equilibrium Analysis of Regional Fiscal Incidence', in H.E. Scarf and J.B. Shoven (eds), *Applied General Equilibrium Analysis*, New York: Cambridge University Press, Section 3.2.

29 Kimbell, L.J. and Harrison, G.W. (1983), 'On the Solution of General Equilibrium Models', Working Paper no. 8301C, Centre for the Study of International Economic Relations, Department of Economics, University of Western Ontario, forthcoming *Economic Modelling*.

30 Harrison, G.W., Jones, R., Kimbell, L.J. and Wigle, R. (1985), 'How Robust is General Equilibrium Analysis?', Working Paper no. 8501, Centre for the Study of International Relations, University of Western Ontario, distinguish between 'conditional' and 'unconditional' systematic sensitivity analyses. The former refers to a series of simulations in which each parameter is perturbed from its point estimate a certain number of times (four in the present case) *conditional* on all other parameters being set *only* to their point estimate value. The latter refers to perturbations of each parameter conditional on all other parameters also being perturbed from their point estimate a certain number of times; thus the set of simulations is 'unconditional'. Clearly the latter type of analysis is more complete than the former, but at a severe cost in terms of the number of required simulations. Given the size of the present model and the large number of parameters subject to perturbations (252 in the seven sector model and 720 in the twenty sector model), we have opted for the conditional systematic sensitivity analysis. We shall consider five values for each parameter, including the point estimate. Thus we have four perturbations for each parameter. Two of these perturbations will be one half of a standard error above and below the point estimate. The exact marginal probabilities for these values depend on the relevant degrees of freedom for the parameter estimate; where we are unable to infer that value from published data it is assumed large enough for asymptotic results to hold. We therefore require 1009 simulations *for each policy change* in the seven sector model (252 relevant parameters times 4 perturbations per parameter, plus one simulation with all parameters equal to their point estimate), and 2881 simulations in the twenty sector model. In all cases reported in Section 3 we initially solved the seven sector model for the given policy change with all parameters set equal to their point estimates. The solution vector of relative factor prices (containing 23 elements) was then employed as starting values for the twenty sector model. Given the solution values for this simulation as starting values for the sensitivity analysis simulations involving a perturbed elasticity, we were able to find the new solution values extremely quickly.

31 Note that there is absolutely no presumption that the policy impacts (welfare impacts in this case) have a Gaussian distribution. One should not therefore assume that the mean and standard deviation are in any way 'sufficient statistics' of that distribution. In fact, many of the policy impact pdf's encountered elsewhere are slightly skewed.

32 Whalley, J. (1984b); and Whalley, J. (1980a), p. 224.

33 Brown, G. and Whalley, J. (1980), Table 14, Experiments 6 and 10.

34 Whalley, J. (1982), p. 360.

35 Harrison, G.W. (1985a); Harrison, G.W. and Kimbell, L.J. (1985).

36 See Cooper, R.W. (1964), 'Tariff Dispersion and Trade Negotiations', *Journal of Political Economy*, vol. 72, pp. 597–603; and Corden, W.M. (1976), 'Customs Union, Theory and the Nonuniformity of Tariffs', *Journal of International Economics*, vol. 6, pp. 99–106.

37 Hamilton, Bob and Whalley, J. (1983), 'Geographically Discriminatory Trade Arrangements', Working Paper no. 8321C, Centre for the Study of International Economic Relations, University of Western Ontario, p. 3 (forthcoming: *Review of Economics and Statistics*).

38 Whalley, J. (1984b).

39 See for example, Allen, R.G.D. and Ely, J.E. (eds), (1953), *International Trade Statistics*, New York: Wiley.

40 Whalley, J. (1984b), Table 11.1.

41 Harrison, G.W. (1984b); Evans, H.D. (1971), 'The Empirical Specification of a General Equilibrium Model of Protection in Australia', in H.E. Grubel and H.G. Johnson (eds), *Effective Tariff Protection*, Geneva, GATT; and Meltzer, B.D. (1980), 'A Comparison of Some Effective Rate of Protection Calculations with the Results of a General Equilibrium Model', General Paper No. G-26, IMPACT Project Research Centre, Melbourne.

42 We compute the ERP for sector j, g_j, using the well known formulae:

$$g_j = \frac{v_j' - v_j}{v_j}$$

$$= \frac{(1 + t_j) - \sum_{i=1}^{n} a_{ij} (1 + t_i) - 1 + \sum_{i=1}^{n} a_{ij}}{1 - \sum_{i=1}^{n} a_{ij}}$$

$$= \frac{t_j - \sum_{i=1}^{n} a_{ij} t_i}{1 - \sum_{i=1}^{n} a_{ij}},$$

where v_j is the free trade (or counterfactual) value added of sector j, v_j' is the distorted (or base case) value added for sector j, t_j is the nominal *ad valorem* tariff equivalent of the trade distortion applying directly to sector j, and a_{ij} are the free trade (counterfactual) input-output coefficients defined as

$$a_{ij} = \frac{X_{ij}p_i}{X_j p_j}$$

where X_{ij} is the amount of commodity i used to produce commodity j, X_j is output of sector j, and p_j is the free trade (counterfactual) price of sector j output. Our 'basic ERP' calculation employs *base case* distorted values for each a_{ij} and p_j. Our 'modified ERP' calculation employs, by courtesy of the GE model, the counter factual free trade values for each a_{ij} and p_j. The GE model simply evaluates g_j directly in terms of the solution values for v_j^1 and v_j. All results are presented in percentage form (i.e., we report g_j times 100). The GE model evaluates g_j directly in terms of the first expression given above, whereas the ERP calculations evaluate it using the third expression.

43 Krueger, A.O. (1974), 'The Political Economy of the Rent-Seeking Society', *American Economic Review*, vol. 64, pp. 291–303.

44 Hamilton, Bob, Mohammad, S. and Whalley, J. (1983), 'Rent Seeking and the North-South Terms of Trade', unpublished manuscript, Department of Eco-

nomics, University of Western Ontario, April (forthcoming, *Journal of Policy Modelling*).

45 Hamilton, C. (1984), 'Voluntary Export Restraints on Asia: Tariff Equivalents, Rents and Trade Barrier Formation', *Weltwirtschaftliches Archiv*, forthcoming.

46 Pearson, S.R. and Ingram, W.D. (1980), 'Economies of Scale, Domestic Divergences, and Potential Gains from Economic Integration in Ghana and the Ivory Coast', *Journal of Political Economy*, vol. 88, pp. 994–1007.

47 Corden, W.M. (1972), 'Economies of Scale and Customs Union Theory', *Journal of Political Economy*, vol. 80, pp. 465–75.

48 Krueger, A.O. (1974).

49 Harris, R. (1985), 'Market Structure and Trade Liberalization: A General Equilibrium Assessment', in T.N. Srinivasan and J. Whalley (eds), *General Equilibrium Trade Policy Modelling*, Cambridge: MIT Press.

50 Also see Harris, R. (1984), 'Applied General Equilibrium Analysis of Small Open Economies with Scale Economies and Imperfect Competition', *American Economic Review*, vol. 74; and Cox, D. and Harris, R. (1984), 'Trade Liberalization and Industrial Organisation: Some Estimates for Canada', *Journal of Political Economy*, vol. 92.

51 See Harrison, G.W. (1985b); and Harrison, Jones, Kimbell and Wigle (1985), for further discussion of this point.

52 Lewis, W.A. (1954), 'Economic Development with Unlimited Supplies of Labor', *The Manchester School*, vol. 22, pp. 139–91.

53 Ranis, G. and Fei, J. (1961), 'A Theory of Economic Development', *American Economic Review (Papers and Proceedings* (11), vol. 51; Fei, J. and Ranis, G. (1964), *Development of the Labour-Supply Economy: Theory and Policy*, Homewood: Irwin; and Dixit, A. (1971), 'Short-Run Equilibrium and Shadow Prices in the Dual Economy, *Oxford Economic Papers*, pp. 384–99.

54 Harris, J.R. and Todaro, M.P. (1970), 'Migration, Unemployment and Development: A Two-Sector Analysis', *American Economic Review*, vol. 60.

55 Bhagwati, J.N. and Srinivasan, T.N. (1973b), 'The Ranking of Policy Interventions Under Factor Market Imperfections: The Case of Sector-Specific Sticky Wages and Unemployment', *Sankhya*, Ser. B; and 'On Reanalyzing the Harris-Todaro Model: Policy Rankings in the Case of Sector-Specific Sticky Wages', *American Economic Review*, vol. 64, June.

56 Blomqvist, A.G. and McMahon, G. (1983), 'Simulating Commercial Policy in a Small, Open Dual Economy with urban Unemployment: A General Equilibrium Approach', Working Paper no. 8403C, Centre for the Study of International Economic Relations, University of Western Ontario (forthcoming: *Journal of Development Economics*).

8 Import quotas and voluntary export restraints: focusing on exporting countries

CARL HAMILTON*
1984

THE objective of this paper is to analyse the effects of preferential import quotas and 'voluntary' export restraints, focusing on the effects on the exporting countries. The presentation is meant to start from the level just above that of the basic texts such as Ethier (1983). It stresses that the effects of bilateral arrangements are not limited to the two contracting parties. These trade barriers are of interest since they are possibly the two most rapidly spreading forms of barriers to trade. Clothing and footwear exports from developing countries are especially affected, as is trade in road motor vehicles, colour television sets, video cassette recorders, and steel.

The typical voluntary export restraint (VER) is an instrument designed to meet the demand of the importing home country to protect its own industry by imposing 'an upper limit' on foreign supply to its market (contraint I), defined by 'source' (constraint II), defined by 'commodity group' (constraint III), and often also defined in terms of import 'volume' rather than value (constraint IV). The voluntary export restraint is administered by the exporting country. It is not voluntary in the normal sense of the word but nevertheless often preferred by the exporting countries to other, alternative, trade barriers that importing countries may use. A country restrained by a VER may or may not have to compete with unrestrained suppliers from other countries.

The preferential import quota is a measure characterised by the same four constraints as an export restraint, but it is administered by

* Carl Hamilton is with the Institute for International Economic Studies, Stockholm. The paper was written while the author was Visiting Fellow at the Economics Department of the Research School of Pacific Studies, Australian National University, Canberra.

the importing home country. Preferential import quotas are used to a lesser extent than export restraints, but Australia, France, Italy, Norway, Spain, and many developing countries employ this form of nontariff barrier.

The literature on VERs and import quotas has been concerned mainly with possible effects in the importing home country and deals only in a limited way with the effects in exporting countries and in the rest of the world.

Throughout this paper it has been assumed that import quotas or export restraints are the only trade barriers and the question of why the nontariff barriers are imposed in the first place has been ignored. The analysis extends only to the case of a single homogeneous product and assumes that protection is vertically integrated, so as to avoid the effective protection complication.[1] There is no discussion of distortions in production and consumption caused by a volume quota or by the fact that several goods are substitutes in production and consumption (constraint IV).[2]

The paper is organised as follows. In the next section the literature is surveyed very briefly. In Section 2 the case of global import quotas is reviewed. In Sections 3, 4, and 5 the preferential import quota, alternatively, the export restraint, is analysed employing a partial equilibrium framework. An issue much discussed in the literature on protection against developing country exports of manufacturers is the possible clash of interests between established exporters and late coming exporters. This issue is analysed generally in Section 6. Section 7 summarises the paper.

1 The literature

In a survey on trade policy aspects of quantitative trade barriers Corden[3] is a natural starting point. Corden analyses primarily the case of a global import quota and its effect in the importing home country. However, Corden also discusses the sharing of the rent from protection between exporters and importers[4] and the possibility of the global import quota being combined with a monopoly on the importing or exporting side. Our analysis differs from Corden's in that its emphasis is on preferential import quotas and export restraints and focuses more on the exporting countries.

The case of export restraints has been discussed in the literature to only a limited extent. Bhagwati and Srinivasan[5] investigate the behaviour of the exporter faced with uncertainty about the size of the next period's export quotas, while Corden takes up voluntary export restraints briefly in a discussion of the terms of trade argument for trade taxes. To our knowledge, Takacs[6] is the only paper that has previously discussed what we will call a 'leaking' supply restriction, a case with which we will be much concerned. Takacs' paper focuses on the home country and analyses the question of nonequivalence of tariffs, (global) import quotas and voluntary export restraints, where a monopolist producer/importer faces an upward sloping import supply curve from all sources of supply. The home country in Takacs' analysis is therefore a 'large' one. Our paper does not assume monopolistic behaviour in the home country but does assume that the home country could be large enough to influence the world price. The paper also uses a customs union theory framework to analyse the home country's policy on quantitative trade barriers.

Cable (1981), Keesing and Wolf (1980, 1981), and Wolf (1983)[7] are useful surveys of empirical work on the regulation of international trade in textiles and clothing. Hamilton (1981) and Jenkins (1980)[8] examine two importing countries (Sweden and Canada, respectively) and estimate the effects of voluntary export restraints on prices, imports, employment, and income distribution in the importing country. Some specific references to empirical studies are made later in the present text.

Customs union theory has been surveyed and contributed to by Krauss (1972), Michaely (1977) and Corden (1982).[9]

2 Global import quotas

A global import quota does not have the source constraint II and is neutral as to the origin of imports. Consider Figure 1, which shows the world market for the commodity in question (Figure 1b). The world market price p_o is determined by the world export supply curve S_wS_w' and the world import demand curved D_wD_w'. Import demand in the home country is D_hD_h' in Figure la and imports are O_hq_2 at price p_o. Suppose the home country imposes a quota reducing imports to O_hq_3. This raises the domestic price to p_1, calls forth the increased domestic production and causes reduced domestic consumption. Provided the cutback on imports in the home country is 'small' in relation to the world market, the world market price will be unchanged at p_o.

Relax the assumption that the home country demand for imports is marginal. In Figure 1a the import quota cuts back imports to O_hq_3 and the new kinked import demand curve becomes D_hGq_3. The demand reduction in the home country is reflected as a kink in the curve for world demand for imports D_wD_w' changing into D_wHD_w'' (Figure 1b). Consequently, the world market price falls, resulting in a term of trade loss (gain) on the part of exporting (importing) countries. The traded quantity on the world market falls by q_1q_o,

Figure 1 The world market

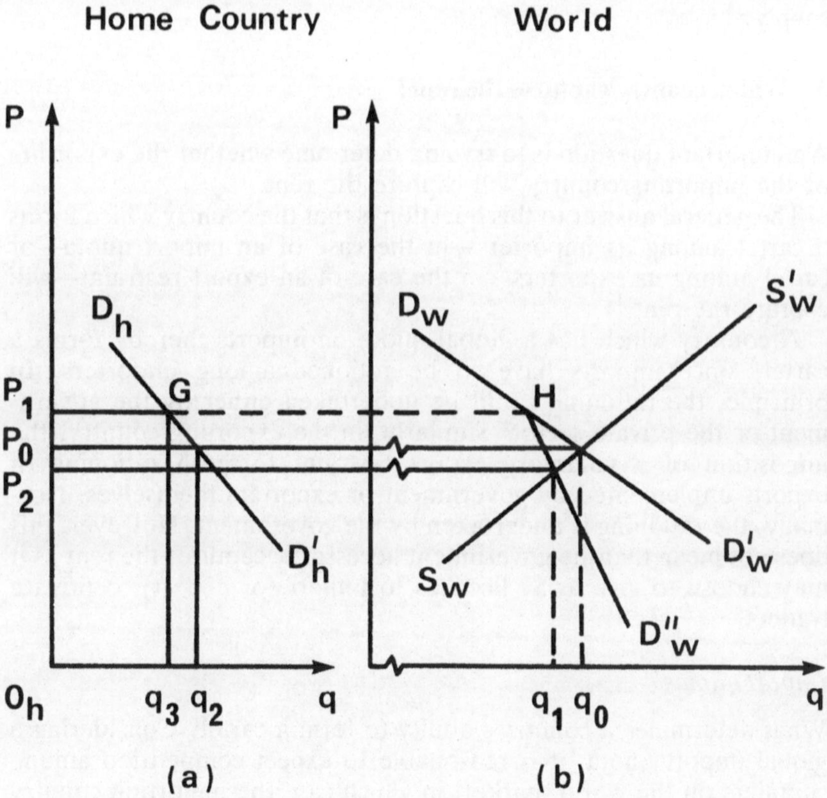

which is smaller than the fall in home country demand, q_3q_2. The cutback in home country demand is partly counteracted by increased imports from the 'rest of the importing world' called forth by the lower world market price.

In the case of a small home country, the imposition of the global quota gives rise to a rent p_1p_0 per unit of the imports. If there is a well functioning market for import licences, the price in that market will equal p_1p_0 per unit of imports. The price of an import licence can be used as a measure of the assistance given to domestic producers.[10] However, in the case of a large home country, the rent revealed in the market for import licences, p_1p_2, will exaggerate the assistance given to domestic producers compared to the free trade situation.

A further complication arises if the government of the home country cannot fully control or predict the supply, owing to smuggling and illegal domestic production. The increase in domestic price—and hence the rent—will then be smaller than p_1p_0 or p_1p_2. This is a very real possibility in many developing countries. In such cases, a distinction has to be made between official and actual domestic supply.[11]

3 Which country captures the rent?

An important question is to try and determine whether the exporting or the importing country will capture the rent.

The general answer to this question is that the country which forms a cartel among its importers—in the case of an import quota—or cartel among its exporters—in the case of an export restraint—will capture the rent.[12]

A country which has a global quota on imports thereby forms a cartel, since imports have to be rationed among importers. In principle, the rationing could be undertaken either by the government or the private sector. Similarly, in the exporting country, the imposition of a voluntary export restraint forces a rationing of exports implemented by government or exporters themselves. Normally, the rationing is undertaken by the government. However, this does not mean that the government necessarily captures the rent as it may choose to give away licences to import (or export) to private traders.

Import quotas

What determines a country's ability to form a cartel? Considering a global import quota, it is reasonable to expect competition among suppliers on the world market, in which case the importing country can capture the whole rent.

Considering a preferential (source specific) import quota, the government limits the number of potential suppliers and increases their ability to form an export cartel. In the extreme case when there is in effect only one preferred supplier, the importing country has given that firm a monopoly position through the source specific import quota and the exporter will extract the whole rent if it is a profit maximiser.[13] Readers familiar with the arguments on tied aid will recognise the situation. The donor country forces the recipient country to have a preferential import quota. The rent comes in the form of donor country firms being able to charge a price to the recipient above the world market one. Experience with tied aid also indicates, as one would expect, that the smaller the donor country and the more sophisticated the product, implying limited substitutability, and the smaller the number of producers in the donor contry, the more significant the rent seems to be.

Export restraints

Considering the export restraint, it is clear that in a static analysis the exporting country captures the whole rent, as the export restraint is always source specific, whereas the import quota can be either global or source specific.

Allowing for several periods, the export cartel's position is unlikely to be as strong as indicated by the static analysis. First, importers—through lobbying—may influence the importing country's decision on the next period's size, and distribution among suppliers, of export restraint volumes.

Second, the importer can buy more from a nonrestrained country. If that happened, a restrained supplier may not only lose rent income but in addition runs the risk of not being able to fill his export quota. He will then typically be penalised by his own government, e.g., by being allocated a smaller export quota next period. To avoid these possible counter actions, the export cartel may well choose to 'bribe' importers by allowing them to capture part of the rent.[14]

4 'Leaking' supply restriction

The preferential import quota or export restraint can be of two types, both of which seem to be of considerable practical interest and which seem to differ in their perceived restrictiveness towards the preferred exporting country. In this section, we consider the leaking preferential import quota and export restraint. The strict case is discussed in Section 5. Henceforth, the term 'supply restriction' will be used as a common term for a preferential (source specific) import quota and a voluntary export restraint.

The leaking supply restriction is characterised by the whole of the home country's import reduction applying to a nonpreferred exporting country (or group of countries) while a preferred exporter (or group of exporters) has free access to the home country market. The supply restriction thus leaks as it permits increased imports from the preferred exporter. A case study of the extent of the trade diverting effect of VERs, in the European textile, clothing, and footwear trade, is reported in the Appendix.

Leaking supply restrictions in the case of a 'very small' home country

Three countries are involved, the home country, the preferred exporter, and the nonpreferred exporter. The home country is comparatively small, which means that export supply curves of the preferred exporter and the nonpreferred exporter are both infinitely

Figure 2 Leaking supply restriction: small home country

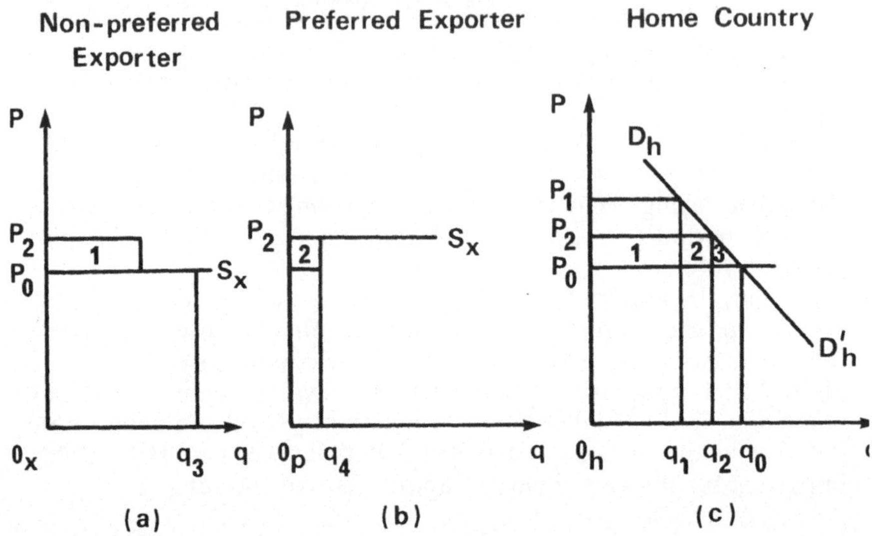

elastic, p_OS_x and p_2S_x respectively (Figure 2). In free trade, all the home country's imports O_hq_O (= O_xq_3) are supplied by the non-preferred exporter at price p_O. Let the home country restrict supply from the nonpreferred exporter to O_hq_1: domestic price increases to p_2 and the preferred exporter becomes competitive, supplying imports q_1q_2 (= O_pq_4). With an export cartel in the nonpreferred country (e.g., through a VER), the rent, area 1, would be captured by the nonpreferred country.[15] Import supply has been increased from the (high cost) preferred exporter, resulting in deadweight loss 2. Deadweight loss 3 will appear under any type of trade barrier.[16]

Leaking supply restriction in the case of a 'not so small' home country

In the case of the 'not so small' home market, the export supply curve of the preferred exporter is upward sloping, but the export supply curve of the nonpreferred exporter is horizontal.

One aspect of agreements on export restraints, preferential import quotas, and orderly market agreements is that their stringency often seems to be determined by the maximum import reduction the home country can get without too much international protest and retaliation. This means the home country may have only limited control

Figure 3 Minimum imports from non-preferred exporter

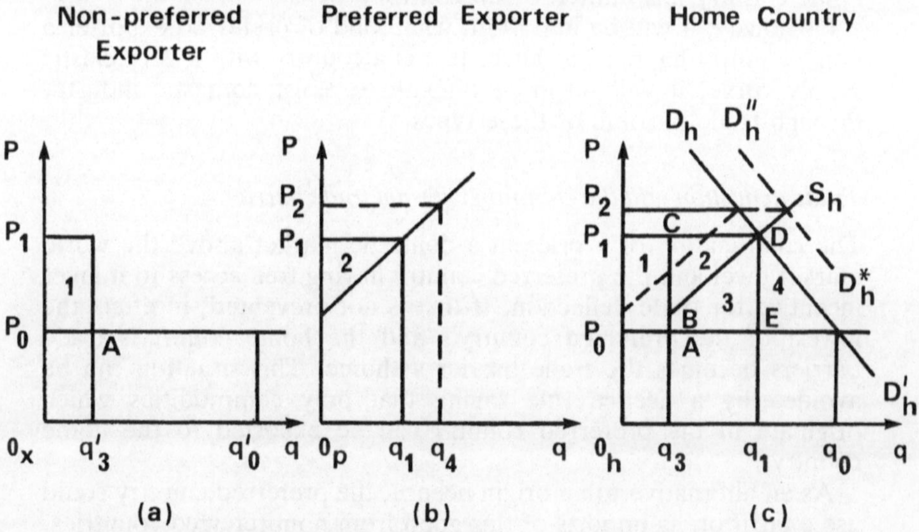

over the extent of import supply restriction. We now examine the implications of this limitation of power. (Note that these pressures may come not only from exporters to the home country but also from other importing countries. The notion that the 'old' industrial countries should share equally the burden of adjusting to the 'new' industrial countries' needs for export outlets is an important example).

Minimum imports from nonpreferred exporter

Suppose the home country can restrict the nonpreferred exporters to O_hq_3 in Figure 3 and is under obligation to allow the preferred exporter to export unrestrictedly to the home country. The domestic price will then settle at p_1. The preferred exporter will increase supply from zero to q_3q_1 ($= O_pq_1$) capturing a surplus equal to area 2. If the exporters can form a cartel, they will capture rent equal to area 1 ($= p_0p_1{}^{CA}$). Areas 3 and 4 are deadweight losses.[17]

The important point is that given (i), the bargaining in which the nonpreferred exporter's supply is determined, the home country's maximum level of assistance to domestic producers, p_1p_0, is determined by (ii), the export supply elasticity of the preferred exporter.

Thus, the level of assistance is fixed by (i) and (ii), over which the home country has limited or no control.

Obviously, it will be important what kind of preferred exporter a home country happens to have. If it is a country with a very elastic supply curve, it will be more difficult to assist domestic industry through trade barriers of these types.

Trade deflection and the common external trade barrier

The combination of a price in a domestic market above the world market level and the preferred country having free access to it gives incentive for trade deflection. If that is not prevented, in effect the lowest of the preferred country's and the home country's trade barriers becomes the trade barrier at home. This situation can be avoided by a licence rule saying that only commodities which originate in the preferred country can be exported to the home country.

As an alternative to an origin licence, the preferred country could use a tariff on its imports of this good from nonpreferred countries. The tariff would make the import price in the preferred country at least as high as its export price to the home country so there would be no incentive for trade deflection. This can be done with an 'accommodating' tariff linked to the domestic price in the home country.

Income growth leads to higher level of protection

Suppose income grows in the home country shifting the demand for imports curve to $D''_h D^*_h$ (Figure 3): if there is no change in the supply restriction on nonpreferred exporters, the domestic price will increase to p_2. The government would welcome that if its previous level of assistance was regarded as too low. However, if it regarded p_1 as the desired domestic price, it should increase supply from nonpreferred exporters to keep the price at p_1. This may not happen, however. First, there is an observation lag of some months before the government may realise that the price has gone up, especially if it is disputed by pressure groups, such as domestic producers and unions. Second, once the price increase is established it may be difficult for the home country government to cut back on preferred country exports (through increased imports from the nonpreferred country). In the preferred country, groups with vested interests in existing levels of production will oppose a cutback in their exports and may alert the preferred country government to prevent this from happening. Depending on the closeness in ties between the two governments, it could be very difficult to cut back a partner's existing exports and possibly risk threatening good relations. Thus, the combination of an observation lag, possibly prolonged by domestic interest groups, and an unwillingness to cut back on a partner's existing exports, can create a ratchet effect pushing the level of protection to ever higher levels as income grows.

Leaking supply restriction in the case of a 'large home' country

In the case of supply to a large home country market, the export supply curves of both the preferred and nonpreferred exporter are upward sloping. Suppose the home country has the import demand curve D_hD_h' in Figure 4. The export supply curve of the nonpreferred exporter is S_xS_x'. In the absence of any restriction, the price is p_0 and the whole of home's import demand ($O_hq_0 = O_xq_0'$) is supplied by that country, by assumption. When the nonpreferred country's supply is restricted to O_hq_3, the aggregated import supply curve becomes p_0ABS_h. Quantity q_1q_3 (= O_pq_4) is supplied by the preferred country. All its production is exported. The domestic price in the home country is p_1, which is the price received for exports by the preferred country. This price exceeds the world price, at which the preferred country can import, and trade deflection is assumed to be prevented by an origin licence.

The nonpreferred country is assumed to have a slightly upward sloping export supply curve. Therefore, commercial policy of the other two countries affects the nonpreferred producers' surplus. Rent income on existing exports to the home country is area 4 (GHJK), provided an export cartel has been formed. Area 5 is lost on exports to the home country. (The world market price is unchanged, by assumption).

Export tax in the preferred country

Suppose the preferred country introduces an export tax, so that the export supply curve changes from S_pS_p' to $S_p''S_p^*$ and the home country's import supply curve becomes p_0AFS_h'. If the home country government, or producers, regard the domestic country price as too low they may be able to raise it by lobbying for higher export taxes in the preferred country. If p_2 is regarded as too high it could be lowered by the home country increasing imports from nonpreferred exporters. That would shift the import supply curve downwards.

The level of assistance and price in the home country depends on the commercial policy—the export tax—as well as on the elasticity of export supply in the preferred country. Again, the home country government has zero or limited influence over these two variables.

Figure 4 Leaking supply restriction: large home country

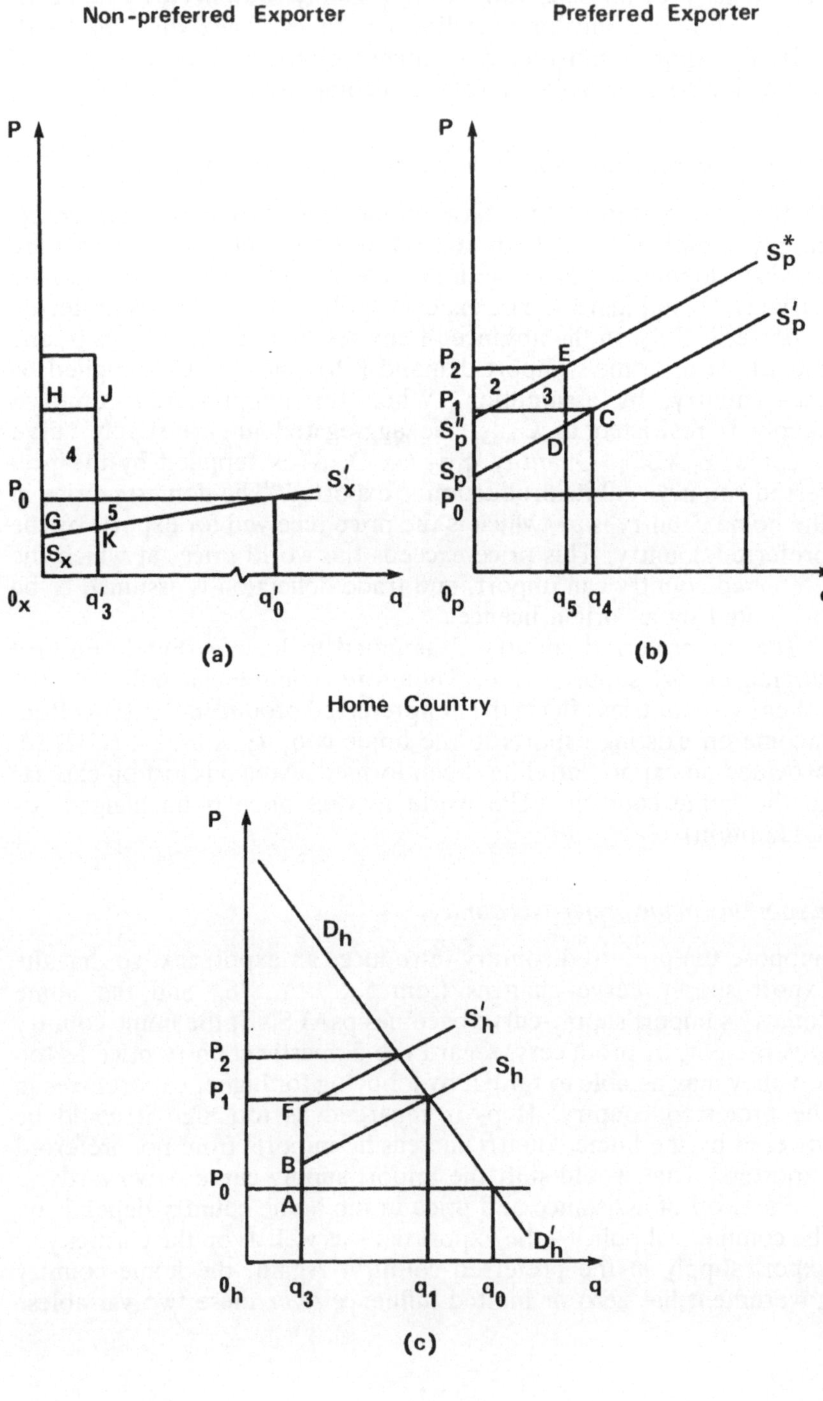

Owing to its limited ability to affect the domestic price through supply restrictions, the government may instead seek to influence its domestic level of assistance by other means, e.g., production subsidies. This would be desirable from a welfare point of view.[18]

Consider the case of a preferred country which exports initially O_pq_4 to the home country at price p_1 and gains surplus p_1CS_p (Figure 4). With the export tax, O_pq_5 is exported and the sum of the tax revenue and producers' surplus is p_2EDS_p. Thus, when the triangle above DC is smaller than area 2 + 3, the preferred exporter gains from the export tax policy. Clearly there is an optimal export tax that will maximise the sum of tax revenue and producers' surplus of the preferred country, given that the home country does not choose to counteract the export tax introduction by increasing supply from non-preferred exporters.

Four cases of leaking supply restriction

Case I: Regional trade area members Suppose the home country is a member of a regional trade area, the purpose of which is to promote free or more liberalised trade between members. The home country may then be prevented from restricting imports from partner countries but could restrict imports from other countries.

Examples are the European Community (EC) and the European Free Trade Area (EFTA), members of which have implemented source specific import quotas and voluntary export restraints against nonmember countries. The EC is supposedly a customs union with a common external trade barrier but some readers may be surprised to know that more than twenty years after the signing of the Rome treaty decisions on import quotas and export restraints were often not taken in Brussels. For example, France and Italy had national import quotas against Japan and Hong Kong specifically, while Benelux, France and the United Kingdom applied voluntary export restraints against developing countries. 'No visible measure is taken by a public body in the EC, but imports are clearly restricted'.[19]

Further, Australia has a free trade agreement with New Zealand which should prevent it from imposing quotas on New Zealand exports. If a Pacific free trade area came into being, Australia as a member would find itself in foreign policy trouble if it gave preference to New Zealand but tried to impose or tighten quotas against the ASEAN countries.[20] However, Australia might still be able to impose quotas against countries outside a Pacific regional grouping.

Case II: The late coming developing country exporter Not all developing countries are subject to source specific import quotas or export restraints on textiles, clothing, electronics, etc. This is especially true of the very latest exporters and, of course, potential exporters. The situation of these countries can, as a simplification, be regarded as that of a preferred nonrestricted exporter with the industrialised countries being the home country.[21] For example, actions taken against Japanese exports have encouraged exports from nonrestrained developing countries. 'The orderly market agreement instituted in 1976, which limited colour TV imports from Japan [to the USA] was a strong stimulus to TV exporters from South Korea and Taiwan' and made it possible for developing countries generally to raise their export prices.[22]

Case III: The United States in Europe The United States textile and clothing industry has proven quite competitive in some commodities (blue jeans!) in Europe. Had the US received the same treatment as many developing countries, US exports would have been subject to import quotas or voluntary export restraints. As far as the author is aware, no such trade barriers *vis a vis* the US exist in Europe. Thus, for foreign policy reasons, Europe as a home country seems to treat the US as a preferred exporter.[23]

Case IV: Exports of cars and steel During recent years Japan has had to agree to restrain its exports of cars to the US and most EC countries. Japan and the EC have also met similar trade barriers against their steel exports to the US. Japanese export of video cassette recorders to France is a very recent example. Other countries are not restrained and the implications of these arrangements could be analysed using the leaking supply restriction model.

5 'Strict' supply restriction

The difference between the 'strict' and the leaking supply restriction is that the preferred exporter in the strict case is not allowed to increase its exports to the home country beyond some initial level. The preferential bias thus seems weaker as compared with the leaking case.

An example would be a home country that for 'historical' or foreign policy reasons does not want to reduce present levels of export from the preferred country but does want to avoid an increase in exports from that country. The group of capital owners and workers, with vested interest in existing production, can presumably apply stronger pressure on the preferred country government—and thus indirectly on the home country—than agents considering potential production.

Suppose the home country imposes a strict supply restriction which reduces total imports, and that source specific arrangements are used to give access to preferred and nonpreferred exporters. Rent per unit of export earned by the preferred country, in the presence of an export cartel, will be higher for a given reduction in nonpreferred exports, as total imports are not allowed to increase. On the other hand, the exported volume is smaller, so that one cannot say *a priori* whether the preferred country will gain more from a strict or a leaking supply constraint. However, the nonpreferred country will unambiguously gain more/lose less from a strict supply restriction for a given cutback in its exports. The reason is the rent per unit of export is higher in the strict case, provided, of course, that there is an export cartel in both cases.

6 The established exporter versus the late coming exporter

A problem discussed especially in relation to the textile and clothing trade is the possible conflict of interest between the established exporters (such as Hong Kong, South Korea, Taiwan, and Singapore) and developing countries that have only recently been able to export (for example Thailand, Malaysia, Indonesia, and the Philippines). The normal procedure in bilateral trade negotiations has been to base access to developed country markets on the export country's historical export performance, possibly with a small increase every year. However, cutbacks have also been frequent in the last few years. There are at least two aspects to this issue.

Income increase in the home country

Suppose income increases in the home country causing an increased demand for imports through a shift to the right in the demand curves. The nonpreferred exporter is not affected by this demand increase at all. The preferred country will enjoy an increased rent from its preferred position.

Practical examples include Sweden and Australia. Increases in Swedish incomes lead to increased import demand which leaks entirely to EC and EFTA partners. If Australian incomes increase, import demand rises, but if the import volume from nonpreferred exporters is fixed, the demand increase leads to increased rent for preferred exporters, which could include New Zealand.

Newcomers being preferred exporters

In the case of EC and EFTA countries, the above description may be reasonably correct in the short run. However, home countries cannot prevent new nonrestrained (developing) countries from increasing their supply and at least temporarily taking advantage of a comparatively high home country price. Given time, these countries can increase their export supply.[24]

In the framework of the leaking case, the established developing country exporters then play the role of a nonpreferred country, being restricted to a fixed export volume. The developing country exporters, not being restrained by a trade barrier, play the role of the preferred exporter, possibly together with the home country's customs union or free trade area partners.

The position of the original preferred exporters is 'eroded' by the latecomers. Their rent is reduced. There is an incentive for existing preferred exporters to press for diversification of the latecomer's exports into other commodities and to put pressure on the home country to impose trade barriers against the latecomers.

As long as the world market price is unaffected there is no conflict between the nonpreferred exporter (the established developing country exporter) and the latecomer. The world market price will stay unaffected when the latecomer sells all exports to the home country where the price is above the world market level. It is not until the home country, by way of trade barriers, forces the latecomer not only to 'freeze' export volume, but actually to require a cutback in export supply that the latecomer may sell to the world market in competition with the nonpreferred exporter. However, for this to happen, the latecomer must be competitive at the ruling world market price.

Thus, there is a clash of interests between the home country producers and partner countries on the one hand, and the latecomers on the other hand. But there is no conflict initially between nonpreferred exporters and latecomers, i.e., no conflict among developing countries.

Another strategy for the latecomer to adopt in a situation of a forced export cutback would be to protect its own domestic market against imports. In this case, a protectionist move by the home country provokes a second protectionist move by the latecomer. Such action would imply a trade barrier against the nonpreferred exporters (the established developing country exporters). Then indeed there would be a clash of interests between the two groups of developing countries.

As we have seen, a latecomer enjoys a potential rent in the leaking case as long as that country has free access to the home country. If there is free entry into the group of latecomers, the nonpreferred country producers will have an incentive to invest in a latecomer country to capture some of the potential rent. In this way the supply curve of the preferred country would shift to the right. Investment of this nature has been observed, for example, in the Pacific ('quota jumping').[25]

7 Summary

The objective of this paper is to analyse effects of two rapidly spreading forms of trade barriers, voluntary export restraints (VER) and import quotas. VERs are, by their nature, always source specific. Import quotas are traditionally analysed as a nondiscriminatory measure but can also be preferential (source specific) in nature. The more source specific the import quota, the greater the ability to form an export cartel. However, VERs are unlikely to result in the whole rent being captured by the exporting country owing to intertemporal considerations. These formal bilateral trade barriers affect 'outside' countries as well, especially if one of the contracting partners is a member of a free trade area or EC. Also, trade restricting agreements, such as those between Japan and the United States, and West Europe and the United States, affect outside countries by increasing trade from other sources of supply. Because of the multicountry nature of the issue one has to distinguish between different situations in which the quantitative supply restrictions are applied (leaking or strict ones), between different sizes of the home country, and also take into account the extent to which trade limitations are acceptable internationally, so that they do not cause excessive welfare losses to the home country owing to retaliation. In the paper, different cases are analysed along these lines. Among several observations, it is noted that because, typically, there are constraints on how much it is politically feasible to reduce imports in the home country, and since the home country has little influence over commercial policy and export supply elasticities in exporting countries, it will have very limited control over the level of assistance given to its domestic producers through the domestic price. The paper ends with an analysis of the possible conflict of interests in trade, such as that in textiles and clothing, between industrialised country producers, established developing country exporters, and latecoming developing country exporters.

Appendix: An example from Europe of how to measure trade diversion effects of VERs

Since the 1960s international trade in textiles, clothing, and footwear (TCF; here defined as International Standard Industrial Classification (ISIC) group 32) has been increasingly regulated by bilateral trade restrictions, mainly VERs. In this appendix we appraise the trade diversion effect on the trade pattern of four members of the EC

Table 8.1A Normalised trade pattern changes in international trades in textile, clothing and footwear (TCF) 1970–1979[a]

(1)	(2) Change in import share of home demand between 1970 and 1979, percentage points	(3) Distribution of increased share of imports among groups of exporters, percent: EC-EFTA members excluding Portugal[b]	(4) Other developed countries	(5) All developed countries[c] (3) + (4)	(6) Established TCF-exporting developing countries[d]	(7) Rest of the world[e]	(8) Total increase in import share of home demand
EC-EFTA members							
France							
Germany	+10.2	47	8	55	20	25	100
Italy	+10.8	5	6	11	54	36	100
Sweden	+12.5	9	20	29	27	44	100
UK	+16.5	40	−2	38	46	16	100
Others:	+8.2	−18	16	−2	63	39	100
Australia							
Canada	+9.7	4	−15	−11	45	66	100
Japan	+0.4	−1.0[f]	3.7[f]	−4.7[f]	2.3[f]	2.8[f]	0.4[f]
USA	+5.1	18	6	24	47[g]	29	100
	+2.2	−30	−80	−110	117	93	100

Notes:

a TCF defined as ISIC group 32.

b Portugal is excluded since supply restrictions have been used against Portugal by EC and EFTA members.

c Australia, Canada, Japan, USA, European Centrally Planned Economies.

d Portugal, Greece, Yugoslavia, Spain, Malaysia, Philippines, Singapore, Thailand, Hong Kong, S. Korea, India, Pakistan, Sri Lanka. Excludes Taiwan, includes S. Europe.

e Includes Taiwan.

f Percentage points and used here.

g Excludes Malaysia.

(France, Germany, Italy and the UK) and one EFTA member (Sweden). The method used is familiar from empirical estimates of integration effects.[26] Since the early 1970s there has been virtually free trade in manufactures between EC and EFTA, and the five countries can, therefore, be regarded as belonging to one free trade area.

In Europe during the 1970s there has been a general trend towards integration in manufactures operating independently of the bilateral restrictions on imports from outside countries. (Such a trend is also observable elsewhere, of course.) When trade creation is measured in the familiar way as changes in partner countries' shares of home demand (production plus imports minus exports), such a trend causes partner countries' shares of home demand to increase even in the absence of bilateral trade restrictions on outside countries.[27] To isolate the trade creation effects due to VERs, in Table 8.1A we have normalised for an underlying integration trend by using other manufactures (ISIC group 3 excluding 32) as the control group. We deducted the percentage point increase in the import share of home demand for other manufactures from that of the VER affected trade in TCF. Data columns 3 to 7 show how the remaining increase in import share of home demand was distributed among sources of supply.

The bilateral trade restrictions have had a comparatively strong trade diverting effect for France and Sweden: in these countries 47 percent and 40 percent, respectively, of the increase in import share of home demand came from partner countries (column 3). For France no less than 55 percent of that increase came from nonrestrained developed countries (column 5). In Sweden and Italy the figures are somewhat lower, 38 and 29 percent, respectively. The German and, in particular, the UK increases of home demand were captured primarily by outside developing countries so the bilateral trade restrictions did not seem to have had much, if any, trade diversion effect. Why the difference? It is often pointed out that France has been more protectionist in TCF policy against outside countries than has Germany, for example.[28] Sweden, traditionally a liberal trade country, pursued a comparatively hard line against LDCs exports of TCF since the mid 1970s,[29] which encouraged a large increase in Swedish home demand for Finnish TCF products, since Finland has a free trade agreement with Sweden. Up to the early 1980s, for example, during the period of this study, the United Kingdom pursued a comparatively liberal import policy, e.g., against Hong Kong exports, but has since shifted to a much more protectionist stance regarding clothing.[30]

Australia, Canada, and the United States took a considerably larger proportion of their import increase of home demand from developing countries than did the Europeans and Japan. However, the European countries have opened up their TCF markets more than did the non-Europeans, apart from Australia, indicating a larger domestic adjustment in Europe. Of the Japanese import increase, a much larger part came from developed countries compared with the other non-European countries.

In summary, until 1980 the bilateral trade restrictions had a trade diverting effect for France, Sweden, and Italy but less so for Germany, and no effect at all for the United Kingdom.

Endnotes

* The author has benefited greatly from discussions with Max Corden, Christopher Findlay, and Kym Anderson. The author is grateful for support from this project, The Swedish Agency for Research Cooperation with Developing Countries (SAREC) and the Tore Browaldh Research Foundation.

1 See Corden, W.M. (1971), *The Theory of Protection*, London: Oxford University Press; Lloyd, P.J. (1982), 'The Australian Textile and Clothing Industry Group: Untoward Effects of Government Intervention', paper presented at Conference

on Structural Adjustment in Trade Dependent Advanced Economies, Stockholm, 2–6 August.

2 See Corden (1971); Falvey, R. (1979), 'The Composition of Trade within Import-restricted Product Categories', *Journal of Political Economy*, Vol. 7, No. 5; Industries Assistance Commission (1981), *Annual Report 1980–81*, Canberra: Australian Government Publishing Service.

3 Corden (1971), Chapters 9, 10.

4 This point is repeated by Bergsten, F. (1975), 'On the Non-equivalence of Import Quotas and 'Voluntary' Export Restraints', in F. Bergsten (ed.), *Towards a New World Trade Policy*, Lexington, Mass.: Lexington Books, who also presents a useful discussion of US experience.

5 Bhagwati, J. and Srinivasan, T.N. (1976), 'Optimal Trade Policy and Compensation under Endogenous Uncertainty: The Phenomenon of Market Disruption', *Journal of International Economics*, 6, pp. 317–36; Corden, W.M. (1974), *Trade Policy and Economic Welfare*, Oxford: Clarendon Press.

6 Takacs, W. (1978), 'The Non-equivalence of Tariffs, Import Quotas and Voluntary Export Restraints', *Journal of International Economics*, vol. 8(4), November, pp. 565–73.

7 Cable, V. (1981), 'An Evaluation of the Multifibre Arrangement and Negotiating Options', *Commonwealth Economic Papers*, No. 15, London: Commonwealth Secretariat; Keesing, D. and Wolf, M. (1980), *Textile Quotas Against Developing Countries*, Trade Policy Research Centre Essay, No. 23, London; Keesing, D. and Wolf, M. (1981), 'Questions on International Trade in Textiles and Clothing', *The World Economy*, 4, pp. 79–101; Wolf, M. (1983), 'Managed Trade in Practice: Implications of the Textile Arrangements', in W.R. Cline (ed.), *Trade Policy in the 1980s*, Institute for International Economics, Washington, DC, Cambridge: MIT Press.

8 Hamilton, C. (1981), 'A New Approach to Estimates of the Effects of Non-tariff Barriers to Trade on Prices, Employment and Import: An Application to the Swedish Textile and Clothing Industry', Weltwirtschaftliches Archiv, 117(2); Jenkins, p. (1980), *Costs and Consequences of the New Protectionism*, Harvard Institute for International Development, Development Discussion paper No. 99, Harvard University, July 1980. Although similar questions are addressed in the two papers the approaches used are quite different.

9 Krauss, M. (1972), 'Recent Developments in Customs Union Theory: An Interpretive Survey', *Journal of Economic Literature*, 10(2); Michaely, M. (1977), *Theory of Commercial Policy*, Oxford: Philip Allan Publishers; Corden, W.M. (1982), 'The Normative Theory of International Trade', in R. Jones and P.B. Kenen (eds.), *Handbook of International Economics*, Vol. 1, Amsterdam: North Holland (forthcoming).

10 The Australian Industries Assistance Commission (IAC) uses the price of import licences as a measure of assistance. Assuming that Australia is a price taker and that importers in Australia actually capture the rent this measure is correct given that the norm of comparison is the ruling world market price. However, for many of the commodities studied by IAC the ruling world market price is likely to be influenced by nonmarginal protectionist measures of the EC and the US, like the VERs on clothing, steel, cars, consumer electronics, and cassava (a foodstuff).

11 There are two recent examples from ASEAN: 'Smugglers are supplying perhaps 10 percent of the Filipino textile market, mostly from the Chinese mainland' (*The Economist*, 12 December, 1981, p. 71). The second refers to Thailand's production and border trade: 'Although official figures are not compiled—since the bulk of the border trade is essentially smuggling—various production and marketing sources told the Review that [border] trade used to account for 30 percent or more of total production. The composition of this trade was roughly 60–65 percent to

Burma, about 30 percent to Indo China and less than 10 percent together for Laos and Malaysia' (*Far Eastern Economic Review*, 7 May, 1982, p. 68). On these kinds of problems in connection with clothing trade see the excellent micro study by Morawetz, D. (1980), *Why the Emperor's New Clothes are Not Made in Colombia*, London: Oxford University Press.

12 There is also a question of the internal distribution of the rent income in the country that captures rent but we leave that issue aside here.

13 However, if there is also a cartel on the importing side there would presumably be a sharing of the rent.

14 Discussion with Swedish importers and Hong Kong exporters of clothing indicates this seems to be the case, not only with regard to the Hong Kong to Sweden trade but also with respect to trade between Hong Kong and other developed countries.

15 This could not happen with a tariff. The rent would be captured by the home country as tariff revenue.

16 These results imply that some care is required in calculation of effective rates of protection. A comparison of the landed price of the import and the domestic price would show zero divergence in the importing country, but if there is a VER in the exporting country, domestic import competing producers would still be receiving some protection, since the price is higher than otherwise. This would be observed if the price comparison was between the domestic prices in the importing and exporting countries. Note that existing producers/exporters and government officials are often in favour of keeping a VER arrangement, as it gives both parties possible rents, from protection and corruption, respectively.

17 We assume that the preferred exporter cannot import at world market prices.

18 Corden, W.M. (1974).

19 Verreydt E. and Waelbroeck J. (1980), 'European Community Protection Against Manufactured Imports from Developing Countries: A Case Study in the Political Economy of Protection', World Bank Staff Working Paper No. 432, Washington, DC.: World Bank, p. 3. See also Cable (1981); Grilli, E. (1980), 'Italian Commercial Policies in the 1970s', World Bank Staff Working Paper No. 428, Washington: World Bank; de Groot, G. (1982), 'The Multi Fibre Arrangement and the Restructuring of the Dutch Textile and Clothing Industry', Occasional Paper No. 13, 1983, Development Research Institute, Tilburg University, Tilburg, Netherlands; Hong Kong Development Council (1981), 'A Summary of Major Restrictions on Hong Kong's Textiles and Clothing Exports', Trade Development mimeo, April 1981, Hong Kong; and Waelbroeck, J. (1982), 'Trends of EC Protection and the Prospects of ASEAN Trade', in N. Akrasanee and H.C. Rieger (eds), *ASEAN-EEC Economic Relations*, Institute of South East Asian Studies, Singapore. To prevent trade deflection within the Community, controls under Article 115 of the Rome treaty are applied by individual member countries. The inclination to use article 115 can be taken as a revelation of member countries' protectionist preferences. See Langhammer, R. (1982), 'ASEAN Manufactured Exports in the EC Markets: An Empirical Assessment of Common and National Tariff and Nontariff Barriers Confronting Them', in N. Akrasanee and H.C. Rieger (eds). EFTA, being a free trade area, is not supposed to have a common external trade barrier. A special case is that West Germany allows free import of certain goods such as textiles and clothing from East Germany, because the latter is part of the pre-World War II German customs union area.

20 According to Anderson, K. and Garnaut, R. (1980), 'ASEAN Export Specialisation and the Evolution of Comparative Advantage in the Western Pacific Region', in R. Garnaut (ed.), *ASEAN in a Changing Pacific and World Economy*, Canberra: Australian National University Press, pp. 374–412, this has already happened. A pertinent example is the following excerpt from the address by the

Malaysian Prime Minister Dr. Mahathir at the dinner given in honour of the visiting Australian Prime Minister, at Kuala Lumpur, 2 August 1982:
'We are glad, Mr. Prime Minister, that you are against protectionism. However, your stand has not yet resulted in a freer flow of Malaysian goods into the Australian market ... We feel strongly that something has to be done about this. We feel strongly that industrialising countries, like Malaysia, should be given the chance to take over the lower technology industries where we would be able to reduce cost, while the advanced countries concentrate on high value added high technology industries. That way there will be an international division of labour which will be beneficial to all. We are glad to see an increasing interest among Australians to invest in Malaysia. However, far too frequently, what they produce here can only be sold in Malaysia and may not be exported to Australia'.

21 However, all developing countries still have to overcome tariffs when exporting to industrialised countries. Allowing for tariffs there are four countries (or groups of countries) to be considered in an analysis: 1 the importing home country; 2 the home country's customs union, or free trade area, partner facing neither tariffs nor supply restrictions when exporting to the home country; 3 a third country facing a tariff but no supply restrictions when exporting to the home country; and 4 the most discriminated against country facing both a tariff and a supply restriction. By assuming away tariffs the number of countries is reduced to three. This simplification does not change the essence of our analysis.

22 Wortzel, L. and Wortzel, H. (1979), 'Marketing Manufactured Exports from LDCs: Progress and Recommendations for Further Progress', Boston, Mass.: School of Management, Boston University, pp. 20, 22.

23 Different reasons for the US competitiveness *vis a vis* the Europeans have been suggested. Keesing, D. and Wolf, M. (1981), pp. 79–101, mention declining relative US wages and controlled energy prices. Yamazawa, I. (1982), 'Renewal of Textile Industry in Developed Countries and World Textile Trade', paper presented at 39th Annual Meeting of Japan Economic Policy Association, 29–30 May, alleges that US firms have been more vigorous in introducing 'labour saving techniques, automation, and speeding up of production processes' (p. 5). OECD (1980), 'Structural Problems and Policies Relating to the OECD Textile and Clothing Industries', DST1/IND/80.36, Paris: mimeo, indicates that US production is localised to poorer regions in the US with comparatively low wage levels.

24 Given some further time the home country may well enforce a supply restriction on the latecomer as well.

25 Keesing, D. and Wolf, M. (1980), p. 124.

26 See for example, Corden (1974) and references there.

27 Note that changes in import shares cannot be used in an analysis of this sort, see eg. Corden (1974).

28 Langhammer, R. (1982), 'ASEAN Manufactured Exports in the EC Markets: An Empirical Assessment of Common and National Tariff and Non Tariff Barriers Confronting Them', in Narongchai Akrasanee and H.C. Rieger (eds), *ASEAN-EEC Economic Relations*; Waelbroeck (1982); Cable (1981).

29 Cable (1981); Hamilton (1981).

30 Cable (1983).

9 ASEAN systems for allocation of export licences under VERs

CARL HAMILTON*
December 1984

THE purpose of this paper is to describe and discuss existing systems in ASEAN for allocating rights to export clothing commodities, when exports are restricted by a voluntary export restraint (VER). To our knowledge this has not been done before. Official documents have been collected but, as these only paint part of the picture, interviews were also undertaken with administrators and exporters in Malaysia, Singapore, Thailand, Philippines and Hong Kong.[1] Some important issues discussed in the final section are the arguments for transferability of quotas, the frequent changes in criteria for allocations, subcontracting, the limitations of the value added criteria often used, value or volume definitions of quotas, and the rationale for under utilised quotas. First, we discuss the creation of rents in standard trade theory models.

The rent from 'voluntary export restraints'

In a country where exports are restricted—i.e., the export volume is lower than exporters would have wished—the right to export carries a value. Alternatively put, to get into a foreign market an 'entrance ticket' is needed and such an entrance ticket carries a value. The impact of a VER on the market for one commodity, say clothing, can be analysed using Figure 5. Importing country demand and exporting country supply schedules are DD′ and SS′ respectively and the resulting price is p_o. In order to satisfy demands for protection of importing country producers, foreigners restrict export supply by q_1q_o by negotiating a VER. This changes the export supply schedule

* Carl Hamilton is with the Institute for International Economic Studies, Stockholm.

Figure 5 The exporting country: one commodity market

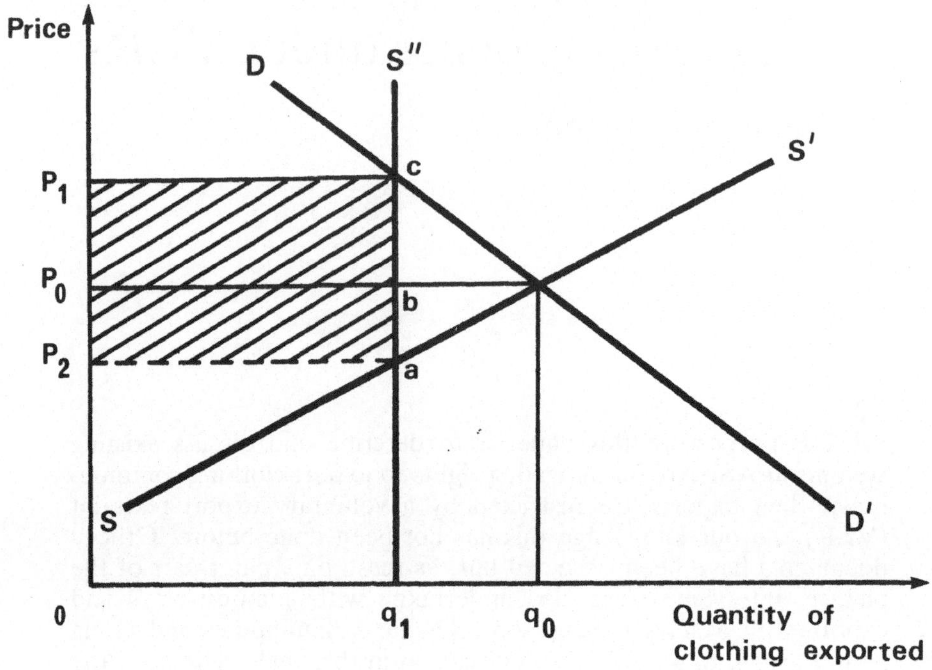

into SaS″. The resulting scarcity of clothing in the importing country market drives up the price to p_1, giving the remaining producers producing Oq_1 an additional income per unit of p_1p_o. This group of producers gains rent income p_obcp_1 and their total rent income is p_2acp_1. The export restriction drives out of business the highest cost producers in the exporting country producing quantity q_oq_1. The lower cost producers producing the restricted volume Oq_1 are prepared to produce this volume at a price of at least p_2 per unit. The government now has several options: first, it could introduce a tax on exports equal to p_1p_2 per unit. That would limit the export volume to Oq_1 and yield export tax revenue p_2acp_1. Second, the government could auction 'entrance tickets'. Exporters would be prepared to pay at least p_1p_2 per unit assuming there is perfect competition among them. The revenue from the auction to the government is again p_2acp_1. Third, the government could hand out as gifts to exporters the set of entrance tickets equal to a value of p_2acp_1. The exported volume is the same in all three cases (Oq_1), but in the third case the government foregoes the rent income which is instead captured by the exporters.

There are more complex cases, such as those with imperfect competition and the reader is referred to Corden.[2] The most important of these cases is when the VER creates an output restricting monopoly in the exporting country. This is discussed as one reason for under utilised quotas later.

Elsewhere, I have estimated import tariff equivalents of VERs on Hong Kong's clothing export to Western Europe and the United States in 1981–1982.[3] It was found that the average EC-EFTA external tariff was 17 percent. On top of this came an import tariff equivalent of VERs of 21 percent, making the total combined European trade barrier 38 percent (1981–1982 average). For the United States, the tariff was 23 percent and the import tariff equivalent also 23 percent (1981–1982). Using these figures, the rent income of Hong Kong 1981–1982 from Western Europe and the United States was estimated to have been US$56 million (1982 prices), i.e., around 1 percent of Hong Kong's GDP 1981–1982.

The next section of this paper contains a description and offers some comments on the present (1983) ASEAN systems for allocating the entrance tickets to foreign clothing markets.[4] In ASEAN there are no legal markets for entrance tickets (quotas).

Country studies

Malaysia

a *Transferability*. No trade in quotas is allowed. Only *bona fide* Malaysian textile manufacturing companies are eligible for the government's free quota allocations.

b *Allocation criteria for existing export quota volumes*. The criterion is past performance during the immediate two preceding years.

The producer chooses the most favourable of two government formulas: either a formula based only on the preceding year's utilisation of quotas, (A), or a formula based on the last two years' utilisations, (B).

$$(A) \quad q_t = [\alpha_{t-1} + \frac{1}{k}(1 - \alpha_{t-1}) \approx q_{t-1} \quad \text{or}$$

$$(B) \quad q_t = [\alpha_{t-1} + \frac{1}{k}(1 - \frac{\alpha_{t-1} + \alpha_{t-2}}{2}) \approx q_{t-1}$$

where k takes on the following values depending on α_{t-1}:

α_{t-1}	k
$\geqslant 0.8$	1.00
[0.70, 0.79]	1.25
[0.60, 0.69]	1.50
[0.50, 0.59]	1.75
[0.40, 0.49]	2.00
$\leqslant 0.39$	2.50

In these expressions, q_t and q_{t-1} are export quotas allocated to a producer in years t and t−1, respectively. The proportions of the two preceding years' allocation actually utilised are denoted by α_{t-1} and by α_{t-2}, where $0 \leqslant \alpha_{t-1} \leqslant 1$ and $0 \leqslant \alpha_{t-2} \leqslant 1$. The Malaysian quota allocation system's relationship between the new quota allocation and that of the previous year is indicated by the ratio q_t/q_{t-1}. There is a degree of penalty for utilising less than 80 percent of the allocated quotas under formula A. An exporter would choose formula (A) if $\alpha_{t-1} < \alpha_{t-2}$, that is, when utilisation is falling.

c *Allocation criteria for increases in Malaysia's export quota volumes*. In the various textile agreements annual increases in export volume have been allowed. Producers whose $\alpha_{t-1} \geqslant 0.8$ are eligible for free increases in volumes with the qualification that if one producer alone already takes more than 40 percent of Malaysia's total growth in that particular category he will receive a reduced increase ('half of the growth factor'). If there are further export volumes available to the government, all producers are eligible to apply to the Ministry of Trade and Industry. The Ministry will then use the following criteria[5] (not necessarily in this order of importance) as the basis for its allocation (free of charge) among applicants:

'(i) higher value of product to be exported;
(ii) manufacturers showing an increasing effort to diversify into exports of nonquota categories;
(iii) manufacturers showing an increasing effort to export to nonquota countries;
(iv) amounts to assist in covering on order;
(v) documentary evidence of confirmed orders (subject to proviso of quota availability);
(vi) *ceteris paribus*, first come first served'.

d *Exit*. A producer who has a quota allocation which cannot be utilised has to hand it back to the government (surrender) before a predetermined day (in the case of Malaysia 'the last day of the seventh month of the quota year'). If this is not done, a penalty will be imposed on the producer equal to the loss of 30 percent of unsurrendered quota volume in the next year's round of allocations. This is in addition to any reduction following from the formulas.

e *Entry*. No newcomers are allowed entry into the export of textile commodity groups unless existing quota holders consistently achieve only $\alpha_{t-1} \leqslant 0.7$ and $\alpha_{t-2} \leqslant 0.7$. Only then are newcomers allowed into exporting the commodity in question but the permitted newcomers must belong to the group of 'already established producers' of other commodity categories. The Malaysian government 'no longer encourages the establishment of new facilities'.

f *Remark*. The past performance of firms is not made public.

Thailand

Two quota systems are applied, one for 'Yarns and fabrics' and one for 'Ready made garments'. The latter is described here.

a *Transferability*. The basic rule is that quotas are allocated free of charge and are not allowed to be traded. There are two qualifications, however. First, two types of firms can hold quotas, producers and 'International Trading Companies'. The trading companies receive 20 percent of the 'residual' quota volume (see below) and they can, in practice, sell the right to export to firms who wish to export through the trading company. Second, there is a complex system of barter of commodities within each country (or EC) agreement on export restraint, requiring the consent of the authorities (substitutability in export commodity mix). However, this option is used by producers only as a last resort. 'Informal' exchange between firms is smoother although illegal and therefore risky.

b *Allocation criteria for existing quota volumes*. Existing volumes are distributed free of charge according to past performance (international trading firms are excluded). This is called the 'principal' or 'basic' quota and is distributed at the beginning of the year. In addition there is a pool from which a potential exporter can apply (residual quotas, see below). An exporter who fails to export at least 90 percent of last year's principal quota will be temporarily barred from participation in the monthly allocations of free residual quotas. This penalty applies from between one and four months.

c *Allocation criteria for increases and residual quota volumes*. Free residual quotas can be applied for once a month. Producers with or without principal quotas and international trade firms can apply. The latter group is guaranteed 20 percent of the residual quota volume. The remaining 8 percent is allocated according to four 'priorities':

1 utilisation of raw materials;
2 price per unit;
3 value added; and
4 time duration between order and delivery dates.

A producer who exports through trading firms acquires the past performance right to export in the next year when the export volume in question is included in the principal quota pool. This is reflected in the price for the right to export through an international trading firm.

d *Exit*. Uncommitted volumes must be surrendered to the government.

e *Entry*. See above under c.

f *Remarks*. Thailand is the only country with free monthly government allocations. From 1983 all firms' past performances as well as the reasons for allocations from the pool of residual quota volumes are planned to be made public. The four criteria given under c above are claimed in practice to discriminate against newcomers.

Philippines

a *Transferability*. Quotas are allocated free of charge. Only manufacturers can hold quotas and quotas are transferable only under a supervised transfer scheme. However, subcontracting is permitted. In this way a small newcomer is said always to be able to enter export markets by being a subcontractor.

b and c *Allocation criteria for existing export quota volumes and increases in quota volumes*. Quota holders who utilised at least 90 percent of their quota allocation during the preceding year will receive the same free quota volume the next year *plus* a free increase equal to the overall rate of growth in quota volume in that particular category. Quota holders who utilise 65 to 90 percent of the allocation (adjusted for any surrender) will lose automatic restoration rights to free quota allocations. However, a holder who pays a fine will be given preference in the allocation of residual free available quotas. A quota holder utilising less than 65 percent and not surrendering will lose privileged access to free quotas and is regarded as a nonperformer in that particular category. Performers are given preference over nonperformers for critical categories. For semicritical categories, applications are treated on a first come, first served basis. Noncritical categories are open to both performers and nonperformers, with priority given to nonperformers.

d *Exit*. Commodities are classified in three categories: critical, semicritical and noncritical. The higher the demand, the more critical the category. Depending on the degree of critical character, and the date when quotas are surrendered, the quota holders' quota allocations are reduced. One month before the quota year starts, definitions of commodity categories' status are given by the government.

e *Entry*. See under a and f.

f *Remarks*. The permitted subcontracting provides for a legal entry possibility. The incentive system for utilising quotas with both penalties and 'bonuses' in the form of better access to additional free quotas has resulted in the most flexible quota system in ASEAN, according to interviews undertaken in Hong Kong. Further, it is claimed in Hong Kong (but denied by officials in Manila) that it is always possible for somebody with 'contacts' to get a quota, i.e., there is illegal trade.

Singapore

The Singapore authorities are more secretive on this subject than the authorities in the other ASEAN countries (possibly with the exception of Indonesia). Quotas are allocated free of charge to manufacturers and some long established specialised exporters.

a *Transferability*. No quota trade is permitted. However, exporters/producers can barter between themselves through the Trade Development Board (TDB).

b *Allocation criteria for existing quota volumes*. Past performance is the basic rule. Existing quota holders receive for free 90 percent of the previous year's quota utilisation.

c *Allocation of increases and surrendered quota volumes*. Specialised traders do not get any free allocations of additional quota volumes. Preference is given to manufacturers of high quality and high value added goods.

d *Exit*. There is no information on exit except that any illegal trade would result in exclusion from allocation of free quotas.

e *Entry*. Newcomers are allowed into the scheme only if they produce categories where quotas are underutilised. There is also a preference for high quality and high value added goods. Subcontracting is permitted only in peak situations with 100 percent capacity utilisation in the quota holder's plant and must be reported to TDB.

f *Remarks*. Quota holdings of individual manufacturers are not made public. The Singapore quota system is regarded as rigid among traders. One reason is that the manufacturer has to apply for quotas on the basis of a nine month forecast, compared for example to Thailand with its monthly quota allocations. The authorities stress that manufacturers should have high certainty regarding the next period's free quota allocation.

Indonesia

a *Transferability*. Free quotas are allocated to manufacturers only and are not transferable.

b and c *Allocation of existing and new quota volumes*. The available quota volume is divided into two parts:
85 percent is allocated to established manufacturers according to rather vague criteria:

	percent
Minimum quota	26
Capacity	20
Past performance	26
Export oriented	9
Small factories	4
	85

15 percent is allocated entirely to newcomers.

d *Exit*. Quotas not used by 30 April must be surrendered to the government for redistribution.

e *Entry*. See above.

f *Remarks*. There is an export incentive system in the form of a subsidy per unit of exports:

	Subsidy per unit, US$ (1983)
Men's shirts	1.07
Men's cotton flannel shirts	1.23
T-shirts	0
Jackets	0
Boys' shirts	0.75
Ladies' blouses	1.15
Men's denim jeans	1.33
Men's polyester trousers	1.00

If export volumes are restricted, an export incentive system only means a cash transfer to quota holders from the government.

Some general remarks

Transferability and efficiency

In the exporting country there must be some sort of mechanism for allocating among potential exporters the valuable export licences. In principle there are two approaches: the market mechanism with market prices for licences, i.e., auctioning off licences and allowing for an ongoing trade in licences or, on the other hand, an administrative decision making process with civil servants deciding according to some criteria which determine, or at least influence, how large an export quota applicants should receive and when.

A constraint imposed on transferability of licences between firms is likely to lead to an inefficient use of resources and quota volumes.

First, if there is auctioning of export licences and firms have different supply prices, the firms with the lowest supply prices will be the ones which are willing to pay most for an export licence. These firms offer to produce at prices below p_2 (Figure 5). The higher cost firms — i.e., the less efficient ones — offering to produce at prices above p_2 will close down and factors of production used in these firms could be employed more productively elsewhere, after allowing for an adjustment period.

Second, suppose there is no auction but past performance criteria are used in a first round of government allocations of export licences. Then transferability after this first round is still desirable. This can be seen by assuming that transferability is nót allowed. Because of the export restriction each firm has to cut back its export volume say, for example, proportionally by 10 percent. If firms originally produced at a cost minimising scale each one of them now has to back up its average cost curve producing at a suboptimal scale. However, had trade in licences been allowed, some firms would have sold their allocations of licences, closed down, and factors of production would be freed for alternative employment. The firms which buy the licences can stay in business producing at the original cost minimising output level. Clearly this argument is always important, but it is more important for industries where economies of scale are substantial, such as automobiles and steel, and less so for clothing.

Third, not permitting trade in licences and using past performance criteria results in new and more efficient producers being discriminated against because they have no past record of export to refer to when applying for free allocation of export licences. However, if trade in licences is allowed, a new firm which is more efficient than existing ones will be prepared to pay a higher price in the market for licences than are existing, less efficient firms. Thus, restricting the possibilities for new entry, like all forms of protection, provides a breeding ground for inefficiency and high cost production, in this case protection against new domestic competition.

Transferability and rents

If the licences are auctioned by the exporting country government, the rent income is captured by this government. If the government allocates the export licences free of charge to a group of exporters, for example, according to past performance, it is of course this group of original receivers of licences which captures the rent income. Because there can be large potential incomes to be gained from receiving licences from the government rather than buying them at an auction, existing exporters form a special interest group which has strong incentives to oppose a change in existing allocation rules based on past performance. Further, the use of administrative criteria can provide huge incentives to bribe civil servants and offer politicians 'gains' in the political marketplace in return for a larger allocation of licences. Prohibition of trade in licences between exporters also reduces their ability to utilise quota volumes fully.

As far as the author is aware, government auctions of licences have been tried nowhere. Just as in ASEAN, the typical approach is that governments hand out licences according to some sort of past performance criteria. In ASEAN countries, as elsewhere, there are usually very complex and discriminating rules regulating entrance of new exporters. However, in Hong Kong, trade in textile and clothing licences is permitted once the original past performance allocation has been made. As was pointed out in the previous section, in ASEAN such trade is illegal.[6]

Instability in rules

A striking impression from studying the rules for allocating export licences (quotas) in ASEAN is that they seem to change almost yearly. This instability in the commercial policy has negative effects: first, firms will devote real resources to lobbying activities for changes in next period's allocation rules. To a firm, the profits gained from this kind of activity in the political market place may well outweigh gains from the same amount of resources spent on efforts to increase productivity. From society's point of view, however, the spending of resources on lobbying is a pure waste. Second, frequent changes discriminate against investment as returns to capital in the licenced activity will have to be correspondingly higher to compensate for the higher uncertainty surrounding the outcome caused by frequent changes in the rules of the system. If the aim is to discourage investment (and newcomers), there are surely less wasteful ways of doing it than increasing uncertainty.

Subcontracting

Subcontracting is typically not permitted officially (except in the Philippines) but occurs in all ASEAN countries because of the difficulty in defining subcontracting. Examples of activities often subcontracted are washing and knitting. In this way there is almost always free entry and competition in some stages of the production process. The original quota holder will retain the rents.

High value added goods

The high value added to goods is sometimes given as a criterion for priority treatment in the quota allocation. This system is open to abuse. The manufacturer/exporter typically presents the authorities a letter of credit showing a high price on that particular commodity category, as proof of high value added. However, as the buyer usually buys several commodity categories from one manufacturer, it is the average price per unit for all goods in the different categories that is important. Consequently, the buyer has little interest in having quoted a correct price for any individual category, and the degree of value added can readily be overstated.

Volume or value quotas

Quotas are typically defined in volume rather than value terms. This is very clearly preferred by *present* exporters, judging from interviews, the argument being that the volume definition provides security. Further, with a volume definition there is an incentive for upgrading. Once a product is upgraded, it is in the interest of established exporters to maintain the volume quota system, an attitude encouraged by official guidelines recommending high value added goods. In this way established exporters are protected against more efficient domestic producers, especially as it is possible to manipulate the proof of high value added as describe above. From the exporting country government's point of view a volume quota is also desirable as it is simpler to administer, since a value quota necessarily involves a judgement of what the correct price of the commodity should be. There would be an incentive for the exporter to quote to the government a low price in his documents, but in effect charge a higher price, and to keep the difference in foreign banks.

Quota utilisation rates

Figures on quota utilisation are available for many exporting countries. However, these figures cannot readily be used as measures of efficiency of the quota allocation system.

There are several rational explanations why a quota may not be filled:[7] an export restriction can result in an export cartel, especially when the government protects existing exporters against competition from new entrants by confining allocation of quotas to existing firms. This cartel may find that its profits are maximised by restricting exports to a volume below the quota volume, i.e., the cartel acts as a profit maximising monopolist. Thus with VERs the export volume can fall for two reasons: first, owing to restrictions imposed by the importing country and, second, owing to a fall due to the export restriction's creation of an exporter's monopoly. This second effect can be an important reason for exporting country governments' observed efforts to penalise exporters for not reaching their historical (past performance) levels of exports. Further, a quota can become redundant if demand is weak, in particular if importing country tariffs are high. Quotas may be given for commodities which are not yet produced and which are potential export goods only. Quotas can be given to unprofitable commodities—for example, the quota may itself be so small that it is unprofitable to try and penetrate the foreign market. To meet these difficulties, agreements on VERs typically contain 'flexibility provisions', which allow substitution possibilities between commodity categories so that over time, quota utilisation rates well over 100 percent for some goods may be feasible, while the rates for other categories or periods are correspondingly lower. Clearly, in all these cases it is rational for exporters not to fill the quota.

Endnotes

1 Interviews and unpublished references: Malaysia: 'Criteria for Quota Allocations', mimeo, Ministry of Trade and Industry, Kuala Lumpur, 1983. Interviews in Kuala Lumpur with, among others, Mr Yap Fung Kong, Ministry of Trade and Industry. Thailand: 'Trade Agreement and Export Regulations of Ready-Made Garments', The Department of Foreign Trade, Bankok, 30.12.1982 (in Thai). Interviews in Bangkok with Mr Pongthian Payakni Hi, Department of Foreign Trade, and Mrs Sansane Yamasmit, Thai Garment Manufacturers Association. Philippines: 'Rules and Regulations to Implement the Provisions of Executive Order No. 537 Creating the Garments and Textiles Board as Amended by Executive Order No. 823', Garments and Textile Board, Ministry of Trade and Industry, Manila. Interviews with Mr Rodolfo Puno, Manila and with traders in Hong Kong. Singapore: Interviews in Singapore with Miss Cheong Choy Hoong, Trade Development Board and with traders in Hong Kong. Indonesia: interviews in Hong Kong only.

2 Corden, W.M. (1971), *The Theory of Protection*, London: Oxford University Press, chapter 9.
3 Hamilton, C. (1984), 'Voluntary Export Restraints on Clothing from Asia: Tariff Equivalents, Rent Incomes and Trade Barrier Formation', Seminar Paper No. 275, Institute for International Economic Studies, Stockholm.
4 See interview details endnote 1.
5 Malaysia, Ministry of Trade and Industry (1983), 'Criteria for Quota Allocations'.
6 For the Hong Kong system, see Hamilton (1984); Hong Kong and Shanghai Banking Corporation, (1978), 'Hong Kong Textile and Garment Export Restrictions. The Quota System: June 1978', Hong Kong: Corporation Secretary's Department; and Morkre, M. (1979), 'Rent-Seeking and Hong Kong's Textile Quota System', *The Developing Economies*, March, pp. 110–18.
7 These possibilities are covered in principle by Corden's discussion of import quota systems: Corden (1971), chapters 9 and 10.

10 Australian protection and the ASEAN countries

PETER G. WARR*
December 1984

1 AUSTRALIA'S bilateral trade with the ASEAN countries is of modest economic significance, whether it is viewed from the ASEAN or from the Australian perspective.[2] But the political importance of the ASEAN countries to Australia and the irritation that Australia's protectionist policies have caused in the ASEAN countries have combined to give the trading relationship an importance which is disproportionate to its size in purely economic terms. Clearly, ASEAN-Australian trade is affected by protection in the ASEAN countries, as well as in Australia, but this paper focuses on the effects of Australian protection on its trade with ASEAN. Since it is the effect of Australia's protection on its imports from ASEAN that has been sensitive politically, the paper is primarily concerned with this issue.[3]

The paper asks three questions. First, does Australia's protection discriminate against imports from the ASEAN countries in particular, relative to imports from Australia's other trading partners? Second, to what extent would Australia's imports from the ASEAN countries increase if Australia's protection policy was liberalised? Finally, of the various import commodities which are subject to tariff and/or quota treatment at the Australian borders, which commodity specific rates of protection have the greatest impact on Australia's imports from the ASEAN countries?

Sections 1 and 2 of this paper attempt, in different ways, to determine whether the differentiation between commodities which is inherent in Australia's system of protection, implies discrimination

* Peter Warr is with the Economics Department, Research School of Pacific Studies, at the Australian National University, Canberra.

Table 10.1 Australia's average implicit protection against ASEAN exports: 1968–69 to 1980–81 (nominal rate of protection, percentage)

	1968–69	1971–72	1974–75	1977–78	1980–81
Indonesia	3.0	4.6	0	2.8	3.0
Malaysia	25.2	23.0	6.5	11.4	11.3
Philippines	19.5	19.8	19.0	25.8	22.6
Singapore	10.2	9.5	4.8	4.7	9.1
Thailand	24.0	33.0	15.1	17.4	14.8
ASEAN	10.8	18.6	6.7	8.8	8.6
LDCs	20.3	24.0	13.1	17.2	16.2
All countries	25.6	26.6	16.9	18.8	18.1

Source: Calculated from data provided by Industries Assistance Commission, Canberra.

against the ASEAN countries in particular. Section 1 does this by calculating implicit rates of protection against imports from the ASEAN countries and other trading partners, using observed import shares as weights. Section 2 approaches the question by simulating the effects that reductions in Australian protection across the board would have on commodity imports, using the ORANI general equilibrium model of the Australian economy. This is then translated into effects on imports from Australia's various trading partners, including the ASEAN countries, by using trade data on Australia's imports by commodity and trading partner. Section 3 extends this use of the ORANI model, in conjunction with the trade data, to examine the effects of reduced protection of individual commodities.

The method of analysis used in the paper draws heavily on an earlier study.[4] This paper extends the analysis presented in that study by looking at the ASEAN countries individually, rather than solely as a block, and also by updating the trade data used in the analysis from 1977–78, as in the earlier paper, to 1980–81. The effects of Australia's trade preferences are not discussed in this paper since the evidence presented in Warr and Lloyd (1983) shows that their importance is small in comparison with the impact of protection.

1 Implicit protection against ASEAN exports

One way to indicate the differential effects of Australian protection on its various trading partners is to construct weighted averages of the nominal rates of protection applying to the various commodities, with the commodity weights used for each trading partner reflecting the share of the commodity concerned in Australia's total imports from that country. The results of this calculation are presented in Table 10.1.[5] The well known difficulty with this procedure is that actual import shares reflect existing rates of protection. Commodities subject to high rates of protection consequently receive too low a weight. Nevertheless, while the use of free trade import shares would be preferable for this purpose, they are unobserved and it is not clear what systematic bias between trading partners might result from this procedure.

These calculations suggest that Australian protection was lighter against exports from the ASEAN countries as a group than against exports from the less developed countries (LDCs) as a group (including ASEAN) throughout the period.[6] Nominal rates of protection applying to ASEAN exports were around half those applying to the LDCs as a whole for most of the period. Moreover, protection against LDC exports was itself consistently lighter than average rates of protection against exports from all countries, although the differences narrowed over time.

For individual ASEAN countries the pattern varies widely. For Indonesia and Singapore, nominal rates have been quite low. Imports from Malaysia and Thailand faced high import barriers early in the period (about the same as the average of all countries and higher than the LDC average), but these rates declined so that they were slightly lower than LDC average rates by 1980–81. The Philippines faced consistently high import barriers in Australia throughout the period. From the mid 1970s onwards these rates have been the highest of the five ASEAN countries and higher than the LDC and 'all country' averages.

To understand these results it is helpful, first, to summarise Australia's system of commodity protection and, then, to look more closely at the pattern of Australia's commodity imports from the ASEAN countries. Nominal rates of protection applying to Australia's imports of primary commodities, including crude petroleum, have been consistently low. Manufacturing protection has been consistently high, but the pattern has changed in a significant way. In 1968–69 these nominal rates averaged around 25 percent in the manufacturing sector. These rates were relatively uniform and consisted primairly of tariffs, rather than quantitative restrictions. There was little change until July 1973. At that time Australia unilaterally reduced tariff protection across the board by 25 percent. This was soon offset within the textiles, clothing, and footwear sector by the progressive introduction of import quotas, beginning in December 1974. By 1979 about 90 percent of clothing and footwear production and 30 percent of textile production were protected by quotas.[7]

In January 1975 the government also introduced import quotas on motor vehicles. This was intended to enforce the policy of reserving 80 percent of the domestic motor vehicle market to Australian producers. By 1980–81 it was possible to say that two groups of industries 'textiles, clothing and footwear' and 'transport equipment' were *more* heavily protected than they were in 1968–69 (nominal rates of 42 percent and 46 percent, respectively, in 1980–81 compared with 32 and 30 percent in 1968–69), while all other manufacturing industries were *less* heavily protected than before (average nominal rates of around 13 percent in 1980–81 and 23 percent in 1968–69).

Turning to the commodity composition of Australia's imports from the ASEAN countries, the distribution of imports by commodity category is summarised for 1980–81 in Table 10.2 and compared with the LDCs as a group and with Australia's imports from all countries.[8] The data in the table (except for the last column) indicate the percentage share of Australia's total imports from each country or group which fall in the commodity categories indicated. The last column indicates average nominal rates of commodity protection in 1980–81. For the ASEAN countries the most important commodity categories to look at are 'petroleum and products' which receive almost no protection in Australia, and 'textiles, clothing and footwear', which receive heavy protection.

For Indonesia and Singapore, the low implicit rates of protection indicated in Table 10.1 are the result of the dominant importance of crude petroleum and refined petroleum products among Australia's imports from those countries. Crude petroleum accounted for 86 and 82 percent of Australia's imports from Indonesia in 1968–69 and 1980–81, respectively, and refined petroleum products accounted for 75 and 73 percent of imports from Singapore in these years. Textiles, clothing and footwear were of minor importance for both countries.

Crude petroleum accounted for an increasing proportion of imports from Malaysia (rising from 1 to 10 percent from 1968–69 to 1980–81), which was partly offset by the rising importance of textiles, clothing, and footwear, following a similar trend. The importance of nonpetroleum primary commodities among Australia's imports from Malaysia diminished dramatically over the period from 51 to 18 percent of imports from that country. Thailand does not export petroleum products to Australia, but the proportion of the heavily protected group, textiles, clothing, and footwear, among these exports declined from 26 percent in 1968–69 to 12 percent in 1980–81. Manufactured food products, with a relatively light average nominal rate of protection of 27 percent, grew from zero to 32 percent of these exports over this period. This accounts for most of the decline in Australia's implicit protection against Thailand's exports.

The composition of the Philippines' exports to Australia is in one respect more typical of the LDCs as a group than of the ASEAN

Table 10.2 Share of ASEAN exports to Australia in particular commodity categories,[a] 1980–81 (percentage)

Commodity category	Indonesia	Malaysia	Philippines	Singapore	Thailand	ASEAN	LDCs	All countries	Average nominal rate of protection
Petroleum and products	83	10	3	73	0	59	31	15	2
Textiles, clothing and footwear	4	10	20	2	12	6	18	8	42
Nonpetroleum primary commodities	9	18	4	1	4	7	8	3	1
Food, beverages and tobacco	0	21	15	2	37	7	6	3	11
Wood and paper products	1	24	25	6	9	8	7	7	9
Chemicals, glass, clay and basic metals	0	3	13	3	11	3	9	15	9
Transport equipment	0	0	7	1	0	1	2	12	46
Other machinery	0	10	2	8	2	5	9	30	15
Leather and plastics	3	4	10	4	24	5	11	8	19

Note: a For identification purposes, the Australian input-output numbers of these commodity groups are: Petroleum and products 12.00 and 27.08; Textiles, clothing and footwear 23.01-24.03; Nonpetroleum primary commodities 01.01-11.02, 14.00, 16.00; Food, beverages and tobacco 21.01-22.01; Wood and paper products 25.01-26.05; Chemicals, glass, clay and basic metals 27.01-31.03; Transport equipment 32.01-32.04; Other machinery 33.01-33.07; Leather and plastics 34.01-34.05.

group as a whole. By 1980–81 textiles, clothing, and footwear accounted for 20 percent of Australia's imports from the Philippines, compared with 18 percent for the LDCs generally and 6 percent for ASEAN. On the other hand, petroleum and petroleum products are of minor importance among imports from the Philippines, but of major importance for the LDCs as a whole. In 1980–81, these commodities comprised only 3 percent of Philippine exports to Australia, but 31 percent of imports from the LDCs in general. Other moderately or lowly protected manufactured commodities, including sheet metal products and sawmill products, were more important among Philippine exports. In addition, highly protected motor vehicle parts comprised 7.2 percent of imports from the Philippines in 1980–1981, compared with 0.6 percent of ASEAN and 1.4 percent for the LDCs.

2 Effects of across the board reductions in protection

The impact of a country's protection on its imports from particular trading partners depends not only on the rates of protection it applies to the commodities exported by those trading partners, but also on the degree to which its imports of those commodities respond to protection. This depends on domestic supply and demand conditions. Countries exporting commodities which have highly elastic import demand within Australia could be more adversely affected by Australian protection than others which export commodities subject to higher rates of protection but for which Australia's import demand is less elastic. Differences among countries in this respect are missed by the sort of analysis presented in Table 10.1, but these differences are clearly central to an assessment of the degree to which imports from particular groups of countries would respond to reduced Australian protection.

This issue is handled in this paper with the aid of a multisectoral general equilibrium model of the Australian economy. This 109 sector model, known as the ORANI model,[9] belongs to the general class of Johansen multisectoral models and is linear in the percentage response of endogenous variables to given percentage changes in exogenous variables. For a given specification of endogenous and exogenous variables, the model computes a matrix of elasticities giving the response of the former to variations in the latter. The structure of this model is particularly well suited to the analysis of the general equilibrium effects of changes in protection.

Two variants of the ORANI model are used. These are referred to as the 1968–69 and 1977–78 variants, and are respectively based on the 1968–69 and 1974–75 Australian input-output tables, 1968–69 and 1977–78 import shares, and 1968–69 and 1977–78 data on tariff equivalent rates of protection by input-output commodity. In other respects, such as in the econometric estimates of domestic consumer demand elasticities, export demand elasticities and other parameters which underlie the model, the two variants are the same. In both variants, real aggregate absorption (consumer spending plus investment plus government spending, all in real terms) is determined endogenously, the balance of payments is fixed exogenously, and in both variants real wages are held fixed and labour markets are therefore slack. All imported commodities are assumed to be available in infinitely elastic supply.[10]

In this exercise, the ORANI model is used in conjunction with Australian trade data to estimate the effects of reduced Australian protection on its imports from various trading partners. The assumption is made that for each commodity the proportion of Australia's imports coming from any one country is unaffected by changes in levels of protection. This assumption is open to criticism. It would seem that the more finely trading partners are disaggregated, the more questionable this assumption will become, and the level of disaggregation used here (individual ASEAN countries) is relatively fine.[11] Nevertheless, it is not obvious what alternative assumption, or analytic procedure, would be more appropriate. Given this assumption, changes in protection will affect individual countries' exports to Australia in different proportions provided the proportional response of Australia's imports to a change in protection differs from one commodity to the other.

The simulated effects that reduced protection has on imports from LDCs are summarised in two ways. First, the dollar value of the increased imports from particular groups of countries that would result from a 30 percent reduction in all rates of protection is estimated. Second, the discriminatory impact of a uniform reduction in protection is captured by computing for each group of countries an *index of discrimination*. This consists of a ratio of that group of countries' *marginal share* to its *average share*. The marginal share is that group of countries' share of the increased volume of imports which enter Australia in response to the uniform reduction in protection. The average share is simply Australia's total imports from that country or group of countries as a proportion of Australia's total imports from all countries.

Clearly, a country or group of countries for which the index of discrimination exceeds unity is one whose share of Australia's imports would *rise* if Australia's protection was reduced, while a country for which the index is less than unity is one whose share of Australia's imports would fall. The higher the index of discrimination, the more heavily Australia's protection discriminates against imports from that country *vis a vis* others. Obviously, the weighted average of all indices of discrimination across trading partners is unity. Therefore, a country whose index is greater (less) than unity is one against which Australia's protection discriminates more (less) heavily than against the average of all other countries.

The results of this exercise are summarised in Table 10.3. These results relate to uniform across the board reductions in protection. The results of this exercise are presented for each of two years, 1968–69 and 1980–81. The 1968–69 and 1980–81 results use the 1968–69 and 1977–78 variants of the ORANI model in conjunction with Australian trade data corresponding to 1968–69 and 1980–81, respectively. As outlined above, the 1968–69 and 1980–81 interval captures significant changes in Australian protection policy. In particular, both the textile, clothing and footwear commodity group and the transport equipment commodity group, which includes motor vehicles, became more heavily protected over this period, while protection was lowered in other areas.

The results presented in Table 10.3 include the estimated effects of a 30 percent reduction in all rates of protection. A '30 percent reduction in all rates of protection' means that the nominal tariff equivalent of all rates of protection, including tariffs themselves and also quantitative restrictions, is lowered by 30 percent. For example, protection on a commodity receiving a 50 percent nominal rate would be lowered by 15 percentage points to 35 percent. A 30 percent rate is lowered by 9 points to 21 percent, and so on.

The estimated effects include an increase in imports from the ASEAN countries of only $A90,000 in 1968–69 and $A13.5 million in 1980–81. These amounts are expressed in current rather than constant prices, but even if inflated in 1980–81 prices the estimated effects for 1968–69 would still be trivial.[12] The estimated effects for 1980–81 are themselves surprisingly small. These results reflect: 1 the fact that Asutralia's imports of the particular commodities exported by the ASEAN countries are less responsive to reduced protection than all other commodities, on average; and 2 the fact that, according to the ORANI model, Australia's total imports would rise by only 2 percent, approximately, in response to a 30 percent across the board cut in protection.[13]

By comparison, the value of Australia's bilateral aid to the ASEAN countries in 1980–81 was $A68 million or five times the estimated effect on imports from ASEAN of a 30 percent across the board reduction in Australian protection. This bilateral aid is in grant form, rather than loans, and even though these funds are frequently tied to the purchase of Australian goods it seems reasonable to

Table 10.3 Australia's total imports from ASEAN and effects of a general cut in Australian protection, 1968–69 and 1980–81

	Actual value of total imports ($m)		Actual share of total imports (%)		Effects of across the board cut in protection			
					Value of increased imports of all commodities from 30% cut ($m)		Index of discrimination	
	1968–69	1980–81	1968–69	1980–81	1968–69	1980–81	1968–69	1980–81
Indonesia	60	416	1.79	2.25	−0.06	2.88	−0.035	0.296
Malaysia	30	183	0.88	0.98	−0.03	2.50	−0.040	0.588
Philippines	3	87	0.09	0.47	0.08	3.74	0.977	1.846
Singapore	12	504	0.37	2.72	0.06	3.46	0.196	0.294
Thailand	2	65	0.05	0.35	0.03	0.94	0.645	0.619
ASEAN total	106	1,255	3.19	6.67	0.09	13.52	0.031	0.461
LDC total	437	3,866	13.05	20.86	6.51	93.22	0.562	1.032
All countries	3,352	18,534	100	100	88.81	432.87	1	1

speculate that $A1 million worth of this aid is worth more, in the short run, to an ASEAN country than $A1 million worth of additional exports to Australia. The social cost of producing these exports has to be considered. Long run considerations involving the possible debilitating effects of dependence on aid and possible dynamic benefits of increased exports could point in a different direction. Economists do not have an impressive record of predicting political trends, but an across the board reduction in protection as large as 30 percent seems extremely unlikely within the foreseeable future. It can reasonably be said that, in the short run at least, the effects on ASEAN of more feasible across the board liberalisations would be quite small relative to the value of Australia's bilateral aid to ASEAN.

The estimated indices of discrimination reveal that the incidence of Australia's protection has shifted towards the ASEAN countries as a group relative to Australia's other trading partners. Nevertheless, although the index rose significantly, it was still well below unity in 1980–81. It cannot be said that Australian protection discriminates against ASEAN as a group, *vis a vis* other countries, or against Indonesia, Malaysia, Singapore, or Thailand. But this statement can be made, in quite a strong form, for the Philippines. The 1980–1981 index of 1.85 is one of the highest estimated for any of Australia's trading partners. It compares with Taiwan 2.43, China 2.18, Japan 1.85, India 1.76, and Korea 1.34. The explanation for this again lies mainly in the fact that of all the ASEAN countries, the Philippines' exports to Australia contain the highest proportion of textiles, clothing, and footwear. In 1980–81 this was 20 percent of Philippine exports, compared with Indonesia's 3.5 percent, Malaysia's 10 percent, Singapore's 2.2 percent and Thailand's 12 percent.

A significant feature of Table 10.3 is the rise between 1968–1980 in the estimated indices of discrimination against imports from the ASEAN countries as a group and each of the individual ASEAN countries except Thailand. Two possible reasons for this are:

1 that the changed structure of protection in Australia and/or other structural changes within Australia imply that imports of the commodities traditionally exported by the ASEAN countries to Australia are now restricted more heavily than before; and
2 that the composition of the ASEAN countries' exports to Australia has shifted towards commodities of which Australia's imports are highly responsive to reduced protection.

It is possible to distinguish between these two explanations by performing the conceptual experiment of combining the 1968–69 variant of the simulation model with the 1980–81 trade data. The estimated index of discrimination for ASEAN as a group is then 0.47. For the individual ASEAN countries, the corresponding estimates are: Indonesia 0.18, Malaysia 0.79, Philippines 1.81, Singapore 0.31, and Thailand 0.94. Clearly, the shift in the commodity composition of ASEAN exports to Australia is the important phenomenon, and not changes in Australian protection or changes in the structure of the Australian economy itself.

3 Effects of commodity specific reductions in protection

It is useful to examine the effects of reduced protection at a more disaggregated level. Of the 109 commodities specified in the Australian input-output tables, 80 are traded goods and 29 are nontraded. Of the traded goods, positive rates of nominal protection have been estimated for 61 items, most of which are manufactured commodities. The ORANI model is capable of estimating the effects of reduced protection on each of the individual items separately, holding the others constant, as well as the effects of reducing all, or groups of items, simultaneously. The results presented in this section of the paper relate to the general equilibrium effect that reductions in the rates of protection of *individual* commodities, holding all other rates constant, are estimated to have on Australia's total imports of *all* commodities from the ASEAN countries.[14]

The effect on imports for ASEAN of reducing protection on individual commodities varies widely. Even the signs of these effects vary. Reduced protection of many items actually causes total imports from ASEAN to decline, owing to the indirect stimulus given to the competitiveness of other import competing manufactures in Australia. This is true for 14 of the 61 tariff items in 1980–81. For individual countries, whose exports to Australia are more specialised, this is even more striking. For the Philippines, 35 of the 61 tariff items have this characteristic. This variation in the response of imports from the ASEAN countries to reduced protection at an individual commodity level is illustrated in Table 10.4. The effects of reduced protection on each of the 61 tariff items have been estimated, and a limited number of these have been selected for inclusion in the table. These items were selected on the basis of the value to the ASEAN group, or to individual ASEAN countries, of a commodity specific reduction in protection.

For ASEAN generally, and then each of the member countries, the commodities listed are those generating the greatest estimated increase in total exports to Austrlia from a 30 percent reduction in the

Table 10.4 Australia: Effects of cuts in protection of selected commodities, 1980–81

Country or group Commodity	Country share of imports of that commodity (%)	Commodity share of imports from that country (%)	Commodity nominal rate of protection (%)	Increase in imports of all commodities from 30% commodity specific cut in protection ($ '000)	Index of discrimination
ASEAN					
Clothing	9.6	1.9	70	4,680	2.00
Footwear	8.1	0.8	57	2,348	1.75
Household appliances	5.7	2.0	25	917	1.24
'Other manufactures'	4.8	2.9	18	857	0.77
Sheet metal products	7.8	0.8	24	680	1.22
Sawmill products	32.0	5.2	3	597	9.22
Motor vehicles and parts	0.4	0.6	56	−1,457	−0.13
Indonesia					
Clothing	3.4	1.9	70	1,820	2.34
Knitting goods	3.5	1.0	53	671	1.48
Footwear	1.3	0.4	57	460	1.03
Motor vehicles and parts	0	0	56	−226	−0.06
Malaysia					
Footwear	2.4	1.6	57	741	3.79
Household appliances	3.8	6.1	25	687	1.08
Margarine, oils and fats	30.0	11.3	9	429	28.46
Cotton, silk, flax yarns	2.3	4.0	40	403	3.99
Motor vehicles and parts	0	0.1	56	−819	−0.09
Philippines					
Clothing	3.7	10.3	70	1,908	11.79
Footwear	3.3	4.5	57	996	10.72
Motor vehicles and parts	0.3	7.2	56	213	0.27
Sheet metal products	1.9	2.8	24	167	4.40
Sawmill products	7.8	18.3	3	148	32.95
Cotton, silk, flax yarns	0.2	0.7	40	−226	−3.15
Singapore					
Clothing	0.8	0.4	70	352	0.375
Furniture, mattresses, brooms	8.9	1.1	21	323	4.24
Household appliances	1.9	1.1	25	323	1.08
Motor vehicles and parts	0.1	0.3	56	−395	−0.09
Thailand					
Clothing	0.8	3.0	57	336	2.78
'Other manufactures'	1.5	17.5	18	247	4.32
Cotton, silk, flax yarns	0.8	3.9	40	130	3.61
Motor vehicles and parts	0	0.1	56	−230	−0.39

For identification purposes, the Australian input-output numbers of the above commodities are, in the order listed: Clothig 24.02; Footwear 24.03; Household appliances 33.03; 'Other manufactures' 34.05; Sheet metal products 31.02; Sawmill products 25.01; Motor vehicles and parts 32.01; Knitted goods 24.01; Margarine, fats and oils 21.04; Cotton, silk, flax yarns 23.03; Funiture, mattresses, brooms 25.04.

protection of that commodity. In addition, one other commodity is selected in each case to illustrate the possibility of negative effects. In these latter cases, the commodity listed is the one for which this negative effect is largest. In all cases except the Philippines, this commodity is motor vehicles and parts. For the ASEAN group, a proportionate reduction in protection of the first six commodities listed generates 75 percent of the value of an across the board reduction in protection of that same proportionate amount. By extending the list to include the top 13 of the 47 commodities for which reduced protection has a positive effect on imports from ASEAN, the combined value would *exceed* the value of an across the board reduction. This concentration of effects is greater for individual countries. For the Philippines, for example, reduced protection of the first five commodities listed generates 92 percent of the value of an equiproportionate across the board reduction.[15]

It is important to understand the mechanism by which reduced protection of some commodities causes estimated imports from ASEAN to decline. In the case of motor vehicles and parts, none of the ASEAN countries, except the Philippines, exports these commodities to Australia in significant amounts. They therefore do not benefit substantially from the increased imports of motor vehicles and parts caused by reduced protection in this area. But reduced protection of motor vehicles and parts stimulates production of other manufactured commodities in Australia which the ASEAN countries do export in significant amounts. This includes, in particular, clothing and footwear. This causes total imports from ASEAN to fall.

The primary reason for the stimulation of production elsewhere derives from the fixed real wage assumption mentioned above in the description of the simulation model. The price of motor vehicles is a major component of the consumer price index (CPI). Reduced motor vehicle protection lowers this price component of the CPI, and the money wage corresponding to a fixed real wage thus falls in terms of the domestic prices of other manufactured goods. A lesser reason is that transport equipment is itself an input into the production of many other manufactured commodities. For both these reasons, reduced protection of motor vehicles improves the competitiveness of other import competing manufacturing industries and thus causes imports of these commodities to decline.

In a similar way, reduced protection of yarns and textiles causes total imports from the Philippines to decline. The reason for this is that the Philippines is a significant supplier of clothing to Australia, but not of yarns and textiles. Reduced protection of yarns and textiles improves the competitiveness of the Australian clothing industry, squeezing out imports of Philippine clothing.

4 Summary and conclusions

It is argued in this paper that over the period 1968–69 to 1980–81 the incidence of Australian protection shifted increasingly against the interests of the ASEAN countries *vis a vis* Australia's other trading partners. But this shift of incidence began from a very low base, and by 1980–81 it was still apparently the case that the proportional reduction in imports from ASEAN due to Australia's protectionist policy was much smaller than the proportional reduction of imports from all other countries. This statement also applies to each of the individual ASEAN countries except the Philippines. Moreover, this shift of the incidence of protection against the interests of the ASEAN countries was due primarily to changes in the composition of ASEAN exports to Australia, rather than to changes in the structure of Australian protection.

The ranking of ASEAN countries in order of the value of Australia's imports from them was, in 1980–81: Singapore, Indonesia, Malaysia, the Philippines, Thailand. But the ranking in the order of the degree to which the estimated value of these imports would rise in response to a uniform proportionate cut in Australian protection was: the Philippines, Singapore, Indonesia, Malaysia, Thailand. The difference is consequent upon the fact that Australia imports commodities from the Philippines which are highly sensitive to changes in rates of protection. These commodities include, in particular, clothing and footwear.

Reductions in rates of protection at the individual commodity level have widely varying effects on Australia's total imports from ASEAN. Reductions in some rates would increase total imports from ASEAN quite considerably, while reductions in others would actually cause them to decline. Negative effects of this type seem paradoxical but they arise from the fact that reduced protection of one commodity can cause imports of other commodities to decline. Of the 61 input-output commodities subject to protection, a given proportional reduction in the rates of protection of only six particular commodities would have 75 percent of the effect on imports from the ASEAN countries of that same proportional reduction applied to all 61 rates. Obviously, a general equilibrium analysis is essential if effects of this sort are to be detected, and the present paper has endeavoured to show how this can be done.

Endnotes

* This paper has benefited from the author's discussions with P.J. Lloyd. The analytic approach used draws in part on our joint paper, Warr and Lloyd (1983).

The paper also benefited from the computational assistance of Prue Phillips. The research makes extensive use of the ORANI general equilibrium model of the Australian economy developed as a component of the IMPACT project and described in Dixon et al. (1977, 1982).[1] Assistance with the ORANI simulations was provided by Robert Butterworth, Tony Lawson and Brian Parmenter.

1 Warr, P.G. and P.J. Lloyd (1983), 'Do Australian Trade Policies Discriminate Against Less Developed Countries?', *Economic Record*, 59, pp. 351–64; Dixon, P.B. et al. (1977), *ORANI, a General Equilibrium Model of the Australian Economy: Current Specification and Illustration of Use for Policy Analysis*, Canberra: Australian Government Publishing Service; Dixon, P.B. et al (1982), *ORANI: A Multisectoral Model of the Australian Economy*, Amsterdam: North-Holland.

2 In 1980–81 trade with the ASEAN countries accounted for 6.8 percent of Australia's imports and 8.4 percent of its exports. In 1962–63 the comparable shares were 4.7 and 4.2 percent, respectively.

3 This sensitivity is reflected in the report of the influential, government appointed (Harries) Committee on Australia's Relations with the Third World (1979), *Australia and the Third World*, Canberra: Australian Government Publishing Service. This report strongly advocates, p. 131, that rather than 'to seek to use our trade policies to isolate ourselves from the consequences of Third World industrialisation', Australia should 'set out deliberately and energetically to facilitate the transition to a more outward-looking Australian industrial structure'.

4 Warr, P.G. and P.J. Lloyd (1983).

5 The nominal rates of protection by input-output category on which these calculations are based represent the *tariff equivalent* of all major forms of protection including tariffs, excise taxes, and import quotas.

6 The LDCs are defined as in World Bank (1981), *World Development Report 1981*, Washington: World Bank, to be those countries with 1979 GNP per capita below US$4,400.

7 Industries Assistance Commission (IAC), (1981), *Trends in Australia's Trade in Manufactures*, Approaches to General Reductions in Production, Information Paper No. 3, Canberra: Australian Government Publishing Service, p. 113.

8 For discussion of trends in Australia's imports from the LDCs, see IAC (1981); and Anderson, K and Garnaut, R. (1982), *Australian Protection and Trade with Developing Countries*, Monograph to be published by the World Bank (forthcoming). For studies dealing with the effects on the LDCs of protection in other OECD countries, see Cline W.R. et al. (1978), *Trade Negotiations in the Tokyo Round: A Quantitative Assessment*, Washington: Brookings Institution; Krueger, A.O. (1978), 'Effects of Exports from New Industrial Countries on US Industries', in W. Kasper and T.G. Parry (eds), *Growth, Trade and Structural Change in an Open Australian Economy*, Centre for Applied Economic Research, University of New South Wales, Kensington, NSW; Birnberg, T.B. (1979), 'Trade Reform Options: Economic Effects on Developing and Developed Countries', in W.R. Cline (ed.), *Policy Alternatives for a New International Economic Order*, New York: Praeger; and Yeats, A.J. (1979), *Trade Barriers Facing Developing Countries*, New York: St. Martins Press.

9 Dixon, P.B. et al. (1977, 1982).

10 Further details on the characteristics of the simulation model used are presented in the Appendix to Warr, P.G. and P.J. Lloyd (1983).

11 There could be a quality response to protection which this approach will miss. This possibility seems strongest in the case of quota protected commodities such as clothing and footwear.

12 The negative value effects and indices of discrimination estimated for Indonesia

and Malaysia in 1968–69 are due to the fact that in the 1968–69 variant of the simulation model the estimated effects of a 30 percent across the board reduction in protection included a 1.13 percent decline in crude petroleum imports. In the 1977–78 variant this becomes a 0.16 percent rise. The estimated 1968–1969 effect occurs because of a 3 percent expansion in Australian crude petroleum, compared with only a 1 percent expansion in the 1977–78 variant.

13 The simulated effect of reduced protection on most endogenous variables is smaller than might have been expected. The reason is that the price elasticity of demand for imports which the parameters of the model imply is relatively low. According to the simulation model, a 30 percent reduction in all rates of protection leads to 1.95 percent rise in aggregate imports. Assuming an average rate of protection of 18 percent this implies a 4.57 percent reduction in the domestic price, implying a price elasticity of demand for aggregate imports of around −0.43. This lies within, but towards the lower end of, the range of estimates for other OECD countries reported in Cline, W.R. et al. (1978).

14 The indices of discrimination reported in Table 10.4 represent the change in imports of *all* commodities from the country specified in response to a cut in the rate of protection applying to a specific commodity relative to the corresponding change in total imports from all countries (the specified country's marginal share) divided by that country's share of imports of *all* commodities (its average share).

15 It should be stressed that these results are based on a static analysis and that dynamic considerations may point in a very different direction. As the comparative advantage of the most advanced LDCs shifts towards more capital and skill intensive goods, this transition might be inhibited by the bias in the present system of more developed country (MDC) protection which favours the most labour intensive goods.

11 Conclusions and suggestions for policy action

CHRISTOPHER FINDLAY AND ROSS GARNAUT*
December, 1985

THE analysis of Chapters 1 to 6 allows us to draw some general conclusions about the origins and prospects for the reform of manufacturing protection in the ASEAN countries and Australia.

There were periods over recent decades in which protection was widely believed to promote the national economic interest in all six countries. The particular benefits believed to derive from protection varied from country to country and over time. The major 'national interest' cases that have been made for protection are examined separately in the second section of this chapter: the need for government revenue; income distribution arguments; employment; balance of payments effects; diversification; and a desire to divert trade amongst trading partners.

Some of the national interest arguments were more influential at times of economic crisis—particularly crisis in the balance of payments, but sometimes in employment. Protection provided in times of crisis was often, even usually, maintained when the crisis receded.

Vested interest groups appeared to play a significant role in policy making in most countries. At very least, they were influential in entrenching any protection that had once been provided, whatever its origin. However, the statistical analysis failed to explain a major part of the inter-industrial variation in protection levels in terms of variations in measures of the power of vested interest groups, in any country other than Australia. These results are discussed in the third section of this chapter.

* Christopher Findlay is with the Economics Department, University of Adelaide and Ross Garnaut is with the Department of Economics, Research School of Pacific Studies, Australian National University.

The third section of the chapter also examines the attempts in recent times to liberalise import trade in Australia and the ASEAN region. There has been some success—most notably in Singapore where the reforms went so far as to positively promote exports at the expense of other, including import competing, production. The most recent history has included relatively little success, however, suggesting the need for new approaches and strategies if trade policy in ASEAN and Australia is to support effectively these countries' ambitions for continued strong economic growth.

The Chapter concludes in section 4 with suggestions for reform, focusing especially on political strategies which have their origins in the earlier analysis. The major suggestions cover the dissemination of information, the redesign of protection policy making institutions, the setting of the wider policy agenda, and the avoidance or at least anticipation of macro-economic dislocation. Finally, it is suggested that the political feasibility of import liberalisation could be enhanced through bilateral or, more likely, regional agreements on trade liberalisation, consistent with the patterns of liberalisation that would have been implemented unilaterally if this had been possible.

National interest arguments for protection

Revenue

A common explanation of tariffs is their role in raising revenue. At early stages of development characterised by low levels of per capita income and a relatively large resource or agricultural sector, it is sometimes cheaper administratively to raise government revenue through trade taxes rather than through income or expenditure taxes. This is because of the structure of employment contracts, the use of family labour, and the low share of goods and services which are exchanged in markets in the domestic economy. An exception to this point can occur when there is a significant resource sector which provides a large base for profit or output taxation.

The revenue motive is reportedly present at some time in every country in this volume. The taxes are also applied to exports. However, the structure of the comparative advantage of countries at that stage of development leads to imports of manufactures and exports of primary based products. As a result the import taxes applied to manufactured goods and export taxes to agriculture. This inevitably has resource allocation effects. In many cases the governments learned to manipulate these taxes, especially the tariff, to achieve particular production effects. Their motivation was also related in these cases to other factors which are discussed below.

Over time cheaper methods of taxation emerge.[1] For example, greater levels of employment in manufacturing industry or service activities facilitate the imposition of an income tax. When, however, cheaper methods of financing government spending emerged the tariff system proved difficult to dismantle. The tariffs had created a group of vested interests in the manufacturing sector who wished the protection system to be retained.

Income distribution

Another common motive was to use industry policy in an attempt to influence the income distribution. The model implicit was that protection would increase the output of the target industry or group of industries, thereby increasing the real incomes of factors used intensively in that industry. The relevance of this model is noted in the Australian paper where it is also pointed out that the analysis applied to a special case and hid from view alternative, more efficient, methods of achieving the income distribution targets.

This income distribution motive emerged in different ways in each country. For example, concern in Australia was to increase the population without causing a fall in the real wage and also to tax the rents of relatively rich landowners. In the ASEAN countries, the interests of groups regarded as economically weak has been a factor in the design of industry policy. In ASEAN, this did not lead directly to the imposition of a tariff, which in some cases was regarded as undesirable from the point of view of the target groups. Instead the policy introduced was a device for escaping a tariff that would otherwise have been applied.

The income distribution motive did not always lead to the adoption of a policy of tariff protection. Attitudes toward the Chinese community had a significant impact on policy design. For example, in Thailand in the late 1940s, the administration deliberately avoided the use of protection policy for fear of promoting the Chinese community. Instead of providing incentives in the form of tariffs, it chose to promote state enterprises. In the Malaysia in the late 1950s, it was felt that protective tariffs would benefit the Chinese and Indians much more than the Malays and this led to a policy focus on income tax measures rather than tariffs. However, these views had changed in Malaysia by the end of the 1960s.

Employment

In Australia, a component of the income distribution motive was to increase population and therefore the size of the work force. Also the argument that protection can be used to reduce high rates of unemployment has been influential in Australia, especially in long periods of unemployment. In Singapore in the early 1960s, the high rate of unemployment led the government to consider an industrialisation strategy that did not depend on entrepot trade. A new industrial policy making body formed in Malaysia in the late 1960s, following the racial riots, had the creation of employment opportunities as one of its objectives. Despite, however, the popular appeal of protection for this purpose, it actually has little effect on overall employment, as argued in the Australian paper, although it will affect employment in particular industries.

Balance of payments

In many countries tariffs or quantitative controls were used as an instrument to attain balance of payments targets. The implication was that failure to attain those targets meant considerable disruption to international trade and to access to imports. For example in the Philippines, combinations of foreign exchange controls, import licensing, multiple exchange rates and tariffs have been used in response to balance of payments problems, with the use of tariffs increasing since the 1960s. Similarly, multiple exchange rate systems, which had protective effects, were introduced in Indonesia in response to such problems. As pointed out in the Australian paper, however, these responses are not efficient. It is efficient to rely on a market for foreign exchange to resolve supplies and demands. Despite that argument, other instruments appear to have some popular appeal in a time of 'crisis' when people approve administrative solutions rather than reliance on a market mechanism. Perhaps the reason is that the market mechanism appears to be unreliable or much more risky than direct action.

Diversification

Another motive was to have a more diversified economy. This is argued to have been influential in the Australian case. It is also reported to have been a consideration in Malaysia in the past when it was aimed to reduce reliance on rubber and tin exports. A sophisticated version of the diversification motive has been developed more recently in Malaysia. The aim is to anticipate the process of capital intensification that is associated with growth and changing comparative advantage. This is reflected in the heavy industry promotion policy in Malaysia.

Another interesting application of this motivation is in Indonesia. Implicitly the argument there is that the oil boom will have placed contractionary pressure on the industrial sector but that when the oil runs out the country will have to industrialise again and this will impose significant adjustment costs. These costs can be reduced if a manufacturing sector is maintained during the resources boom and these costs are less than the costs of distorting the pattern of output in the economy.

The responses to these types of argument are twofold. One is empirical; the questions are, in the Indonesian case, how long will the oil last, or more generally, what is the likelihood of a disruption of supplies of a particular good or service, how long will that disruption last and what are the adjustment costs involved. Also, it is not clear why there are not natural forces which will sustain an industrial sector during a short term resources boom. For example, the adjustments required may be greater in the financial sector, and in the term structure of debt, for example, than in the structure of production.

A common characteristic of tariff structures is escalation. That is, the rate of nominal tariff tends to rise with the degree of processing so as to preserve or increase the effective rate of protection. The effective rate structure therefore appears to be biased in favour of finished or more highly processed goods, and there is a divergence between effective rates for intermediate and finished goods. This is reported in the Indonesian, Malaysian, Philippine and Thai country studies in this volume. A related strategy is to apply taxes on natural resource exports to encourage local processing.

The presence of tariff escalation could be related to the diversification motive. Countries which in the absence of protection would export raw materials and import processed products may prefer to have a wider range of industrial activities because that appears to leave them less exposed to reliance on world markets. However, the application of this strategy is not straighforward. The development of a processing industry may simply expose the country to new forms of insecurity, for example, about supplies of complementary materials needed for processing: energy is an example.

Furthermore, the unilateral adoption of a protectionist policy does not tackle the fundamental cause of insecurity in trade. This is the fear that other countries will themselves unilaterally act in a manner which is detrimental to the home country. The adoption of a protectionist policy only serves to reinforce these expectations. A method of breaking out of this spiral is to develop mutual confidence among trading partners, especially concerning the stability of each other's outward looking trade policies.

A focus on the behaviour of vested interests offers an explanation of tariff escalation. The resistance to protection tends to fall as group size increases. The size of the buyer group tends to fall as the product moves down the production chain, and reaches its largest when the finished product is sold to consumers. Therefore, the lack of political resistance could contribute to the escalating rates of protection.

Preferential trade policy

Another motive in the operation of protection policy has been to facilitate preferential access to domestic markets for particular countries or to retaliate against policy in other countries. For example, protection policy in Indonesia and Malaya in the 1930s was designed to discriminate in favour of their colonial administrations and against the Japanese. Singapore operated import licensing for imports from Japan and also used tariffs in retaliation to tariffs imposed by Malaysia in 1959. Malaysia, Singapore and Australia were party to Commonwealth country preferential arrangements and all the ASEAN countries are party to the ASEAN preferential trading arrangements (PTA) although the volume of trade under these is relatively small.

Inter-industry variation in protection and the role of vested interest groups

Typically there is very wide dispersion in rates of assistance to industries yet statistical attempts to explain this variation performed poorly both in this volume and for a sample of industrialised countries.[2] The method used in all these studies was to extend the model of supply and demand for protection in a political market to the industry level. A number of variables can be identified which are likely to influence the levels of protection. The variables reflect reasons why the losers from protection might be willing to supply it and reasons why the gainers operating solely in their own interests might be more successful in satisfying their demands. These variables include the share of exports or imports in total consumption, the industry's labour intensity, and a group of variables which reflect the returns to private interest lobbying.

In many cases a significant statistical relationship between the hypothesised variable and the level of protection was found. However, the set of variables overall tended to explain a low share of the total variation in protection, especially in the ASEAN countries. The Australian equation performed better than the others, suggesting that the importance of the role of vested interests is a less important part of the story in the ASEAN group. A related point is that the review of the history of protection policy in each country suggests that each was responding to a particular set of circumstances, or crises, and this, combined with the ratchet effects at work, may explain the poor performance of the statistical work. In other words, there are a lot of country specific variables which have been excluded from the statistical work.

The statistical analysis does suffer from other problems of specification. It is sometimes not clear whether the level of the hypothesised variable is a cause or a result of the protection. For example, highly protected industries are often found to have low levels of import penetration but this could reflect a high nominal tariff. The specification needs to be extended to allow for some of these simultaneity problems. There is also an identification problem in the equations. This is because some variables may have influences on both the supply and demand sides. For example, as an industry's size increses, it becomes more influential but also there are greater returns to consumers from becoming involved in a counter lobby.[3]

While it is difficult to do powerful tests of the impact of vested interests, it is possible to use this approach to explain particular events. For example the continuing high levels of protection for motor vehicles and TCF industries in Australia can be rationalised this way. It is related to the established lobbying capacity of the industries (a ratchet variable), the high numbers of workers employed, the regional location of the industries and the size of the crisis they faced due to the tariff cut of 1973 and, in the TCF case, due also to emerging competition from an industrialising Asia accompanied by increases in real wages in Australia.

Attempts at tariff reform

All the countries in the volume have also attempted across the board reforms of their protection policies. In some cases, this took the form of a switch from an active import substitution policy to a more explicit export promotion policy. For example in Thailand, export promotion policy was adopted after 1982. Singapore switched out of an import subsitution policy by the late 1960s. In 1973 in Australia, the Labor government cut all tariffs by 25 percent. In 1981 in the Philippines, a major tariff reform was announced. The Malaysian government made explicit in the Third Malaysia plan (1976–1980) its intention to reduce the level of effective protection for manufacturing.

Broadly there appear to be two sets of factors that contribute to these liberalisation moves. One is the economic conditions of the time, either good or bad. In Thailand, the commodities boom of the mid 1970s was argued to make it easier to shift to export promotion.

In Australia, the government used the tariff cut to reduce inflationary pressures. The policy change in Singapore appears to be in response to the accelerated withdrawal of the British. Crises therefore appear to change the political calculus and to switch it in favour of using protection policies, particularly unemployment or balance of payments crises, but there are also examples of crises stimulating a shift to more open policies, as in Singapore.

The other set of factors contributing to reform was the more widespread understanding in the community of the impact of protection. This knowledge was spread in a number of ways, either by local academics, by government officials (many trained overseas) or by the international agencies. The ASEAN paper writers stress the role of a coalition of these groups and not the dominance of any one of them. In particular, they stress that policy making has not been dominated by the international agencies. In Australia, contributions to the change in the climate of opinion are argued to have been made by academics, by the financial press, by officials and, as a result of lobbying, by primary sector interest groups which were hurt by protection.

One point that emerges is that the influence of the international agencies has not always been in favour of reduced levels of protection. A World Bank Mission to Malaysia in 1955 argued in favour of the use of tariffs as an instrument for industrialisation. However, this prescription was not accepted by the new government of Malaysia. Nevertheless, it is noted that a World Bank Mission to Malaysia in 1963 could have been influential in arguing for economic diversification based on manufacturing. Also in Singapore in 1960 a UN Industrial Survey Mission argued for infant industry protection, and this recommendation was accepted.

Impact of tariff reforms

In Singapore, the move to an open economy policy that began in the late 1960s was sustained during the period examined in the paper. In other countries the results of reform are mixed. In the Philippines, the protection structure after reform is argued to continue to favour materials and capital goods. However, it is also argued that the levels of protection are more even than before. In Thailand the policy change appears to have led to a more escalated protection system rather than an across the board liberalisation and also to greater use of administrative forms of protection, all without a significant reduction in the bias against the export sector. The author of the Malaysian paper reports that protection for the manufacturing sector did decline over the late 1970s and that the variation in effective rates also declined. He is concerned, however, that there is a new move toward protection of inefficient heavy industries in Malaysia. In Australia, while the average rate of effective protection for manufacturing declined, the variation in effective rates increased, principally because of increases in protection for motor vehicles, clothing, textiles and footwear, while rates for other industrial groups declined.

The Singapore country paper identifies a number of factors contributing to the stability of the policy in Singapore. These factors appear to be peculiar to Singapore. One is the size of the country and its more obvious inability to sustain growth without an open policy, especially in the entrepot activities in which Singapore had a traditional advantage because of its location. Singapore's location also gave it an advantage in later years when the economy was able to diversify into a wider range of internationally traded services. Furthermore the small area reduced the adjustment costs associated with the structural change as the economy grew, thereby reducing the demand for protection. No one parliamentary constituency could be identified with the interests of any particular industry in Singapore, so that it was difficult for any political entrepreneurs to marshall support through a protection campaign. Finally, the extraordinary economic success of Singapore during the period of free trade policies worked to entrench support for the maintenance of these policies.

In other countries, there are a number of factors which contributed to the mixed results of reform. First there is a ratchet effect. Once protection is granted to an industry or sector for whatever reason (revenue, balance of payments etc.) it generates considerable benefits for groups in that industry. These groups then have the incentive to lobby to maintain the policy. Other groups have difficulty in organising to work against the policy. Therefore the policy tends to take on a 'life of its own', even though rhetoric of the policy makers may be stressing efficiency of resource allocation and the value of reductions in protection. That is, the experience of the effects of protection tends to shift the political calculus.

Second, governments tend to pursue a number of objectives simultaneously. They may use trade policy to pursue a number of targets, and they may use a variety of instruments under the control of different agencies. This means that the resource allocation motives for reducing protection can be undone by actions on behalf of other objectives or actions by other agencies. For example, if tariffs are cut, a government agency in charge of promoting domestic investment will pursue its target with another policy, such as tax incentives, so that the effective rate of assistance may not change. Both the conflicting objectives and the problem of pursuing too many targets with the one instrument are highlighted in the Indonesian paper.

The interactions between policies can be potentially very important for influencing the politics of protection. For example, the earlier discussion indicates how public finance policy or exchange rate policy affects the willingness of the community to support protection. Also at the level of the individual, taxation policies which affect the forms in which people save tend to exaggerate the costs of adjustment that must occur in an open economy by tying their wealth to the wealth of a particular industry or location. Similarly tax policy affects the willingness of industry to take risks and therefore its unwillingness to operate in what appears to be a more risky open economy.

Strategies for reform

While hypotheses about the causes of protection and of the choice of policy instrument have not been subject here to powerful testing, it is possible to make some suggestions about strategies for more extensive and longer lasting reform. The aim of these suggestions is to strengthen Government's capacity to resist pressures for protection and to reduce pressures for protection from private vested interests.

Unilateral action

The standard theoretical result is that unilateral action to reduce trade barriers improves the welfare of the liberalising country. There is a number of variants of unilateral action. One is an across the board and complete liberalisation. A less drastic version is to put a ceiling on rates of protection, as suggested in the Malaysian paper, or to put a freeze on current rates. The latter implies no decreases, thereby lowering the uncertainties for industries already protected, but also implies granting no increases either. It means no introduction of new categories for tariffs. This has been called the 'Do Little' strategy for reform.[4] Obviously time will wear down the significance of protection more quickly where specific rates of duty are applied and there is inflation. A difficulty is that it should ideally be a comprehensive policy, covering all forms of assistance. In the case of quotas, it could be interpreted as allowing imports to take up any market growth, which is the intention under the Australian TCF plan. Monitoring forms of assistance other than tariffs which the country studies show are prevalent may be difficult. Also the freeze needs to be complemented by other actions which will permit governments to sustain it in the face of new 'crises'.

With the exception of Singapore, unilateral tariff reform has had limited effects in the countries studied. This was so even if the reductions were across the board. The 1973 tariff cut in Australia had a substantial permanent effect on lowering average rates of manufacturing protection, but it can be argued to have generated new forms and higher levels of protection for TCF and motor vehicles.

Information

An obvious further step is to continue to generate information about the levels of assistance and their effects. The relevance of collecting and reporting information by either government agencies or academics is supported by the general view that their previous work has had an impact on industry policy, through its effects on the climate of opinion.

The questions arise also of the methods of generating information about the impact of protection and the relevance of NRPs, ERAs and computable GE models. In the Introduction to this volume we outlined some of the criteria for choosing between these approaches. This topic is taken up in detail by Harrison and Rutström in their chapter.

They compare ERP estimates derived using the standard procedure, which they call basic ERPs, with estimates of changes in value added from the results of a GE model in which global trade liberalisation is simulated. They argue that the GE estimates are generally lower in absolute terms than the basic ERPs. They attribute this to the items excluded from the basic ERP calculation. These are the effects of tariffs in trading partners, equivalent to export taxes on final goods, and substitutability between intermediate goods which their simulation incorporates. They also attribute the difference to the constraints on growth of highly protected sectors by the changes in input prices and conversely the benefits to more lightly protected sectors. In some cases, the measure of the resource pull effect even changes sign.

While the GE modelling techniques need further refinement, Harrison and Rutström make a case for attempting to measure these effects through GE modelling. As the authors argue their case, a large amount of effort and of information is required to compute an ERP with strong predictive power. Instead they argue, effort on that scale is more productively directed to GE computations. Our own view is that the standard, and simpler, ERP analysis can generate operationally useful results in many circumstances, but that it is worthwhile developing GE modelling further with the aim of obtaining even more powerful guidance for policy.

Work on the impact of protection should also stress the problem of aligning targets and instruments. It should stress the availability of alternative policy instruments for publicly supported objectives such as income distribution, balance of payments or employment targets. In the ASEAN countries trade policy is used alongside investment incentives as an instrument of industry policy. It would be interesting to study in more detail some of the debates, for example in Malaysia, on the choice between these instruments.

The review in this volume of the history of industry policy and the failure of the econometric studies to explain a large part of the variation in industry protection levels, suggests there are important country-specific factors which, combined with ratchet mechanisms, influence levels of protection. The best approach to understanding these factors is to refine the historical work that has been begun here and to attempt to explain in more detail some of the turning points in industry policy in each country.

Redesign of protection policy making institutions

Some of the country papers suggest the need for new institutions for reviewing industry assistance policy in a rational and systematic way. The Indonesian paper argues for an institution that will overhaul the tariff system, make decisions on tariffs and other protection policy and set the timetable for protection policy changes in order to reconcile conflicting objectives of industry policy in Indonesia. The Malaysian paper points out the value of having more information on which to base policy and argues that this will require more resources than the industry policy authority now has.

A model for these sorts of proposals might be the Australian Industries Assistance Commission (IAC) which is an advisory body and not a final decision making authority. Decisions on policy remain with the government. There are different views in Australia about the contribution of the IAC. One is that the Commission makes a substantial contribution to reducing protection in two ways: its reports and public enquiries have contributed to changing public opinion on protection, and most of the IAC's recommendations during its systematic and general review of the tariff in the 7 year period after 1974 were accepted by governments, both Labor and conservative.

The other view is that the IAC has had little effect. The industries affected by its general review of the tariff accounted for only 10 percent of manufacturing value added and the average effective rate for those industries was only halved. Protection for TCF and motor vehicles increased enormously during the same period and reports by the IAC on those industries, as well as iron and steel, were rejected by the governments of the time (in the case of steel, in favour of an outcome that provided less assistance than had been recommended by a majority of Commissioners in the IAC's draft report). The contribution of the IAC to public opinion is difficult to identify. Finally, the IAC can be interpreted as simply a political device designed to act as an umpire between the conflicting claims of various interest groups. The IAC in this view represents the institutional form of the political market for protection. It reduces the costs of putting political pressures, and reduces the rewards to a counter lobby by consumers, by using a case by case approach to its policy reviews. It also reduces lobbying costs by providing a stable environment for policy making and it legitimises the granting of protection because of the enquiry that precedes the decision.[5]

Of course, a difficulty with an assessment of the latter view is that it is impossible to tell what the outcome would otherwise have been. It may be that in the absence of this umpire the outcome would have been even higher protection. The staff of the commission, by their analytical and empirical work, and the public hearings process constrained the claims that the private interests could make. Also the IAC consistently provided a systematic view of the levels of protection and their impacts. These are the important contributions organisations of this type can make, even if their recommendations are not always accepted, and this should be borne in mind in their design.

In general, we would say that the public information role of the IAC has been of substantial assistance to the cause of trade liberalisation through its effects on the general climate of opinion on protection, but that its case by case approach has severely limited its role as a more direct agent for reduction in protection.

Agenda setting

Following the work on the levels and impacts of protection, another important component of a liberalisation strategy is to set an agenda for reform. This is because of the interaction between policies not specifically directed to industry targets and the demands for protection. For example, a flexible exchange rate regime and supporting institutions for trading risks and for financing firms across longer run fluctuations in exchange rates seems to be a substantial aid in building support for abandoning or reducing import controls. A public finance system that does not rely heavily on trade taxes as a revenue source and an income tax system which does not encourage demands for protection because of its distorting effects are also important to the building of an environment favourable to trade liberalisation. It is also important to design policy toward inter-regional relationships within a country, for example, Federal-State relations in Australia, so that it does not exaggerate demands for protection. The relevance of policy interactions appears to be well understood in Singapore whose manpower, education and welfare, as well as trade, policies have been nominated as contributors to its growth.

Individual politicians can be important in the development of agendas for reform. In economic models the importance of individuals in events is usually discounted. For example, in the political processes that were examined statistically in several of the country papers, there is no distinction between the intermediaries in the political process. However, an important feature of the political process may be the many coalitions of support that a politician can marshall in order to stay in power. In these circumstances, the choice of groups to approach will depend on the preferences of the political entrepreneur. In these conditions, the background, interests, and training of a politician can influence the outcome. Whether the ruling party builds a coalition of open economy forces or protectionist forces for example may depend on the party leadership. There is some evidence that policy in Australia, especially the 1973 tariff cut, in Thailand, and in Singapore has depended on the role of particular people.

Anticipation of crises

A theme of the historical review of industry policy was its use as a reaction to 'crises', such as balance of payments problems for the whole country or massive import penetration at the industry level. It would be useful to have a longer term view of the economy, and of the inevitable changes in the competitiveness of trading partners as they grow, so that crises of this sort can be easily anticipated. Sudden crises during which policy is vulnerable to protectionist pressures might then be avoided.

Bilateral and multilateral action

There are commercial trade policies which can facilitate reform. These lead to simultaneous liberalisation in two or more countries. Their aim is to reduce demands for protection by promoting growth and therefore reducing the costs of adjustment due to the removal of trade barriers. The opportunity for export growth also raises the opposition to protection, that is, it motivates the counter lobbies.

This strategy could be implemented through bilateral or multilateral action. Bilateral action would focus on countries with whom the complementarity in import and export composition was high but where there were severe resistances to trade. An example of this strategy is the 1957 trade agreement between Japan and Australia.

Multilateral action could operate through GATT or through regional trade liberalisation.

Some estimates of the impact of multilateral trade liberalisation are provided by Harrison and Rutström. Their GE estimates of welfare gains are small as a proportion of national output, for example, up to about 5 percent of GDP for global liberalisation of trade in all sectors or just over 3 percent of GDP for global liberalisation of trade in manufactures. The Harrison and Rutström model includes each of the ASEAN countries and Australia, plus Korea, Japan, Canada, the USA and the EEC while remaining countries are combined into a 'rest of the world' aggregate. However, though small in percentage terms, the gains are large in absolute terms. Furthermore, they are an annual gain, not just a one-off event, so that the accumulated gain is significant. The GE methodology is limited in that it incorporates no dynamic effects of protection, for example, on innovation and managerial performance, and no scale effects. Harrison and Rutström canvas the latter as an important extension to their work along with the incorporation of rent seeking activity. Taking these effects into account would increase the size of the estimated gain from liberalisation.

Information on the impact of protection on the pattern of trade is useful for the design of commercial strategies. The chapter by Peter Warr in this volume applies a GE model to measure the impact of Australian protection on its imports from the ASEAN countries. As he shows, this depends on the commodity composition of ASEAN exports and the inter-industry structure of Australian protection. However, the impact also depends on the elasticity of import demand. An efficient way to intergrate information on protection, import volumes and demand elasticities is in a GE model. Warr finds that Australian protection policy does not discriminate against ASEAN as a group although it does discriminate against the Philippines. He accounts for this by noting that of all the ASEAN countries, the Philippines' exports to Australia contain the highest proportion of textiles, clothing and footwear. At the individual commodity level, the GE analysis produces some paradoxical results. For example, reduced protection for some commodities can cause imports from ASEAN to decline. This is because the reduction in protection increases the competitiveness of other Australian activities which causes imports to decline. The GE analysis can also identify commodity markets for which liberalisation can have relatively greater effect.

In all cases, the aim of the commercial policy action should be to increase the volume of trade without discrimination. Carl Hamilton in his chapter examines the impacts and highlights the problems of discriminatory trading arrangements. He examines analytical questions concerning the impact of voluntary export restraints and preferential import quotas, compares them to global import quotas and looks at the actual practice of the administration of VERs in the ASEAN countries.

From the political economy point of view, an important result is that the VERs transfer rents from protection to exporting countries, thereby buying some support for protection from exporters. However, at the same time they create conflicts among exporters, particularly the way VERs rigidify market shares and deny 'latecomers' access. Hamilton also points out that from the importing country point of view the VER offers less control over the level of protection than directly administered policies. He also stresses the long run implications of using discriminatory policies: which is the retaliation that they will attract from the excluded suppliers.

By promoting a simultaneous improvement in market access, commercial policies can also reduce those demands for protection which are derived from the perceived insecurities associated with reliance on world markets. The presence of bilateral and multilateral strategies, however, does not diminish the case in principle for unilateral action. Indeed, some unilateral action may be a necessary component, to illustrate a willingness to put 'our own house in order', of the broader strategy.[6]

Endnotes

1 It is possible to design a test for the statistical significance of the revenue motive. All taxes impose deadweight costs and if the tax system is optimal from the efficiency point of view these costs at the margin should be equal for all taxes. In that case, the tax system will achieve its revenue target at minimum deadweight cost. The information required for this test could be derived quickly from a GE model, if available, for the country of interest.

2 Anderson, K. and Baldwin, R. (1986), 'The Political Market for Protection in Industrial Countries: New Empirical Evidence', in S. Odano (ed.), *The State of the World Economy: Essays in Honour of Kiroshi Kitamura*, London, Macmillan.

3 The functional form of these econometric equations has also been criticised. Dougan, W.R. (1984), 'Tariffs and the Economic Theory of Regulation', in *Research in Law and Economics*, Vol. 6, pp. 187–210, argues that if the tariff rate is used as a dependent variable in the equations, then the factors which influence the political power of various groups will enter the equation in a nonlinear way, not via a linear function which has been used in the analysis to date. Dougan also argues that the tariff rate is not the appropriate dependent variable if the problem is to test hypotheses about political power. He points out that the observed tariff will also depend on the elasticities of supply and demand. That is, the observed tariffs will differ between industries even if relative levels of political power of either consumer groups or owners of fixed factors are the same. His technique is to derive the implied indices of political power, using observations of tariff rates and of

supply and demand elasticities, and then attempt to explain variation in these indices, and not variation in tariff rates, using industry characteristics.

4 Pincus, J.J. (1982), 'Industry Assistance—Prospects for Long-Run Reform of Protection', Summary of talk given to Australian Agricultural Economics Society, Canberra, 19 August, 1982.

5 ibid.

6 Anderson, K. et al. (1985), 'Pacific Economic Growth and Prospects for Australian Trade', Research Paper No. 122, Canberra: Australia-Japan Research Centre.

Bibliography

Akrasanee, N. (1973), 'The Manufacturing Sector in Thailand: A Study of Growth, Import Substitution and Effective Protection 1960–1969', PhD Dissertation, Johns Hopkins University.

——(1975), *The Structure of Effective Protection in Thailand: A Study of Industry and Trade in the Early 1970s*, March.

——(1977), *Industrial Development in Thailand*, Report to the World Bank, September.

Akrasanee, N. and Jongtanasarnsombat, S. (1980), 'Protection and Control on Industry', Thai University Research Association: Research Report No. 3, NESDB, November.

Akrasanee, N. et al. (1982), 'The Electrical and Electronics Industry in Thailand', Bangkok: National Economic and Social Development Board, (mimeograph).

Alikhani, I. and Havrylyohyn, O. (1982), 'The Political Economy of Protection in Developing Countries: A Case Study of Colombia and South Korea', Division Working Paper, No. 1982–4, World Bank, July.

Allen, G.C. and Donnithorne, A.G. (1954), *Western Enterprise in Indonesia and Malaya*, London: George Allen and Unwin.

Allen, R.G.D. and Ely, J.E. (eds), (1953), *International Trade Statistics*, New York: Wiley.

Anderson, Karl. (1938), 'Protection and the Historical Situation: Australia', *Quarterly Journal of Economics* 53(4): 86–104.

——(1939), 'The Australian Case for Protection Re-examined: Comment', *Quarterly Journal of Economics* 54(4): 149–51.

Anderson, Kym. (1980), 'The Political Market for Government Assistance to Australian Manufacturing Industries', *Economic Record* 56(153): 132–44.

——(1981), 'Northeast Asian Agricultural Protection in Historical and Comparative Perspective: The Case of South Korea', Research Paper No. 82, Australia-Japan Research Centre, Australian National University, May.

——(1986), 'Tariffs and the Manufacturing Sector', in R. Maddock and I. McLean (eds), *The Australian Economy Since 1900: Performance and Policies*, Cambridge: Cambridge University Press, (forthcoming).

Anderson, K. and Baldwin, R. (1986), 'The Political Market for Protection in Industrial Countries: New Empirical Evidence', in S. Odano (ed.), *The State of the World Economy: Essays in Honour of Hiroshi Kitamura*, London, Macmillan.

Anderson, K. and Garnaut, R. (1980), 'ASEAN Export Specialisation and the Evolution of Comparative Advantage in the Western Pacific Region', in R. Garnaut (ed), *ASEAN in a Changing Pacific and World Economy*, Canberra, Australian National University Press, pp. 374–412.

Anderson, K. and Garnaut, R. (1982), *Australian Protection and Trade with Developing Countries* (forthcoming volume in a series to be published for the World Bank).

Anderson, Kym et al. (1985), 'Pacific Economic Growth and Prospects for Australian Trade', Research Paper No. 122, Canberra: Australia-Japan Research Centre.

Ariff, K.A.M. (1975), 'Protection for Manufactures in Peninsula Malaysia', *Hitotsubashi Journal of Economics*, vol. 15(2), February.

Ariff, K.A.M. and Hill, H. (1986), *Export-oriented Industrialisation: the ASEAN Experience*, Sydney, Allen and Unwin.

Arndt, H.W. (1965), 'Australia-developed, Developing or Midway?', *Economic Record* 41(95): 318–40.

Asher, M.G. and Booth, A. (1983), *Indirect Taxation in ASEAN*, Singapore: Singapore University Press.

Australian Development Assistance Bureau, (1982), 'Australian Official Development Assistance to Developing Countries, 1981–82: Statistical Summary', Canberra: mimeo.

Balassa, B. (1965), 'Tariff Protection in Industrial Countries: An Evaluation', *Journal of Political Economy*, vol. 73(6), December.

Balassa, B. (ed), (1971), *The Structure of Protection in Developing Countries*, Baltimore: Johns Hopkins University Press.

Balassa, B. and Associates (1982), *Development Strategies in Semi-industrial Economies*, World Bank.

Baldwin, R.E. (1969), 'The Case Against Infant Industry Protection', *Journal of Political Economy* 77(3): 295–305.

Baldwin, R. (1975), *Foreign Trade Regimes and Economic Development: The Philippines*, New York: NBER Inc.

Barber, A. (1939), 'Six Years of Economic Planning in Netherlands India', *Far Eastern Survey*, 16 August.

Bautista, R.M. (1978), 'Domestic Resource Costs in Philippine Manufacturing Industries', *Philippine Economic Journal*, SVII, 3 November.

Bautista, R.M., Poer, J. and Associates (1980), *Industrial Promotion Policies in the Philippines*, Philippine Institute for Development Studies.

Bautista, R.M. and Teeson, G.R. (1978), 'Plant Characteristics and Export Orientation in Philippine Manufacturing', *Philippine Review of Business and Economics*, 15, September.

Benham, F.C. (1926), 'The Australian Tariff and the Standard of Living: A Reply', *Economic Record* 2: 20–42.

——(1927), 'The Australian Tariff and the Standard of Living: A Restatement', *Economic Record* 3: 239–48.

Bergsten, F. (1975), 'On the Non-Equivalence of Import Quotas and 'Voluntary' Export Restraints', in F. Bergsten (ed.), *Towards a New World Trade Policy*, Lexington, Mass.: Lexington Books.

Bhagwati, J.N. (1964), 'The Pure Theory of International Trade: a Survey', *Economic Journal*, 74 (213), March, pp. 1–84.

——(1982), *Import Competition and Response*, Chicago: University of Chicago Press.

Bhagwati, J.N. and Srinivasan, T.N. (1973a), 'The General Equilibrium Theory of Effective Protection and Resource Allocation', *Journal of International Economics*, vol. 3, August, pp. 259–282.

Bhagwati, J.N. and Srinivasan, T.N. (1973b), 'The Ranking of Policy Interventions Under Factor Market Imperfections: The Case of Sector-Specific Sticky Wages and Unemployment', *Sankhya*, Ser, B.

Bhagwati, J.N. and Srinivasan, T.N. (1974), 'On Reanalyzing the Harris-Todaro Model: Policy Rankings in the Case of Sector-Specific Sticky Wages', *American Economic Review*, vol. 64, June.

Bhagwati, J.N. and Srinivasan, T.N. (1975), 'Alternative Policy Rankings in a Large, Open Economy with Sector-Specific, Minimum Wages', *Journal of Economic Theory*, vol. 11, December, pp. 356–371.

Bhagwati, J. and Srinivasan, T.N. (1976), 'Optimal Trade Policy and Compensation under Endogenous Uncertainty: The Phenomenon of Market Disruption', *Journal of International Economics*, 6, 317–36.

Birnberg, T.B. (1979), 'Trade Reform Options: Economic Effects on Developing and Developed Countries', in W.R. Cline (ed.), *Policy Alternatives for a New International Economic Order*, New York: Praeger, pp. 217–83.

Blomqvist, A.G. (1979), 'Urban Unemployment and Optimal Tax Policy in a Small, Open Dual Economy', *Journal of Development Studies*, vol. 15(2).

Blomqvist, A.G. and McMahon, G. (1983), 'Simulating Commercial Policy in a Small, Open Dual Economy with Urban Unemployment: A General Equilibrium Approach', Working Paper No. 8403C, Centre for the Study of International Economic Relations, the University of Western Ontario, October, (forthcoming, *Journal of Development Economics*).

Boadway, R. and Treddenick, J. (1978), 'A General Equilibrium Computation of the Effects of the Canadian Tariff Structure', *Canadian Journal of Economics*, vol. 11, pp. 424–446.

Boediono, (1972), 'Foreign Exchange Policies for Stabilization: The Indonesian Experience 1950–1965', MEc. thesis, Department of Economics, Monash University.

Boeke, J.H. (1946), *The Evolution of the Netherlands Indies Economy*, New York: Institute of Pacific Relations.

Booth, A. (1980), 'The Economic Impact of Export Taxes in ASEAN', *Malay Economic Review*, April, 25(1), pp. 36–61.

——(1983), 'Foreign Trade and Economic Growth in the Colonial Economy', Paper given at the Conference on Indonesian Economic History in the Dutch Colonial Period.

Borrie, W.D. (1963), 'The Peopling of Australia', in H.W. Arndt and W.M. Corden (eds), *The Australian Economy: A Volume of Readings*, Melbourne: Cheshire.

Breton, A. (1974), *The Economic Theory of Representative Government*, Chicago: Aldine.

Brigden, J.B. (1925), 'The Australian Tariff and the Standard of Living', *Economic Record* 1(1): 29–46.

——(1927a), 'The Australian Tariff and the Standard of Living: A Rejoiner', *Economic Record* 3: 102–16.

——(1927b), 'Comment on Mr. Benham's Restatement', *Economic Record* 3: 249–51.

Bridgen, J.B. et al. (1929), *The Australian Tariff: An Economic Enquiry*, Melbourne: Melbourne University Press.

Broek, J.O.M. (1942), *Economic Development of the Netherland Indies*, New York: Institute of Pacific Relations.

Brown, F. and Whalley, J. (1980), 'General Equilibrium Evaluations of Tariff-Cutting Proposals in the Tokyo Round and Comparisons with More Extensive Liberalization of World Trade', *Economic Journal*, V. 90, pp. 838–866.

Buchanan, J. and Tullock, G. (1962), *The Calculus of Consent*, Ann Arbor: University of Michigan Press.

Cable, V. (1981), 'An Evaluation of the Multifibre Arrangement and Negotiating Options', *Commonwealth Economic Papers*, no. 15, London: Commonwealth Secretariat.

——(1983), *Protectionism and Industrial Decline*, London: Hodder and Stoughton.

Cain, N. (1973), 'Political Economy and the Tariff: Australia in the 1920s', *Australian Economic Papers* 12: 1–20.

Carmody, A.T. (1952), 'The Level of the Australian Tariff: a Study in Method', *Yorkshire Bulletin of Economic and Social Research* 4(1): 51–65.

Chee, P.L. (1983), 'The Malaysian Car Industry at the Crossroads: Time to Change Gear?', paper presented at the Seventh Malaysian Economic Convention, Kuala Lumpur, January 1983.

Cline, W.R., Kawanabe, N., Kronsjo, T.O.M. and Williams, T. (1978), *Trade Negotiations in the Tokyo Round: A Quantitative Assessment*, Washington, D.C.: Brookings Institution.

Committee on Australia's Relations with the Third World (Harries Committee) (1979), *Australia and the Third World*, Canberra: Australian Government Publishing Service.

Cooper, R.N. (1974), 'Tariffs and Smuggling in Indonesia', in J. Bhagwati (ed), *Illegal Transactions in International Trade*, Amsterdam: North Holland.

Cooper, R.N. and Glassburner, B. (1973), 'Foreign Exchange Policy and the Foreign Trade Sector: Suggestions for Indonesia's Second-Five-Year-Plan', monograph, August.

Cooper, R.N. and White, L.J. (1970), 'Some Proposals Regarding Indonesia's Tariff System', monograph, August.

Cooper, R.W. (1964), 'Tariff Dispersion and Trade Negotiations', *Journal of Political Economy*, vol. 72, 1964, pp. 597–603.

Copland, D.B. (1931), 'A Neglected Phase of Tariff Controversy', *Quarterly Journal of Economics*, 45(1): 289–303.

——(1935), 'Notes on Tariff Theory with Special Reference to the Australian Tariff', *Economic Record*, 11 (Supplement): 33–39.

Corden, W.M. (1963), 'The Tariff', Chapter 6 in A. Hunter (ed.), *The Economics of Australian Industry*, Melbourne: Melbourne University Press.

——(1966), 'The Structure of a Tariff System and the Effective Protective Rate', *Journal of Political Economy*, vol. 74, June, pp. 221–237.

——(1967), 'The Exchange Rate System and the Taxation of Trade', in T.H. Silcock (ed.), *Thailand: Social and Economic Studies in Development*, Australian National University Press, pp. 151–169.

——(1971), *The Theory of Protection*, London: Oxford University Press.

——(1972), 'Economies of Scale and Customs Union Theory', *Journal of Political Economy* vol. 80, pp. 465–475.

——(1974), *Trade Policy and Economic Welfare*, Oxford: Clarendon Press.

——(1976), 'Customs Union Theory and the Nonuniformity of Tariffs', *Journal of International Economics*, vol. 6, pp. 99–106.

——(1979), 'Tell Us Where the Extra Jobs Will Be', *Bank of New South Wales Review* 30: 1–5, October.

——(1981), 'Exchange Rate Protection', in R.N. Cooper (ed.), *The International Monetary System Under Flexible Exchange Rates*, Cambridge, Mass: Ballinger.

——(1982), 'The Normative Theory of International Trade', in R. Jones and P.B. Kenen (eds), *Handbook of International Economics*, vol. 1, Amsterdam: North Holland Publishing Co.; forthcoming.

——(1984), 'Booming Sector and Dutch Disease Economics: A Survey', *Oxford Economic Papers*, forthcoming.

Corden, W.M. and Mackie, J.A.C. (1962), 'The Development of the Indonesian Exchange Rate System', *Malayan Economic Review*, vol. 7, no. 1, April.

Corden, W.M. and Neary, J.P. (1982), 'Booming Sector and Deindustrialization in a Small Open Economy', *Economic Journal*, 92 (368).

Corden, W.M. and Warr, P.G. (1981) 'The Petroleum Boom and Exchange Rate Policy in Indonesia'. *EKI*, 29(3).

Cox, D. and Harris, R. (1984), 'Trade Liberalization and Industrial Organization: Some Estimates for Canada', *Journal of Political Economy*, vol. 92.
Crawford, J.G. (1934), 'Tariff Level Indices', *Economic Record*, 10: 213–21.
CSIS, (1972), *Industrialisasi Dalam Rangka Pembangunan Nasional*, Jakarta: February 1982.
Currie, J.M., Murphy, J.A. and Schmitz, A. (1971), 'The Concept of Economic Surplus and Its Uses in Economic Analysis', *Economic Journal*, December, 81 (324), pp. 741–99.
de Groot, G. (1982), 'The Multi Fibre Arrangement and the Restructuring of the Dutch Textile and Clothing Industry', Occasional Paper No. 13, 1983, Development Research institute, Tilburg University, Tilburg, The Netherlands.
Dixit, A. (1971), 'Short-Run Equilibrium and Shadow Prices in the Dual Economy', *Oxford Economic Papers*, pp. 384–399.
Dixon, P.B., Parmenter, B.R. and Powell, A.A. (1983), 'Trade liberalization and labour market disruption', General Paper No. G-46, IMPACT Project, University of Melbourne.
Dixon, P.B., Parmenter, B.R., Ryland, G.J. and Sutton, J. (1977), *ORANI, A General Equilibrium Model of the Australian Economy: Current Specification and Illustration of Use for Policy Analysis*, Canberra: Australian Government Publishing Service.
Dixon, P.B., Parmenter, B.R., Sutton, J. and Vincent, D.P. (1982), *ORANI: A Multisectoral Model of the Australian Economy*, Amsterdam: North-Holland.
Dougan, W.R. (1984), 'Tariffs and the Economic Theory of Regulation', in *Research in Law and Economics*, vol. 6, pp. 187–210, Greenwich, Conn., JAI Press.
Downs, A. (1957), *An Economic Theory of Democracy*, New York: Harper and Row.
Edwards, C.B. (1975), *Protection, Profits and Policy: An Analysis of Industrialisation in Malaysia*, 2 Volumes, Norwich: University of East Anglia.
Edwards, C.T. (1965), 'The Structure of Import and Excise Duty Taxation in the States of Malaya and Singapore', *Malayan Economic Review*, 10(2), October.
Ethier, W. (1983), *Modern International Economics*, New York and London: Northon and Company.
Evans, H.D. (1971), 'The Empirical Specification of a General Equilibrium Model of Protection in Australia' in H.E. Grubel and H.G. Johnson (eds), *Effective Tariff Protection*, Geneva, GATT.
Falvey, R. (1979), 'The Composition of Trade within Import-restricted Product Categories', *Journal of Political Economy*, 87(5).
Federal Industrial Development Authority (1970), *Annual Report, 1970*, Kuala Lumpur.
Fei, C. and Ranis, G. (1964), *Development of the Labour-Supply Economy: Theory and Policy*, Homewood: Irwin.
Fong, L.S. (1969), 'The TAB and Industrial Development', Unpublished graduation exercise, Faculty of Economics and Administration, University of Malaya.
Freebairn, J.W. (1978), Projections of Australia's World Trade Opportunities: Mid and Late Nineteen Eighties', IMPACT Working Paper No. I-07, Industries Assistance Commission, Melbourne.
Furnival, J.S. (1944), *Netherlands India*, New York: Macmillan.
Giblin, L.F. (1927), 'The Australian Tariff and the Standard of Living: A Note on Mr. Benham's Statistics', *Economic Record* 3: 148–56.
——(1936), 'Some Economic Effects of the Australian Tariff', The Joseph Fisher Lecture in Commerce, University of Adelaide, 23 June.
Glassburner, B. (ed.), (1971), *The Economy of Indonesia: Selected Readings*, Ithaca: Cornell University Press.
——(1973), 'The January 1973 Tariff Revision', *BIES*, November.
——(1978), 'Political Economy and the Soeharto Regime', *BIES*, November.

Goh Keng Swee (1977), *The Practice of Economic Growth*, Singapore: Federal Publications.

Grilli, E. (1980), 'Italian Commercial Policies in the 1970s', World Bank Staff Working Paper No. 428, Washington, D.C.: World Bank.

Hagen, E. (1958), 'An Economic Justification of Protectionism', *Quarterly Journal of Economics*, November, 72, pp. 496–514.

Hamilton, Bob, Mohammad, S. and Whalley, J. (1984), 'Rent Seeking and the North-South Terms of Trade', unpublished manuscript, Department of Economics, University of Western Ontario, April, (forthcoming, *Journal of Policy Modelling*).

Hamilton, Bob and Whalley, J. (1983), 'Geographically Discriminatory Trade Arrangements', Working Paper No. 8321c, Centre for the Study of International Economic Relations, University of Western Ontario, November, (forthcoming, *Review of Economics and Statistics*).

Hamilton, C. (1981), 'A New Approach to Estimation of the Effects of Non-Tariff Barriers to Trade on Prices, Employment and Import: An Application to the Swedish Textile and Clothing Industry', *Weltwirtschaftliches Archiv*, 117(2).

——(1984), 'Voluntary Export Restraints on Clothing from Asia: Tariff Equivalents, Rent Incomes and Trade Barrier Formation', *Weltwirtschaftliches Archiv*, forthcoming.

Harris, J.R. and Todaro, M.P. (1970), 'Migration, Unemployment and Development: A Two-sector Analysis', *American Economic Review*, vol. 60, March.

Harris, R. (1984), 'Applied General Equilibrium Analysis of Small Open Economies with Scale Economies and Imperfect Competition', *American Economic Review*, Vol. 74.

——(1985), 'Market Structure and Trade Liberalization: A General Equilibrium Assessment', in T.N. Srinivasan and J. Whalley (eds), *General Equilibrium Trade Policy Modelling*, Cambridge: MIT Press.

Harrison, G.W. (1984a), 'Economic Interdependence Between ASEAN and Australia: A General Equilibrium Approach', Working Paper, ASEAN-Australia Economic Relations Research Project, Australian National University.

——(1984b), 'Structural Interdependence Between ASEAN and Australia: An Input-Output Approach', Working Paper, ASEAN-Australia Economic Relations Research Project, Australian National University.

——(1985a), 'A General Equilibrium Analysis of Tariff Reductions', in T.N. Srinivasan and J. Whalley (eds), *General Equilibrium Trade Policy Modelling*, Cambridge: MIT Press.

——(1985b), 'Sensitivity Analysis and Trade Policy Modelling', in T.N. Srinivasan and J. Whalley (eds), *General Equilibrium Trade Policy Modelling*, Cambridge: MIT Press.

Harrison, G.W., Jones, R., Kimbell, L.J. and Wigle, R. (1985), 'How Robust Is Applied General Equilibrium Analysis?' Working Paper No. 8501, Centre for the Study of International Economic Relations, University of Western Ontario.

Harrison, G.W., Kimbell, L.J. (1985), 'Economic Interdependence in the Pacific Basin: A General Equilibrium Approach', in J. Piggott and J. Whalley (eds), *New Developments in Applied General Equilibrium Analysis*, New York: Cambridge University Press.

Harrison, G.W. and Manning, R. (1984), 'Best Approximate Aggregation of Input-Output Systems', Working Paper No. 8401, Centre for the Study of International Economic Relations, Department of Economics, University of Western Ontario, January.

Heilleiner, G.K. (1977), 'The Political Economy of Canada's Tariff Structure: An Alternative Model', *Canadian Journal of Economics*, 10(2), May.

Higgins, B. (1957), *Indonesia's Economic Stabilization and Development*, Institute of Pacific Relations, New York: American Book—Stratford Press Inc.

Hill, H. (1984), 'Survey of Recent Developments', *Bulletin of Indonesian Economic Studies*, (BIES), vol. 20(2), April.

Hoffman, L. and Tan, S.E. (1980), *Industrial Growth, Employment, and Foreign Investment in Peninsular Malaysia*, Kuala Lumpur: Oxford University Press.

Hollinger, W.C. (1970), 'The Foreign Exchange Rate as a Policy Instrument, The Indonesian Case: 1966–70', Paper presented at the Dubrovnik conference of the Development Advisory Service of Harvard University, 20–26 June.

Hong Kong and Shanghai Banking Corporation (1978), 'Hong Kong Textile and Garment Export Restrictions. The Quota System: June 1978', Corporation Secretary's Department, Hong Kong.

Hong Kong Development Council (1981), 'A Summary of Major Restrictions on Hong Kong's Textiles and Clothing Exports', Trade Development Mimeo, April, Hong Kong.

Howe, H.J. (1975), 'Development of the Extended Linear Expenditure System', *European Economic Review*, vol. 6, pp. 305–310.

Hughes, H. (1985), 'The Singapore Economy: The Next 25 Years', Lim Tay Boh Memorial Lecture, Singapore, December 1984 (revised January, 1985).

Hughes, H. and You Poh Seng (eds), (1969), *Foreign Investment and Industrialization in Singapore*, Canberra: Australian National University Press.

IBRD (1955), *The Economic Development of Malaya*, Baltimore: Johns Hopkins University Press.

——(1959), *A Public Development Program for Thailand*, Baltimore: Johns Hopkins University Press.

——(1963), *Report on the Economic Aspects of Malaysia*, Kuala Lumpur: Government Press.

——(1981), 'Malaysia's Manufacturing Sector: Development Issues and Policy Options', Washington DC: IBRD.

Industries Assistance Commission (1976), *Assistance to Manufacturing Industries in Australia, 1968–69 to 1973–74*, Canberra: Australian Government Publishing Service.

——(1978), *Annual Report 1977–78*, Canberra: Australian Government Publishing Service.

——(1979), *Annual Report 1978–79*, Canberra: Australian Government Publishing Service.

——(1980a), *Trends in the Structure of Assistance to Manufacturing*, Canberra: Australian Government Publishing Service.

——(1980b), *Tariffs as Taxes*, Canberra: Australian Government Publishing Service.

——(1981a), *Annual Report 1980–81*, Australian Government Publishing Service, Canberra.

——(1981b), *The Regional Implications of Economic Change*, Canberra: Australian Government Publishing Service.

——(1981c), *Trends in Australia's Trade in Manufactures*, Approaches to General Reductions in Protection, Information Paper No. 3, Canberra: Australian Government Publishfing Service.

——(1982), *Annual Report 1981–82*, Canberra: Australian Government Publishing Service.

——(1983), *Annual Report 1982–83*, Canberra: Australian Government Publishing Service.

Ingram, J.C. (1971), *Economic Change in Thailand 1850–1970*, London: Oxford University Press.

Institute of Developing Economies, (1982), *International Input-Output Table for ASEAN Countries: 1975*, IDE Statistical Data Series No. 39, Tokyo: Asian Economic Press.

Jayakamur, S. (ed.), (1982), *Our Heritage and Beyond*, Singapore: Singapore National

Trade Union Congress.

Jenkins, P. (1980), *Costs and Consequences of the New Protectionism*, Harvard Institute for International Development, Development Discussion Paper No. 99, Harvard University, July.

Johnson, H.G. (1965), 'An Economic Theory of Protectionism, Tariff Bargaining and the Formation of Customs Unions', *Journal of Political Economy* 73(3): 256–83.

Johnson, H.G. (1971), *Aspects Theory of Tariff*, London: George Allen and Unwin Press.

Kanesa-thasan, S. (1966), 'Multiple Exchange Rates: The Indonesian Experience', *IMF Staff Papers*, July.

Keesing, D. and Wolf, M. (1980), *Textile Quotas Against Developing Countries*, Trade Policy Research Centre Essay, No. 23, London.

——(1981), 'Questions on International Trade in Textiles and Clothing', *The World Economy*, 4, 79–101.

Khoo, H.A. (1983), 'The Malaysian Car: Finding a Head Start', *Malaysian Business*, March.

Khoo Saik Liew, (1970/71), 'A Study of the Singapore Footwear Industry and Its Protection', unpublished academic exercise, University of Singapore.

Kimbell, L.J. and Harrison, G.W. (1983), 'On the Solution of General Equilibrium Models', Working Paper No. 8301C, Centre for the Study of International Economic Relations, Department of Economics, University of Western Ontario, January, forthcoming, *Economic Modelling*.

Kimbell, L.J. and Harrison, G.W. (1984), 'General Equilibrium Analysis of Regional Fiscal Incidence', in H.E. Scarf and J.B. Shoven (eds), *Applied General Equilibrium Analysis*, New York: Cambridge University Press.

Kmenta, Tan (1971), *Elements of Econometrics*, New York: Macmillan.

Koomsup, Praiphol (1973), 'The Protection and Industralization in Thailand: A Case Study of the Textile Industry', MA Thesis, Thammasat University.

Krauss, M. (1972), 'Recent Developments in Customs Union Theory: An Interpretative Survey', *Journal of Economic Literature*, 10(2).

Krueger, A.O. (1974), 'The Political Economy of the Rent-Seeking Society', *American Economic Review*, vol. 64, pp. 291–303.

——(1978), 'Effects of Exports from New Industrial Countries on US Industries', in W. Kasper and T.G. Parry (eds), *Growth, Trade and Structural Change in an Open Australian Economy*, Centre for Applied Economic Research, University of New South Wales, Kensington, N.S.W.

Kuznets, S.S. (1966), *Modern Economic Growth: Rate, Structure and Spread*, New Haven: Yale University Press.

——(1971), *Economic Growth of Nations: Total Output and Production Structure*, Cambridge: Harvard University Press.

Landon, K.P. (1941), *The Chinese in Thailand*, New York, Russell and Russell, 1973 reprint of 1941 edition published by OUP, London.

Langhammer, R. (1982), 'ASEAN Manufactured Exports in the EC Markets: An Empirical Assessment of Common and National Tariff and Non-Tariff Barriers Confronting Them', in N. Akrasanee and H.C. Rieger (eds), *ASEAN-EEC Economic Relations*, Institute of South East Asian Studies, Singapore.

Lewis, W.A. (1954), 'Economic Development with Unlimited Supplies of Labor', *The Manchester School*, vol. 22, pp. 139–191.

Lim Chong-Yah (1984), 'Economic Development Experience in Singapore Since Self-Government in 1959', paper presented at the Asian Development Consultation, Christian Conference of Asia, Singapore, 26 November.

Lim, D. (1973), *Economic Growth and Development in West Malaysia, 1947–1970*, Kuala Lumpur: Oxford University Press.

Lim, T.S., Nair, S. and Razak, S.M. (1974), *The Impact of Customs*, Singapore: Singapore National Printers Ltd.

List, F. (1885), *The National System of Political Economy*, London: Longmans, Green. (Translated from the German text first published in 1841).

Little, I., Scitovsky, T. and Scott, M. (1970), *Industry and Trade in Some Developing Countries: A Comparative Study*, London: Oxford University Press for the OECD.

Lloyd, P.J. (1973), *Non-Tariff Distortions of Australian Trade*, Canberra: Australian National University.

——(1978), 'Protection Policies', in F.H. Gruen, *Surveys of Australian Economics*, Sydney: George Allen and Unwin.

——(1982a), 'The Australian Textile and Clothing Industry Group: Untoward Effects of Government Intervention', paper presented at Conference on Structural Adjustment in Trade Dependent Advanced Economies, Stockholm, 2–6 August.

——(1982b), 'Protection for Whom?' *Economic Papers* 1(2), pp. 1–16.

Lluch, C. (1973), 'The Extended Linear Expenditure System', *European Economic Review*, vol. 4, pp. 21–32.

Lluch, C., Powell, A.A. and Williams, R.A. (1977), *Patterns in Household Demand and Saving*, New York: Oxford University Press.

Lo, S.Y. (1972), *The Development Performance of West Malaysia: 1955–1967*, Kuala Lumpur: Heinemann Eductaional Books.

McCawley, P. (1978), 'Industrialisation in Indonesia—Developments, Policies, Prospects', (Paper presented at Asian Studies Association of Australia, Second National Conference, UNSW, May 15–19, 1978).

——(1979), 'Industrialisation in Indonesia: Developments and Prospects', *Occasional Paper No. 13*, Development Studies Centre, Australian National University.

MacPhee, C.R. (1974), *Restrictions on International Trade in Steel*, Lexington, Mass.: Lexington Books.

McPherson, C.P. (1972), 'Tariff Structures and Political Exchange', Unpublished Ph.D. dissertation, University of Chicago.

Maddison, A. (1982), *Phases of Economic Development*, London: Oxford University Press.

Maddock, R. and McLean, I. (1984), 'Supply-Side Shocks: The Case of Australian Gold', *Journal of Economic History*, 44(4), pp. 1047–67.

Malaya, Government of, (1957), *Report of the Industrial Development Working Party*, Kuala Lumpur: Government Printer.

Malaya, State of, (1930), *Estimate of State Revenue and Expenditure 1930*, Kuala Lumpur: Government Printing Office.

Malaysia, Government of, (1966), *First Malaysia Plan, 1966–1970*, Kuala Lumpur: Government Printer.

Malaysia, Government of, (1971), *Second Malaysia Plan, 1971–1975*, Kuala Lumpur: Government Printer.

Malaysia, Government of, (1975), 'Effective Protection and Industrialization Policies: Report of a Research Project', Kuala Lumpur: Mimeographed, Economic Planning Unit.

Malaysia, Government of, (1976), *Third Malaysia Plan, 1970–1980*, Kuala Lumpur: Government Printer.

Malaysia, Government of, (1981), *Fourth Malaysia Plan, 1981–1985*, Kuala Lumpur: National Printing Department.

Malaysia, Government of, (1984), *Mid-Term Review of the Fourth Malaysia Plan, 1971–1975*, Kuala Lumpur: National Printing Department.

Manger, G.J. (1981a), 'The Australian Case for Protection Reconsidered', *Australian Economic Papers*, 20(37): 193–204.

——(1981b), 'Summary Upon the Australian Case for Protection: Comment',

Quarterly Journal of Economics, 96(1): 161–7.

Mansur, A. and Whalley, J. (1984), 'Numerical Specification of Applied General Equilibrium Models: Estimation, Calibration, and Data', in H.E. Scarf and J.B. Shoven (eds), *Applied General Equilibrium Analysis*, New York: Cambridge University Press.

Marshall, A. (1890), *Principles of Economics*, London: Macmillan.

Meltzer, B.D. (1980), 'A Comparison of Some Effective Rate of Protection Calculations with the Results of a General Equilibrium Model', General Paper no. G-26, IMPACT Project Research Centre, Melbourne.

Michaely, M. (1977), *Theory of Commercial Policy*, Oxford: Philip Allan Publishers Ltd.

Moffatt, G.G. (1970), *Import Control and Industrialization: A Study of The Australian Experience*, Melbourne: Melbourne University Press.

Morawetz, D. (1980), *Why the Emperor's New Clothes Are Not Made in Colombia*, London: Oxford University Press.

Morkre, M. (1979), 'Rent-Seeking and Hong Kong's Textile Quota System', *The Developing Economies*, March, pp. 110–18.

Mousney, A. (1964), *The Economy of Thailand: An Appraisal of a Liberal Exchange Policy*, Bangkok: Social Science Association of Thailand.

Muscat, R. (1966), *Development Strategy of Thailand*, New York: Praeger.

Nartsupha, C. and Prasartsetl, S. (1980), *The Political Economy of Thailand 1910–1932*, Bangkok: Social Science Association of Thailand.

Nogues, J., Olechowski, A. and Winters, L.A. (1985), 'The Extent of Non-tariff Barriers in Industrial Countries', World Bank Development Research Department Discussion Paper No. 115, Washington, DC.

OECD (1980), 'Structural Problems and Policies Relating to the OECD Textile and Clothing Industries' DSTI/IND/80.36, Paris, mimeo.

Paauw, D. (1960), *Financing Economic Development: The Indonesian Case*, Glencoe, Illinois: Free Press.

——(1963), 'From Colonial to Guided Economy', in McVey, R., *Indonesia*, New Haven: Hraf Press.

——(1978), 'Labour-Intensive Exports: Cost and Market Distortions', monograph, Jakarta.

Panchamukhi, V.R. (1972), 'Effective Protection and Intra-Regional Trade: A Case Study of Malaysia', Bangkok: Mimeographed, ECAFE Secretariat.

Patmasiriwat, D. (1981), *Industrial Growth and Employment*, Bangkok: TURA Report to NESDB, November.

Pearson, R. and Ingram, W.D. (1980), 'Economies of Scale, Domestic Divergences, and Potential Gains from Economic Integration in Ghana and the Ivory Coast', *Journal of Political Economy*, vol. 88, pp. 994–1007.

Phraya Suriyanuvat, *Political Economy*, Book 3, p. 4.

Piggott, J. and Whalley, J. (1985), *Economic Effects of U.K. Tax-Subsidy Policies: A General Equilibrium Appraisal*, New York: Cambridge University Press.

Pincus, J.J. (1975), 'Pressure Groups and the Pattern of Tariffs', *Journal of Political Economy*, 83(4), August.

——(1977), *Pressure Groups and Politics in Antebellum Tariffs*, New York: Columbia University Press.

——(1982), 'Industry Assistance—Prospects for Long-Run Reform of Protection', Summary of a talk given to the Australian Agricultural Economics Society, Canberra, 19 August.

Pincus, J.J. and G.A. Withers (1983), 'Economics of Regulation', in F.H. Gruen (ed.), *Surveys of Australian Economics*, Vol. 3, Sydney: George Allen and Unwin.

Pitt, M. (1977), 'Economic Policy and Agricultural Development in Indonesia', PhD dissertation, University of California, Berkeley.

——(1981), 'Alternative Trade Strategies and Employment in Indonesia', in Krueger, A.O., et al (ed.), *Trade and Employment in Developing Countries*, vol. 1, Chicago: NBER.

Power, J.H. (1967), 'Import Substitution as Industrialization Strategy', *Philippine Economic Journal*, vol. 5 (Spring).

——(1971), 'The Structure of Protection in West Malaysia', in B. Balassa and Associates, *The Structure of Protection in Developing Countries*, Baltimore: Johns Hopkins University Press.

Power, J. and Sicat, G. (1971), *The Philippines: Industrialisation and Trade Policies*, London: Oxford University Press.

Puthucheary, J.J. (1960), *Ownership and Control in the Malayan Economy*, Singapore: Eastern Universities Press.

Ranis, G. and Fei, J. (1961), 'A Theory of Economic Development', *American Economic Review (Papers and Proceedings)*, vol. 5, pp. 433–565.

Reddaway, W.B. (1937), 'Some Effects of the Australian Tariff', *Economic Record*, 13: 22–30 and 249–50.

Reitsma, A.J. (1960), *Trade Protection in Australia*, Brisbane: University of Queensland Press.

Samuelson, M.C. (1939), 'The Australian Case for Protection Re-examined', *Quarterly Journal of Economics* 54(4): 143–49.

Samuelson, P.A. (1981a), 'Summing Up on the Australian Case for Protection', *Quarterly Journal of Economics* 96(1): 147–60.

——(1981b), 'Justice to the Australians', *Quarterly Journal of Economics* 96(1): 169–70.

Sanittanont, S. (1967), *Thailand: Rice Export Tax: Its Effects on the Rice Economy*, Bangkok.

Saxon, E. and Anderson, K. (1982), 'Japanese Agricultural Protection in Historical Perspective', Research Paper No. 92, Australia-Japan Research Centre, Research School of Pacific Studies, Australian National University, July.

Shepherd, G. (1980), 'Policies to Promote Industrial Development', in K. Young, W.C.F. Bussink and P. Hasan (eds), *Malaysia: Growth and Equity in a Multiracial Society*, Baltimore: Johns Hopkins University Press.

Singapore, Colony of, *Annual Report of the Customs and Excise for the Year 1947, 1949.*

Singapore, *Annual Report 1960, 1962*, Singapore: Government Printing Office, 1964.

Singapore, *Annual Report of Customs and Excise Department*, for years 1965, 1969, 1973, 1977, Singapore:

Singapore, (1971), *Trade Division Annual Report, 1971*, Singapore:

Singapore, Statistics Department of, *Economic and Social Statistics, Singapore, 1960–1982.*

Singapore, *Economic Survey of Singapore 1984.*

Sitsen, P.H.W., (1942), 'The Industrial Development of the Netherland Indies', paper submitted at the Eight Conference of the Institute of Pacific Relations, Mont Tremblant, Quebec, Canada, December.

Snape, R.H. (1984), 'Australia's Relations with GATT', *Economic Record* 60(168): 16–27.

Soo, W.K. (1969), 'The Tariff Advisory Board and the Malaysian Common Market', unpublished graduation exercise, Faculty of Economics and Administration, University of Malaya.

Stahl, K.M. (1951), *The Metropolitan Organisation of British Colonial Trade*, London: Faber and Faber.

Stolper, W.F. and Samuelson, P.A. (1941), 'Protection and Real Wages', *Review of Economic Studies* 9(1): 58–73.

Suhartono, R.B. (1981), *Pola dan Struktur Industri Indonesia Sekitar Akhir Pelita II*, monograph, Department of Industry.

Sutarno, F.S. (1981), 'Peraturan dan Prosedur Impor', *Bulletin Ekonomi*, BAPINDO, March.

Suwankiri, T. (1970), 'The Structure of Protection and Import Substitution in Thailand', MA Thesis, University of Philippines, April.

Szenberg, M., Lombardi, J.W. and Lee, E.Y. (1977), *Welfare Effects of Trade Restrictions: A Case Study of the US Footwear Industry*, N.Y., Academic Press.

Takacs, W. (1978), 'The Non-equivalence of Tariffs, Import Quotas and Voluntary Export Restraints', *Journal of International Economics*, vol. 8(4), November, pp. 565–573.

Tan, A.H.H. (1970), 'Differential Tariffs, Negative Value-Added and the Theory of Effective Protection', *American Economic Review*, Vol. 60, pp. 107–116.

Tan, N.A. (1979), 'The Structure of Protection and Resource Flows in the Philippines', unpublished PhD dissertation, School of Economics, University of Philippines.

Tata Niaga Impor, (1983), Jakarta: Jaya Prasada, September.

Teh, K.P. (1977), 'Protection, Fiscal Incentives and Industrialization in West Malaysia Since 1957', *Monograph Series on Malaysian Economic Affairs*, Kuala Lumpur: Faculty of Economics and Administration, University of Malaysia.

Thailand, Board of Investment, (1971), *Investor's Manual*, Bangkok.

Thailand, National Economic and Social Development Board, *Five-Year Plans, First (1961–1966) and Second (1967–1971)*, Bangkok: NEDB, 1963 and 1968.

Thomas, K.D. and Panglayleim, J. (1973), 'Indonesia—The Effect of Past Policies and President Suharo's Plans for the Future', *Committee for Economic Development of Australia*, P. Series no. 11, November 1983.

Tyers, R. and Chisholm, A. (1982), 'Agricultural Policies in Industrialised and Developing Countries and International Food Security', unpublished manuscript, Research School of Pacific Studies, Australian National University, February.

United Nations Industrial Survey Mission, (1963), *A Proposed Industrialization Programme for the State of Singapore*, United Nations, Commissioner of Technical Assistance, Department of Economic and Social Affairs, 1 September, 1963.

Van Eeglien, M. (1937), 'The Beginnings of Industrialization in Netherlands India', *Far Eastern Survey*, June 9, 1937.

Van Gelderen, J. (1939), *The Recent Development of Economic Foreign Policy in the Netherland Indies*, London and New York.

Verreydt, E. and Waelbroeck, J. (1980), 'European Community Protection Against Manufactured Imports from Developing Countries: A Case Study in the Political Economy of Protection', World Bank Staff Working Paper No. 432, Washington, D.C.: World Bank.

Viner, J. (1929), 'The Australian tariff', *Economic Record* 5(9): 306–15.

von Rabenau, K. (1975), 'Trade Policies and Industrialization in a Developing Country: The Case of West Malaysia', *Economics Discussion Paper*, no. 55, Regensburg: University of Regensburg. An abridged version can be found in *Malayan Economic Review*, vol. 21(1), April.

Waelbroeck, J. (1982), 'Trends of EC Protection and the Prospects of ASEAN Trade', in N. Akrasanee and H.C. Rieger (eds), *ASEAN-EEC Economic Relations*, Institute of South East Asian Studies, Singapore.

Warr, P.G. (1983), 'Structural Effects of Increasing Australia's Imports from Less Developed Countries', invited paper for presentation at 7th World Congress of International Economic Association, Madrid, to be published by International Economic Association.

——(1984), 'Exchange Rate Protection in Indonesia', *BIES*, August.

Warr, P.G. and Lloyd, P.J. (1983), 'Do Australian Trade Policies Discriminate Against Less Developed Countries?', *Economic Record*, 59, 351–64.

Weisberg, S. (1980), *Applied Linear Regression*, New York: John Wiley & Sons.

White, L.J. (1972), 'The Indonesian Foreign Exchange System, 1966–1971', Paper presented at NBER Conference on Exchange Control, Liberalization and Economic Development, 24–26 February.

Whalley, J. (1980a), 'General Equilibrium Analysis of U.S.-EEC-Japanese Trade and Trade Distorting Policies', *Economie Applique*, vol. 33, 1980, pp. 191–230.

——(1980b), 'Discriminatory Features of Domestic Factor Tax Systems in a Goods Mobile-Factors Immobile Trade Model: An Empirical General Equilibrium Approach', *Journal of Political Economy*, vol. 88, 1980, pp. 1177–1202.

——(1982), 'An Evaluation of the Tokyo Round Trade Agreement Using General Equilibrium Computational Methods', *Journal of Policy Modelling*, vol. 3, pp. 341–361.

——(1984a), 'The North-South Debate and the Terms of Trade: An Applied General Equilibrium Approach', *Review of Economics and Statistics*, vol. 66.

——(1984b), *Trade Liberalization Among Major World Trading Areas: A General Equilibrium Approach*, Cambridge: MIT Press.

——(1985), 'Impacts of a 50% Tariff Reduction in an Eight-Region Global Trade Model', in T.N. Srinivasan and J. Whalley (eds), *General Equilibrium Trade Policy Modelling*, Cambridge: MIT Press.

Whalley, J. and Wigle, R. (1983), 'Are Developed Country Multilateral Tariff Reductions Necessarily Beneficial for the U.S.?', *Economics Letters*, vol. 12, pp. 61–67.

Whalley, J. and Wigle, R. (1984), 'Adjustment to Trade Policy Changes: Alternative Formulations and Initial Calculations', in C. Jungenfelt (ed.), *Structural Adjustment in Trade Dependent Advanced Economies*, New York: St. Martin's Press and IEA.

Wheelwright, E.L. (1963), 'Industrialization in Malaya' in T.H. Silcock and E.K. Fisk (eds), *The Political Economy of Independent Malaya*, Singapore: Eastern Universities Press.

Wolf, M. (1983), 'Managed Trade in Practice: Implications of the Textile Arrangements', in W.R. Cline (ed.), *Trade Policy in the 1980s*, Institute for International Economics, Washington, D.C., Cambridge/London: MIT Press.

Wongwuthiwat, P. (1975). 'The Structure of Differential Incentives in the Manufacturing Sector: A Study of Thailand's Experience During 1945–1974', MA Thesis, Thammasat University, June.

World Bank (1980), *World Tables (1980 Edition)*, New York: Oxford University Press.

World Bank (1981), 'The Foreign Trade Regime', Annex 2 in *Indonesia: Selected Issues of Industrial Development and Trade Strategy*, Washington, D.C.: World Bank.

World Bank (1981), *World Development Report 1981*, Washington, D.C.: World Bank.

World Bank (1985), *World Development Report 1985*, New York: Oxford University Press for World Bank.

Wortzel, L. and Wortzel, H. (1979), 'Marketing Manufactured Exports from LDCs: Progress and Recommendations for Further Progress', Boston University, School of Management, Boston, Mass.

Yamazawa, I. (1982), 'Renewal of Textile Industry in Developed Countries and World Textile Trade', paper presented at 39th Annual Meeting of Japan Economic Policy Association, 29–30 May.

Yang, S.C. (1967), *A Multiple Exchange Rate System*, Madison, Wisconsin.

Yeats, A.J. (1979), *Trade Barriers Facing Developing Countries*, New York: St. Martins Press.

Index